Lecture Notes in Artificial Intelligence   6304

Edited by R. Goebel, J. Siekmann, and W. Wahlster

Subseries of Lecture Notes in Computer Science

W0193313

Darina Dicheva   Danail Dochev (Eds.)

Artificial Intelligence: Methodology, Systems, and Applications

14th International Conference, AIMSA 2010
Varna, Bulgaria, September 8-10, 2010
Proceedings

 Springer

Series Editors

Randy Goebel, University of Alberta, Edmonton, Canada
Jörg Siekmann, University of Saarland, Saarbrücken, Germany
Wolfgang Wahlster, DFKI and University of Saarland, Saarbrücken, Germany

Volume Editors

Darina Dicheva
Winston-Salem State University
601 S. Martin Luther King Drive, Winston Salem, NC 27110, USA
E-mail: dichevad@wssu.edu

Danail Dochev
Bulgarian Academy of Sciences, Institute of Information Technologies
29A Acad. Bonchev Str., Sofia 1113, Bulgaria
E-mail: dochev@iinf.bas.bg

Library of Congress Control Number: 2010933081

CR Subject Classification (1998): I.2, H.3, H.4, H.5, I.4, I.5

LNCS Sublibrary: SL 7 – Artificial Intelligence

ISSN   0302-9743

ISBN-10   3-642-15430-1 Springer Berlin Heidelberg New York
ISBN-13   978-3-642-15430-0 Springer Berlin Heidelberg New York

springer.com

© Springer-Verlag Berlin Heidelberg 2010
Printed in Germany

Typesetting: Camera-ready by author, data conversion by Scientific Publishing Services, Chennai, India
Printed on acid-free paper   06/3180

Preface

The 14th International Conference on Artificial Intelligence: Methodology, Systems, Applications (AIMSA 2010) was held in Varna, Bulgaria, during September 8–10, 2010. The AIMSA conference series has provided a biennial forum for the presentation of artificial intelligence research and development since 1984. The conference covers the full range of topics in artificial intelligence (AI) and related disciplines and provides an ideal forum for international scientific exchange between Central/Eastern Europe and the rest of the world. The 2010 AIMSA edition continued this tradition.

For AIMSA 2010, we decided to place special emphasis on the application and leverage of AI technologies in the context of knowledge societies where knowledge creation, accessing, acquiring, and sharing empower individuals and communities. A number of AI techniques play a key role in responding to these challenges. AI is extensively used in the development of systems for effective management and flexible and personalized access to large knowledge bases, in the Semantic Web technologies that enable sharing and reuse of and reasoning over semantically annotated resources, in the emerging social Semantic Web applications that aid humans to collaboratively build semantics, in the construction of intelligent environments for supporting (human and agent) learning, etc. In building such intelligent applications, AI techniques are typically combined with results from other disciplines such as the social sciences, distributed systems, databases, digital libraries, information retrieval, service-oriented applications, etc. AIMSA 2010 reflected this plethora of avenues and the submitted and accepted papers demonstrated the potential of AI for supporting learning, sharing, and amplifying of knowledge.

A total of 93 papers were submitted to the conference coming from 35 different countries. Each submission was reviewed by three independent reviewers. The Program Committee accepted 26 full papers and 13 posters for presentation at the conference, which are included in this volume.

In addition to the selected papers, the conference featured a tutorial "Semantic Technologies and Applications: Web Mining, Text Analysis, Linked Data Search and Reasoning."

AIMSA 2010 had three outstanding keynote speakers: John Domingue (the Open University, UK) outlined the key role that Semantic and Web service technologies will play in the next-generation global network infrastructure, giving examples from a number of running EU projects, Wolfgang Wahlster (German Research Center for AI) presented his view on sharing digital lifelogs of people and objects in instrumented environments, and Atanas Kyriakov (Ontotext AD, Sirma Group, Bulgaria) discussed the emergence of a Web of linked data, the challenges for their adoption, and how reasonable views can make it easier to use linked open data for specific purposes.

Many people contributed to the success of AIMSA 2010. First, we would like to thank all the authors for providing such an excellent set of papers. Second, we would like to thank all members of the AIMSA 2010 Program Committee and the external reviewers for their dedication in the review process and for working hard to meet sometimes impossible deadlines. Last, but not least, we would like to offer our thanks to the Local Organizing Committee recruited from the Institute of Information Technologies, Bulgarian Academy of Sciences, for making the conference as convenient as possible for all participants.

September 2010   Darina Dicheva
                                                                                      Danail Dochev

Organization

Program Chair

Darina Dicheva   Winston-Salem State University, USA

Local Organization Chair

Danail Dochev   Bulgarian Academy of Sciences, Bulgaria

Program Committee

Gennady Agre   Bulgarian Academy of Sciences, Bulgaria
Galia Angelova   Bulgarian Academy of Sciences, Bulgaria
Grigoris Antoniou   ICS-FORTH, Heraklion, Greece
Annalisa Appice   University of Bari, Italy
Sören Auer   University of Leipzig, Germany
Franz Baader   Technical University Dresden, Germany
Roman Barták   Charles University, Czech Republic
Petr Berka   University of Economics, Prague
Mária Bieliková   Slovak University of Technology, Slovakia
Guido Boella   University of Turin, Italy
Paulo Bouquet   University of Trento, Italy
Diego Calvanese   Free University of Bozen-Bolzano, Italy
Valerie Camps   IRIT, Paul Sabatier University, France
Yves Demazeau   CNRS, LIG Laboratory, France
Christo Dichev   Winston-Salem State University, USA
Ying Ding   Indiana University, USA
Danail Dochev   Bulgarian Academy of Sciences, Bulgaria
Peter Dolog   Aalborg University, Denmark
Ben du Boulay   University of Sussex, UK
Stefan Edelkamp   TZI, Bremen University, Germany
Floriana Esposito   University of Bari, Italy
Jérôme Euzenat   INRIA Rhône-Alpes, France
Dragan Gasevic   Athabasca University, Canada
Chiara Ghidini   FBK, Center for Information Technology, Italy
Enrico Giunchiglia   University of Genova, Italy
Vania Dimitrova   University of Leeds, UK
Martin Dzbor   Open University, UK
Michael Fisher   University of Liverpool, UK

Harry Halpin	University of Edinburgh, UK
Dominikus Heckmann	Saarland University, Germany
Pascal Hitzler	Wright State University, USA
Geert-Jan Houben	Delft University of Technology, The Netherlands
Irena Koprinska	University of Sydney, Australia
Atanas Kyriakov	Ontotext Lab, Sirma Group Corp., Bulgaria
H. Chad Lane	USC/Institute for Creative Technologies, USA
Ruben Lara	Telefonica R&D, Spain
Dominique Longin	IRIT, Paul Sabatier University, France
Pierre Marquis	University of Artois, France
Erica Melis	German Research Institute for Artificial Intelligence (DFKI), Germany
Michela Milano	University of Bologna, Italy
Tanja Mitrovic	University of Canterbury, New Zealand
Riichiro Mizoguchi	Osaka University, Japan
Marco Pistore	FBK, Center for Information Technology, Italy
Allan Ramsay	University of Manchester, UK
Zbigniew Ras	University of North Carolina, Charlotte, USA
Ioannis Refanidis	University of Macedonia, Greece
Francesca Rossi	University of Padova, Italy
Paolo Rosso	Polytechnic University of Valencia, Spain
Giovanni Semeraro	University of Bari, Italy
Luciano Serafini	FBK, Center for Information Technology, Italy
Pavel Shvaiko	TasLab, Informatica Trentina, Italy
Umberto Straccia	Institute of Information Science and Technologies - CNR, Italy
Valentina Tamma	University of Liverpool, UK
Annette ten Teije	Free University Amsterdam, The Netherlands
Dan Tufis	Research Institute for AI, Romanian Academy, Romania
Petko Valtchev	University of Montréal, Canada
Julita Vassileva	University of Saskatchewan, Canada
Johanna Voelker	University of Mannheim, Germany

External Reviewers

Carole Adam (Australia)
Anastasios Alexiadis (Greece)
Edouard Amouroux (France)
Pierpaolo Basile (Italy)
Fernando Bobillio (Spain)
Stefano Bortoli (Italy)
Donatello Conte (Italy)
Tiago de Lima (France)
Gerry Dozier (USA)
Albert Esterline (USA)

David John (USA)
Yevgeny Kazakov (UK)
Petar Kormushev (Italy)
Daniel Le Berre (France)
Sanjiang Li (Australia)
George Markou (Greece)
Cataldo Musto (Italy)
Preslav Nakov (Singapore)
Jeff Pan (UK)
Gennaro Percannella (Italy)

Stefano Ferilli (Italy)
Daniel Fleischhacker (Germany)
Dariusz Frejlichowski (Poland)
Benoit Gaudou (Vietnam)
Valerio Genovese (Italy)
Bjoern Gottfried (Germany)
Christophe Gueret (The Netherlands)
Evert Haasdijk (The Netherlands)
Yusuke Hayashi (Japan)
Norman Heino (Germany)
Sebastian Hellmann (Germany)
Shahid Jabbar (Germany)

Kiril Simov (Bulgaria)
Karolina S. Jabbar (Germany)
Dimitrios Sklavakis (Greece)
Martin Stommel (Germany)
Jan Suchal (Slovakia)
Damian Sulewski (Germany)
Stefan Trausan-Matu (Romania)
Michal Tvarozek (Slovakia)
Mario Vento (Italy)
Guandong Xu (Denmark)
David Young (UK)
Bozhan Zhechev (Bulgaria)

Local Organizing Committee

Danail Dochev
Gennady Agre
Kamenka Staykova
Ivo Marinchev

Sponsoring Institutions

Bulgarian Artificial Intelligence Association
Institute of Information Technologies at the Bulgarian Academy of Sciences
Ontotext Lab, Sirma Group Corp.

Table of Contents

Machine Learning, Data Mining, and Information Retrieval

AI in Education

Applications

Posters

Deduction in Existential Conjunctive First-Order Logic: An Algorithm and Experiments

Khalil Ben Mohamed, Michel Leclère, and Marie-Laure Mugnier

LIRMM, CNRS - Université Montpellier 2, France
{benmohamed,leclere,mugnier}@lirmm.fr

Abstract. We consider the deduction problem in the existential conjunctive fragment of first-order logic with atomic negation. This problem can be recast in terms of other database and artificial intelligence problems, namely query containment, clause entailment and boolean query answering. We refine an algorithm scheme that was proposed for query containment, which itself improves other known algorithms in databases. To study it experimentally, we build a random generator and analyze the influence of several parameters on the problem instance difficulty. Using this methodology, we experimentally compare several heuristics.

Keywords: Deduction, Negation, Conjunctive Queries with Negation, Graphs, Homomorphism, Algorithm, Heuristics, Experiments.

1 Introduction

We consider deduction checking in the fragment of first-order logic (FOL) composed of *existentially closed conjunctions of literals* (without functions). The deduction problem, called DEDUCTION in this paper, takes as input two formulas f and g in this fragment and asks if f can be deduced from g (noted $g \vdash f$). DEDUCTION is representative of several fundamental artificial intelligence and database problems. It can be immediately recast as a *query containment* checking problem, which is a fundamental database problem. This problem takes two queries q_1 and q_2 as input, and asks if q_1 is contained in q_2, i.e. if the set of answers to q_1 is included in the set of answers to q_2 for all databases (e.g. [1]). Algorithms based on query containment can be used to solve various problems, such as query evaluation and optimization [2][3] or rewriting queries using views [4]. So-called (positive) *conjunctive queries* form a class of natural and frequently used queries and are considered as basic queries in databases [2][5] and more recently in the semantic web. Conjunctive queries with negation extend this class with negation on atoms. Query containment checking for conjunctive queries with negation is essentially the same problem as DEDUCTION, in the sense that there are natural polynomial reductions from one problem to another, which preserve the structure of the objects. Another related problem in artificial intelligence is the *clause entailment* problem, i.e. a basic problem in inductive logic programming [6]: given two clauses C_1 and C_2, does C_1 entail C_2? If we consider first-order clauses, i.e. universally closed disjunctions of literals, without function symbols, by contraposition we obtain an instance of DEDUCTION.

D. Dicheva and D. Dochev (Eds.): AIMSA 2010, LNAI 6304, pp. 1–10, 2010.

Query answering is a key problem in the domain of knowledge representation and reasoning. Generally speaking, it takes a knowledge base and a query as input and asks for the set of answers to the query that can be retrieved from the knowledge base. When the query is boolean, i.e. with a yes/no answer, the problem can be recast as checking whether the query can be deduced from the knowledge base. When the knowledge base is simply composed of a set of positive and negative factual assertions, i.e. existentially closed conjunctions of literals (possibly stored in a relational database), and the query is a boolean conjunctive query with negation, we again obtain DEDUCTION. Integration of an ontology in this knowledge base is discussed in the conclusion of this paper.

If the considered fragment is restricted to positive literals, deduction checking is "only" NP-complete and this has been intensively studied from an algorithm viewpoint, in particular in the form of the equivalent constraint satisfaction problem [7]. In contrast, when atomic negation is considered, deduction checking becomes Π_2^p-complete[1](e.g. [8]) and very few algorithms for solving it can be found in the literature. Several algorithms have been proposed for the database query containment problem [9][10][11]. They all use homomorphism as a core notion. In this paper, we refine the algorithm scheme introduced in [11] for query containment checking, which itself improves other algorithms proposed in databases. To study it experimentally, we build a random generator and analyze the influence of several parameters on the problem instance difficulty. Then we experimentally compare several heuristics using this methodology. A research report with more figures, algorithms, as well as preliminary results on the comparison with logical provers is available [12][2].

Paper layout. Section 2 introduces the framework. In Section 3, we present our experimental methodology and choices. Section 4 is devoted to the comparison of several heuristics, which leads to refine the algorithm. Section 5 outlines the prospects of this work.

2   Framework

We note $FOL\{\exists, \wedge, \neg_a\}$ the fragment of FOL composed of existentially closed conjunctions of literals, with constants but without other function symbols. A formula in $FOL\{\exists, \wedge, \neg_a\}$ can also be seen as a *set* of (positive and negative) literals.

In [11], queries (i.e. formulas in the present paper) are seen as labeled graphs. This allows us to rely on graph notions that have no simple equivalent in logic (such as pure subgraphs, see later). More precisely, a formula f is represented as a bipartite, undirected and labeled graph F, called *polarized graph (PG)*, with two kinds of nodes: term nodes and predicate nodes. Each term of the formula is associated to a term node, that is unlabeled if it is a variable, otherwise it is labeled by the constant itself. A positive (resp. negative) literal with predicate r is associated to a predicate node labeled $+r$ (resp. $-r$) and it is linked to the nodes assigned to its terms. The labels on edges correspond to the position of each term in the literal (see Figure 1 for an example). For simplicity, the subgraph corresponding to a literal, i.e. induced by a predicate node and

---

[1] $\Pi_2^p = (co\text{-}NP)^{NP}$.

[2] http://www.lirmm.fr/~benmohamed/Publications/RR.pdf

its neighbors, is also called a *literal*. We note it $+r(t_1, \ldots, t_n)$ (resp. $-r(t_1, \ldots, t_n)$) if the predicate node has label $+r$ (resp. $-r$) and list of neighbors $t_1, \ldots, t_n$. The notation $\sim r(t_1, \ldots, t_n)$ indicates that the literal with predicate r may be positive or negative. Literals $+r(t_1, \ldots, t_n)$ and $-r(u_1, \ldots, u_n)$ with the same predicate but different signs are said to be *opposite*. Literals $+r(t_1, \ldots, t_n)$ and $-r(t_1, \ldots, t_n)$ with the same list of arguments are said to be *contradictory*. Given a predicate node label (resp. literal) l, $\bar{l}$ denotes the complementary predicate label (resp. literal) of l, i.e. it is obtained from l by reversing its sign. Formulas are denoted by small letters (f and g) and the associated graphs by the corresponding capital letters (F and G). We note $G \vdash F$ iff $g \vdash f$. A PG is *consistent* if it does not contain two contradictory literals (i.e. the associated formula is satisfiable).

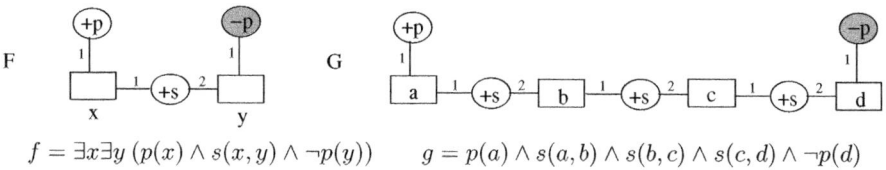

$$f = \exists x \exists y\ (p(x) \wedge s(x, y) \wedge \neg p(y)) \qquad g = p(a) \wedge s(a, b) \wedge s(b, c) \wedge s(c, d) \wedge \neg p(d)$$

Fig. 1. Polarized graphs associated with f and g

Homomorphism is a core notion in this work. A *homomorphism* h from a PG F to a PG G is a mapping from nodes of F to nodes of G, which preserves bipartition (the image of a term - resp predicate - node is a term - resp. predicate - node), preserves edges (if rt is an edge with label i in F then $h(r)h(t)$ is an edge with label i in G), preserves predicate node labels (a predicate node and its image have the same label) and can instantiate term node labels (if a term node is labeled by a constant, its image has the same label, otherwise the image can be any label). On the associated formulas f and g, it corresponds to a *substitution* h from f to g such that $h(f) \subseteq g$. When there is a homomorphism h from F to G, we say that F *maps* to G by h. F is called the *source* graph and G the *target* graph.

Definition 1 (Complete graph and completion). *Let G be a consistent PG. It is com-plete w.r.t. a set of predicates $\mathcal{P}$, if for each $p \in \mathcal{P}$ with arity k, for each k-tuple of term nodes (not necessarily distinct) $t_1, \ldots, t_k$ in G, it contains $+p(t_1, \ldots, t_k)$ or $-p(t_1, \ldots, t_k)$. A completion G' of G is a PG obtained from G by repeatedly adding new predicate nodes (on term nodes present in G) without yielding inconsistency of G. Each addition is a completion step. A completion of G is called total if it is a complete graph w.r.t. the set of predicates considered, otherwise it is called partial.*

If F and G have only positive literals, $G \vdash F$ iff F maps to G. When we consider positive and negative literals, only one side of this property remains true: if F maps to G then $G \vdash F$; the converse is false, as shown in Example 1.

Example 1. See Figure 1: F does not map to G but $g \vdash f$. Indeed, if we complete g w.r.t. predicate p, we obtain the formula g' (equivalent to g): $g' = (g \wedge p(b) \wedge p(c)) \vee (g \wedge \neg p(b) \wedge p(c)) \vee (g \wedge p(b) \wedge \neg p(c)) \vee (g \wedge \neg p(b) \wedge \neg p(c))$. Each of the four conjunctions of g' is a way to complete g w.r.t. p. F maps to each of the graphs associated with them. Thus f is deductible from g'.

One way to solve DEDUCTION is therefore to generate all total completions obtained from G using predicates appearing in G, and then to test if F maps to each of these graphs.

Theorem 1. *[11] Let F and G be two PGs (with G consistent), $G \vdash F$ iff for all G^c, total completion of G w.r.t. the set of predicates appearing in G, F maps to G^c.*

We can restrict the set of predicates considered to those appearing in opposite literals both in F and in G [11]. In the sequel, this set is called the *completion vocabulary* of F and G and denoted $\mathcal{V}$.

A brute-force approach, introduced in [9], consists of computing the set of total completions of G and checking the existence of a homomorphism from F to each of them. However, the complexity of this algorithm is prohibitive: $\mathcal{O}(2^{(n_G)^k \times |\mathcal{V}|} \times hom(F, G^c))$, where n_G is the number of term nodes in G, k is the maximum arity of a predicate, $\mathcal{V}$ is the completion vocabulary and $hom(F, G^c)$ is the complexity of checking the existence of a homomorphism[3] from F to G^c.

Two types of improvements of this method are proposed in [11]. First, let us consider the space leading from G to its total completions and partially ordered by the relation "subgraph of". This space is explored as a binary tree with G as root. The children of a node are obtained by adding, to the graph associated with this node (say G'), a predicate node in positive and negative form (each of the two new graphs is thus obtained by a completion step from G'). The aim is to find a set of partial completions covering the set of total completions of G, i.e. the question becomes: "Is there a set of partial completions $\{G_1, \ldots, G_n\}$ such that (1) F maps to each G_i for $i = 1 \ldots n$; (2) each total completion G^c of G is covered by a G_i (i.e. G_i is a subgraph of G^c) ?" After each completion step, we check whether F maps to the current partial completion: if yes, this completion is one of the sought G_i, otherwise the exploration continues.

Figure 2 illustrates this method on the very easy case of Example 1. Two graphs G_1 and G_2 are built from G, respectively by adding $+p(c)$ and $-p(c)$. F maps to G_1, thus there is no need to complete G_1. F does not map to G_2: two graphs G_3 and G_4 are built from G_2, by adding $+p(b)$ and $-p(b)$ to G_2. F maps to G_3 and to G_4, respectively. Finally, the set proving that F is deductible from G is $\{G_1, G_3, G_4\}$ (and there are four total completions of G w.r.t. p). Algorithm 1 implements this method (the numbers in the margin are relative to the refinements studied in Section 4).

The second kind of improvement consists of identifying subgraphs of F for which there must be a homomorphism to G when $G \vdash F$. Such a subgraph, say F', can be used as a filter to detect a failure before entering the completion process: if F' does not map to G, then $G \nvdash F$. In [10], this property is exhibited for F^+, which is the set of all positive literals in F. This result is generalized in [11] with the notions of *pure subgraph* and *compatible* homomorphism.

Definition 2 (pure subgraph). *A PG is said to be* pure *if it does not contain opposite literals (i.e. each predicate appears only in one form, positive or negative). A* pure

---

[3] Homomorphism checking is NP-complete. A brute-force algorithm solves it in $\mathcal{O}(n_G^{n_F})$, where n_F is the number of term nodes in F.

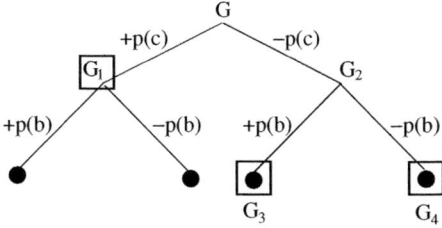

Fig. 2. The search tree of Example 1. Each black dot represents a G^c and each square a G_i.

Algorithm 1. recCheck(G)

Input: a consistent PG G
Data: $F, \mathcal{V}$
Result: true if $G \vdash F$, false otherwise
begin
    if *there is a homomorphism from F to G* **then return** *true* ;
    if *G is complete w.r.t. $\mathcal{V}$* **then return** *false* ;
(3)   *** *Filtering step* ***\
(1)   Choose $r \in \mathcal{V}$ and $t_1, \ldots, t_n$ in G such that $+r(t_1, \ldots, t_n) \notin G$ and
    $-r(t_1, \ldots, t_n) \notin G$;
    Let G' be obtained from G by adding $+r(t_1, \ldots, t_n)$;
    Let G'' be obtained from G by adding $-r(t_1, \ldots, t_n)$;
(2)   **return** *recCheck(G') AND recCheck(G'')* ;
end

---

subgraph *of F is a subgraph of F that contains all term nodes in F (but not necessarily all predicate nodes)*[4] *and is pure.*

We will use the following notations for pure subgraphs of F: F^{max} denotes a pure subgraph that is maximal for inclusion; F^+ is the F^{max} with all positive predicate nodes in F; F^- is the F^{max} with all negative predicate nodes in F; F^{Max} denotes a F^{max} of maximal cardinality.

Moreover, a homomorphism from a pure subgraph of F to G has to be "compatible" with a homomorphism from F to a total completion of G.

Definition 3 (Border, Compatible homomorphism). *Let F and G be two PGs and F' be a pure subgraph of F. The predicate nodes of $F \setminus F'$ are called* border predicate nodes *of F' w.r.t. F. A homomorphism h from F' to G is said to be* compatible *w.r.t. F if, for each border predicate node inducing the literal $\sim p(t_1, \ldots, t_k)$, the opposite literal $\overline{\sim p}(h(t_1), \ldots, h(t_k))$ is not in G.*

Theorem 2. *[11] If $G \vdash F$ then, for each pure subgraph F' of F, there is a compatible homomorphism from F' to G w.r.t. F.*

The following filtering step can thus be performed before the recursive algorithm: select some F^{max}; if there is no compatible homomorphism from F^{max} to G then return *false.*

---

[4] Note that this subgraph does not necessarily correspond to a set of literals because some term nodes may be isolated.

3 Experimental Methodology

Due to the lack of benchmarks or real-world data available for the studied problem, we built a random generator of polarized graphs. The chosen parameters are as follows:

- the number of *term* nodes (i.e. the number of terms in the associated formula)[5];
- the number of distinct *predicates*;
- the *arity* of these predicates (set at 2 in the following experiments);
- the *density* per predicate, which is, for each predicate p, the ratio of the number of literals with predicate p in the graph to the number of literals with predicate p in a total completion of this graph w.r.t. $\{p\}$.
- the *percentage of negation* per predicate, which is, for each predicate p, the percentage of *negative* literals with predicate p among all literals with predicate p in the graph.

An instance of DEDUCTION is obtained by generating a pair (F, G) of polarized graphs. In this paper, we chose the same number of term nodes for both graphs. The difficulty of the problem led us to restrict this number to between 5 and 8 term nodes. In the sequel we adopt the following notations: nbT represents the number of term nodes, $nbPred$ the number of distinct predicates, SD (resp. TD) the Source (resp. Target) graph Density per predicate and neg the percentage of negation per predicate.

In order to discriminate between different techniques, we first experimentally studied the influence of the parameters on the "difficulty" of instances. We measured the difficulty in three different ways: the running time, the size of the search tree and the number of homomorphism checks (see [12] for more details). Concerning the negation percentage, we checked that the maximal difficulty is obtained when there are as many negative relation nodes as positive relation nodes.

One can expect that increasing the number of predicates occurring in graphs increases the difficulty, in terms of running time as well as the size of the searched space. Indeed, the number of completions increases exponentially (there are $(2^{n_G^2})^{nbPred}$ total completions for $nbPred$ predicates). These intuitions are only partially validated by the experiments: see Table 1, which shows, for each number of predicates, the density values at the difficulty peak. We observe that the difficulty increases up to a certain number of predicates (3 here, with a CPU time of 4912 and a Tree size of 72549) and beyond this value, it continuously decreases. Moreover, the higher the number of predicates, the lower SD which entailed the greatest difficulty peak, and the higher the difference between TD and SD at the difficulty peak.

For each value of the varying parameter, we considered 500 instances and computed the mean search cost of the results on these instances (with a timeout set at 5 minutes). The program is written in Java. The experiments were performed on a Sun fire X4100 Server AMD Opteron 252, equipped with a 2.6 GHz Dual-Core CPU and 4G of RAM, under Linux. In the sequel we only show the CPU time when the three difficulty measures are correlated.

---

[5] We do not generate constants; indeed, constants tend to make the problem easier to solve because there are fewer potential homomorphisms; moreover, this parameter does not influence the studied heuristics.

Table 1. Influence of the number of predicates: *nbT*=5, *neg*=50%

NbPred	SD	TD	CPU time (ms)	Tree size
1	0.24	0.28	20	68
2	0.08	0.16	3155	53225
3	0.08	0.4	4912	72549
4	0.08	0.68	3101	44089
5	0.08	0.76	683	8483

4   Refinement of the Algorithm

In this section, we analyze three refinements of Algorithm 1, which concern the following aspects: (1) the choice of the next literal to add; (2) the choice of the child to explore first; (3) dynamic filtering at each node of the search tree.

1. Since the search space is explored in a depth-first manner, the choice of the next literal to add, i.e. $\sim r(t_1, \ldots, t_n)$ in Algorithm 1 (Point 1), is crucial. A brutal technique consists of choosing r and $t_1, \ldots, t_n$ randomly. Our proposal is to guide this choice by a compatible homomorphism, say h, from a F^{max} to the current G. More precisely, the border predicate nodes $\sim r(e_1, \ldots e_n)$ w.r.t. this F^{max} can be divided into two categories. In the first category, we have the border nodes s.t. $\sim r(h(e_1) \ldots h(e_n)) \in G$, which can be used to *extend* h; if all border nodes are in this category, h can be extended to a homomorphism from F to G. The choice of the literal to add is based on a node $\sim r(e_1, \ldots e_n)$ in the second category: r is its predicate symbol and $t_1, \ldots, t_n = h(e_1) \ldots h(e_n)$ are its neighbors (note that neither $\sim r(h(e_1) \ldots h(e_n))$ nor $\overline{\sim r}(h(e_1) \ldots h(e_n))$) is in G since $\sim r(e_1, \ldots e_n)$ is in the second category and h is compatible). Intuitively, the idea is to give priority to predicate nodes potentially able to transform this compatible homomorphism into a homomorphism from F to a (partial) completion of G, say G'. If so, all completions including G' are avoided.

Figure 3 shows the results obtained with the following choices:

- *random choice*;
- *random choice + filter*: random choice and F^+ as filter (i.e. at each recCheck step a compatible homomorphism from F^+ to G is looked for: if none exists, the false value is returned);
- *guided choice*: F^+ used both as a filter and as a guide.

As expected, the guided choice is always much better than the random choice (with or without filter): on the guided choice peaks (*TD*=0.15 and *TD*=0.2), it is almost 11 and 8 times better than the random choice with filter. The greatest difference is for *TD*=0.25 with the guided choice almost 116 times better than the random choice with filter.

2. Experiments have shown that the order in which the children of a node, i.e. G' and G'' in Algorithm 1 (Point 2), are explored is important. Assume that Point 1 in Algorithm 1 relies on a guiding subgraph. Consider Figure 4, where F^+ is the guiding subgraph (hence the border is composed of negative predicate nodes): we see that it is

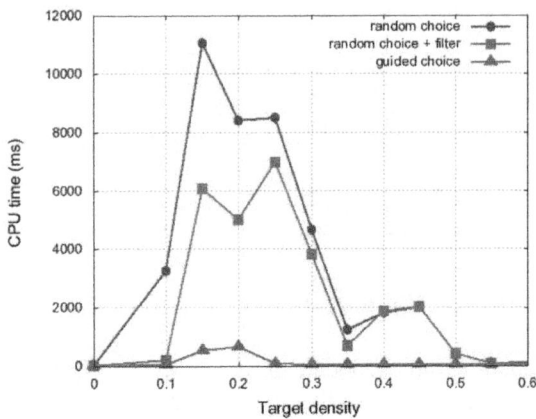

Fig. 3. Influence of the completion choice: nbT=7, $nbPred$=1, SD=0.14, neg=50%

always better to explore G' before G''. If we take F^- as the guiding subgraph, then the inverse order is better. More generally, let $\sim r(e_1 \ldots e_n)$ be the border node that defines the literal to add. Let us call *h-extension* (resp. *h-contradiction*) the graph built from G by adding $\sim r(h(e_1) \ldots h(e_n))$ (resp. $\overline{\sim r}(h(e_1) \ldots h(e_n))$). See Example 2. It is better to first explore the child corresponding to the *h-contradiction*. Intuitively, by contradicting the compatible homomorphism found, this gives priority to failure detection.

Example 2. See Figure 1. $F^+ = \{+p(x), +s(x,y)\}$. Let F^+ be the guiding subgraph. The only border node of F^+ w.r.t. F is $-p(y)$. $h = \{(x,a),(y,b)\}$ is the only compatible homomorphism from F^+ to G. The h-extension (resp. h-contradiction) is obtained by adding $+p(b)$ (resp. $-p(b)$).

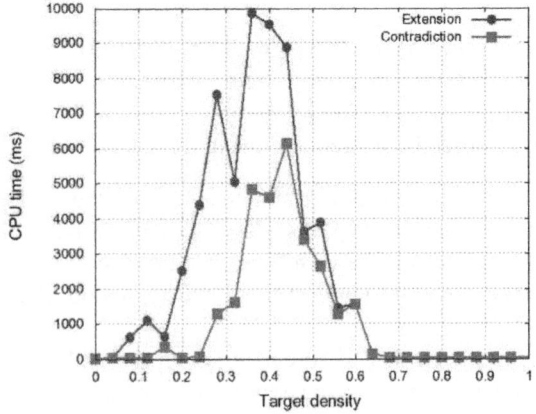

Fig. 4. Influence of the exploration order: nbT=7, $nbPred$=3, SD=0.06, neg=50%

Table 2. Influence of the dynamic filtering: nbT=8, $nbPred$=3, SD=0.03, TD=0.16, neg=50%

Configuration	CPU time (ms)	Tree size	Hom check
Max	3939	2423	3675
$Max\text{-}\overline{Max}$	3868	2394	3858
$Max\text{-}all$	3570	1088	6338

3. The last improvement consists of performing dynamic filtering at each node of the search tree. Once again, the aim is to detect a failure sooner. More precisely, we consider a set of F^{max} and check if there is a compatible homomorphism from each element in this set to the newly generated graph. Table 2 shows the results obtained at the difficulty peak with the following configurations: Max: F^{Max} as guide and no filter; $Max\text{-}\overline{Max}$: F^{Max} as guide and $F^{\overline{Max}}$ (the subgraph on the predicate nodes in $F \setminus F^{Max}$) as filter; $Max\text{-}all$: F^{Max} as guide and all other F^{max} as filters.

Unsurprisingly, the stronger the dynamic filtering, the smaller the size of the search tree. The CPU time is almost the same for all configurations (and all TD values) though $Max\text{-}all$ checks much more homomorphisms than the others. Since our current algorithm for homomorphism checking can be improved, these results show that $Max\text{-}all$ is the best choice. The algorithm finally obtained is given in [12].

5   Perspectives

Our experiments show that the problem is really complex in practice in the difficult area. This may be an argument in favor of restricting conjunctive queries with negation (to pure queries for instance) or using alternative kinds of negation. Closed-world negation is often used. However, even in this case, good algorithms are required for comparing queries (cf. the query containment problem, which is the basis of many mechanisms in databases) and we again find the deduction problem with classical negation.

Ontologies play a central role in knowledge bases, and this role is increasing in databases. A lightweight ontology, in the form of a partial order, or more generally a preorder, on predicates (i.e. on the concepts and predicates of the ontology) can be taken into account without increasing complexity. Note that, in this case, we obtain exactly the deduction problem in a fragment of conceptual graphs [13][14]. Homomorphism is extended in a straightforward way to take the partial order into account. The heuristics studied here still work with an extension of opposite literals: $+r(t_1, \ldots, t_n)$ and $-s(u_1, \ldots, u_n)$ are opposite if $r \geq s$ (then, a pure subgraph is defined as previously with this extended definition). How this work can be extended to more complex ontologies is an open issue.

On the experimental side, further work includes precise comparison with techniques used by logical solvers. We have not found logical algorithms dedicated to the DEDUCTION problem addressed in this paper. For the moment, we have considered the free tools *Prover9* (the successor of the Otter prover [15]) and *Mace4* [16], which consider full first-order logic. Prover9 is based on the resolution method and Mace4 enumerates models by domains of increasing size. These tools are complementary: Prover9 looks for a proof and Mace4 looks for a model that is a counterexample. On preliminary results, we show that our refined algorithm is better than the combination Prover9-Mace4. However the comparison needs to be refined and other provers have to be considered.

References

1. Abiteboul, S., Hull, R., Vianu, V.: Foundations of Databases: The Logical Level. Addison-Wesley, Reading (1995)
2. Chandra, A.K., Merlin, P.M.: Optimal Implementation of Conjunctive Queries in Relational Databases. In: 9th ACM Symposium on Theory of Computing, pp. 77–90 (1977)
3. Aho, A.V., Sagiv, Y., Ullman, J.D.: Equivalences Among Relational Expressions. SIAM J. Comput. 8(2), 218–246 (1979)
4. Halevy, A.Y.: Answering Queries Using Views: a Survey. VLDB Journal 10(4), 270–294 (2001)
5. Ullman, J.D.: Principles of Database and Knowledge-Base Systems, vol. 2. Computer Science Press, Rockville (1989)
6. Muggleton, S., De Raedt, L.: Inductive Logic Programming: Theory and Methods. J. Log. Program. 19/20, 629–679 (1994)
7. Rossi, F., van Beek, P., Walsh, T.: Handbook of Constraint Programming. Elsevier, Amsterdam (2006)
8. Farré, C., Nutt, W., Teniente, E., Urpí, T.: Containment of Conjunctive Queries over Databases with Null Values. In: Schwentick, T., Suciu, D. (eds.) ICDT 2007. LNCS, vol. 4353, pp. 389–403. Springer, Heidelberg (2007)
9. Ullman, J.D.: Information Integration Using Logical Views. In: Afrati, F.N., Kolaitis, P.G. (eds.) ICDT 1997. LNCS, vol. 1186, pp. 19–40. Springer, Heidelberg (1997)
10. Wei, F., Lausen, G.: Containment of Conjunctive Queries with Safe Negation. In: Calvanese, D., Lenzerini, M., Motwani, R. (eds.) ICDT 2003. LNCS, vol. 2572, pp. 343–357. Springer, Heidelberg (2003)
11. Leclère, M., Mugnier, M.-L.: Some Algorithmic Improvments for the Containment Problem of Conjunctive Queries with Negation. In: Schwentick, T., Suciu, D. (eds.) ICDT 2007. LNCS, vol. 4353, pp. 401–418. Springer, Heidelberg (2007)
12. Ben Mohamed, K., Leclère, M., Mugnier, M.-L.: Deduction in existential conjunctive first-order logic: an algorithm and experiments. RR-10010, LIRMM (2010)
13. Kerdiles, G.: Saying it with Pictures: a Logical Landscape of Conceptual Graphs. Univ. Montpellier II, Amsterdam (2001)
14. Mugnier, M.-L., Leclère, M.: On Querying Simple Conceptual Graphs with Negation. Data Knowl. Eng. 60(3), 468–493 (2007)
15. McCune, W.: OTTER 3.3 Reference Manual. CoRR, vol. cs.SC/0310056 (2003)
16. McCune, W.: Mace4 Reference Manual and Guide. CoRR, vol. cs.SC/0310055 (2003)

Use-Based Discovery of Pervasive Services

Raman Kazhamiakin[1], Volha Kerhet[2], Massimo Paolucci[3],
Marco Pistore[1], and Matthias Wagner[3]

[1] Fondazione Bruno Kessler, Trento TN 38050, Italy
{raman,pistore}@fbk.eu
[2] Faculty of Computer Science, Free University of Bozen-Bolzano
kerhet@inf.unibz.it
[3] DoCoMo Euro-Labs, Landsberger Strasse 312, 80687 Munich, Germany
{paolucci,wagner}@docomolab-euro.com

Abstract. Pervasive services accomplish tasks that are related with common tasks in the life of the user such as paying for parking or buying a bus ticket. These services are often closely related to a specific location and to the situation of the user; and they are not characterized by a strong notion of goal that must be achieved as part of a much broader plan, but they are used to address the contingent situation. For these reasons, these services challenge the usual vision of service discovery and composition as goal directed activities. In this paper we propose a new way to look at service discovery that is centered around the activities of the user and the information that she has available rather than the goals that a given service achieves.

Keywords: Mobile&Telco Services, Pervasive Services, Service Discovery.

1 Introduction

The last few years have seen the deployment of an increasing number of pervasive services that support everyday tasks of our life. Such services range from SMS-based services for paying for parking in a number of European cities such as Milan[1] and Vienna to train ticketing in Germany[2].

Pervasive services, like the ones listed above, provide two major challenges to service oriented computing. First, they are closely related to specific locations and context of the user, because they are provided through short range protocols, or because their use is intrinsically related to the environment in which the user operates (e.g., SMS-based parking). Second, these services challenge the idea that service discovery and composition are goal oriented processes; rather users exploit these services without ever creating an explicit notion of goal. Consider for example a performance ticket: to go to the performance, the user may take advantage of a number of services such as navigation services to go to the show, "Point of Interest" service to suggest restaurants where to eat nearby; on-line social services to share comments with friends; contextual services such as parking services which depend on the exact location of the parking: on

---

[1] http://www.atm-mi.it/ATM/Muoversi/Parcheggi/SOSTA_MILANO_SMS.html (l.v. 15.12.08).
[2] http://www.bahn.de/p/view/buchung/mobil/handy_ticket.shtml (l.v. 22.03.09).

D. Dicheva and D. Dochev (Eds.): AIMSA 2010, LNAI 6304, pp. 11–20, 2010.

the street side vs. in a private parking lot. Crucially all these services are of interest for the user and provide added value to the performance ticket, and yet the ticket itself is neither the input of any of them, nor these services contribute to explicitly stated goals of the user. Since there is no explicit notion of goal, discovery algorithms that are based on input/output matching (see [9,12] among the many others) or on precondition and effect matching [1], and possibly quality of service [11,4] hardly apply. Instead, there is a need of different algorithms that organize the available services around the tasks that the user is pursuing. To this extent, in this paper we propose *use-based service discovery* as a new framework for service discovery that aims at matching services with the user's tasks. The resulting discovery framework *recommends* services that may be useful at any given time, leaving up to the user the task of deciding whether to use them or not.

The intuition behind Use-based discovery is to rely on a trend that sees the mobile phone as a collector of very important pieces of information about our life. Already now objects like Deutsche Bahn's "handy ticket", a train ticket delivered to the mobile phone[3], and electronic boarding passes provided by airlines (e.g., airCanada[4]), are evidence of this trend. Crucially, these objects are evidence of (possibly future) user activities. The objective is to envision a service discovery mechanism that given a new object, selects the known services that are relevant for it; and, conversely, given new services selects the objects to which they apply.

The results of Use-based discovery is a contextualized discovery mechanism that reacts to changes in the life of the user trying to support her through her daily life by offering services that may address problems of the user. Ultimately, Use-based discovery is an essential piece to transform the mobile phone from a collector of the users data to an organizer of such data.

The paper is organized as follows. We address the technological problems related with Use-based discovery, namely the representation of services (section 2) and the matching algorithms (section 3). We then proceed with an evaluation in section 4, discuss the open problems in section 5, and conclude in section 6.

2   Specification of Service Advertisement

The first problem to address when facing a new discovery algorithm is the process of service advertisement and the description of the services. Mobile phones have different ways to discover available services: UPnP protocols [3], Bluetooth discovery[5], the reading of 2D QR codes[6], Near Field Communication transmission [8], P2P interactions [13], or simple typing of URLs of service advertisements. Whereas, these algorithms concentrate in finding services, there is still very limited support for organizing the services so that they can be automatically associated to the user activities and invoked.

---

[3] http://www.bahn.de/p/view/buchung/mobil/handy_ticket.shtml (l.v.4.4.2009).

[4] http://www.aircanada.com/en/travelinfo/traveller/mobile/mci.html (lv 24.06.2009).

[5] See www.bluetooth.org (lv 25.06.2009).

[6] See http://www.denso-wave.com/qrcode/index-e.html (lv 25.06.2009).

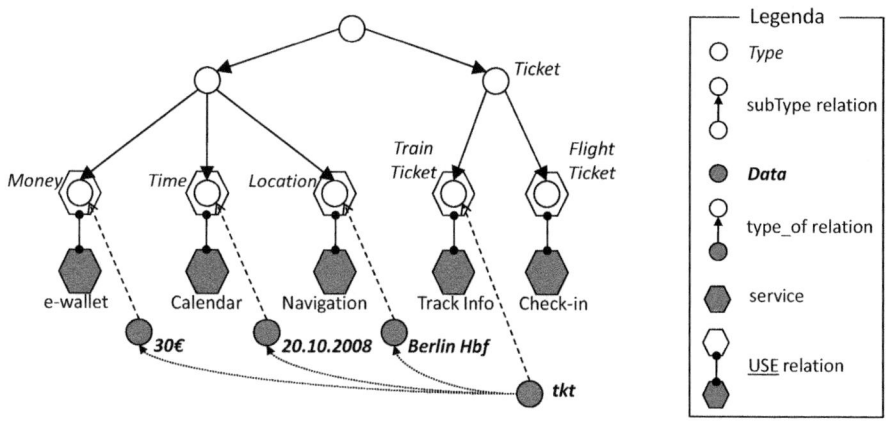

Fig. 1. A representation of service and objects in the ontology

The intuition behind Use-based discovery is that services, exploiting the channels reported above, advertise themselves to mobile phones by declaring for which type of objects a mobile user would use such service. More formally, we define advertisements with the relation <u>USE</u> : $S \times T$ that maps services (S) to types (T), where types are classes in an ontology[7] that are used to specify for which types of objects a service may be used. For example, a navigation service described as to be "used" for locations, means that the service is expected to handle location information.

Fig. 1 shows more in details the relation between types of data and services that are advertised. The circles represent data types organized in a hierarchy as they are known to a system: *Money*, *Time*, *Location*, *Ticket*, *Train Ticket*, and *Flight Ticket*. In addition to types, Fig. 1 shows the available services represented as filled gray labeled hexagons. As the figure shows, services advertise themselves through the types to which they associate themselves. These types are decorated by empty hexagons overlapping on the corresponding types: the e-wallet service is used in conjunction with data of type *Money*, the Calendar with *Time*, the Navigation with *Location*, the Track Info with *Train Ticket*, and the Check-in with *Flight Ticket* respectively.

Whereas services associate to fundamental classes, service advertisements can exploit the whole expressive power of the ontology language, and therefore, other restrictions can be expressed. That is, the navigation service nav may specify that it works only in specific locations by specifying that the relations nav <u>USE</u> *US* (only in USA) or nav <u>USE</u> *ShoppingMall* (only in the shopping mall), where of course *US* and *ShoppingMall* are all restrictions or subtypes of *Location* that are supported by the ontologies. Similarly, depending on the availability of type conjunction and disjunction it is possible to say that the navigation service can be used for Location and Routes meaning that it is good for both types.

---

[7] Although our implementation is based on the OWL descriptions and relates the use to OWL classes, the proposed algorithms could also be modeled in other ways. Using OWL still takes advantage of richer expressivity and of the logic inference in OWL inference engines.

3   Matching Algorithms

In Use-based discovery, the matching process should satisfy two requirements: first, the discovery should be bidirectional, and, second, the discovery should support partial matching of objects and services. Bi-directionality allows the matching process to be triggered by both objects and services. Triggering the process by objects leads to the discovery of services which apply to the object; triggering the discovery from services discovers the objects to which the service applies.

Partial matching relates services to only some of the aspects of objects. For example, in Fig. 1, the navigation relates but only to the location component of the ticket. To this extent, we need to compute an extension of the USE relation that relates services to some properties of objects. We compute this extension through an *association* function that relates the data to the property types. The discovery can then follow associations to find the different parts of the data to which services may apply.

Definition 1. *Given some data d with properties $k_1, \ldots, k_n$, associate function defines the set of types $A \subseteq T$ that can be associated to d (T being the set of all types):*

a) $\forall t \in Type(d) \Rightarrow t \in A$,
b) *if* $d = < k_1, \ldots, k_n >$, $\forall k_i \in k_1, dots, k_n \wedge \forall t \in associate(k_i) \Rightarrow t \in A$.

The first condition specifies that a data instance d is associated with its own types; the second condition specifies that if a data has a set of properties $k_1, \ldots, k_n$, then it is associated to these properties. As an example consider the train ticket `tkt` in Fig. 1 to the destination `BerlinHbf`, at the date 20.10.2008, with a cost 30 €. That is, $tkt = <$`BerlinHbf`, 20.10.2008, 30€$>$. Using the formalisms outlined above we define: $associate(tkt) = \{TrainTicket, Location, Time, Money\}$.

The example also highlights the difference between the classification and association. The ticket is not a location, so it cannot be classified as belonging to that type, but it is associated with it by virtue of its *destination* property. Similarly, the ticket is neither a cost nor a moment, but it is associated to *Time* and *Money* through its properties *departure date* and *cost*.

3.1   Discovery Services from Objects

The objective of the algorithm presented in Fig. 2 (left) is to find all the services that can be used in conjunction with some given data. The algorithm is essentially a search across the taxonomy of types to find all services that can be utilised in conjunction with some given data. More precisely, line 1 specifies the data that is the input parameter, line 2 specify a variable to store the services found (Found); line 3 specifies the set of types from which to start the search: the types that are associated to the data d on the bases of the *associate* function. Line 5 starts the search through the type hierarchy. Line 6 selects all services that can be used to process data of type t by searching its subclasses and super-classes; line 7 applies filters that can weed out services on the bases of different criteria, and line 8 records the services. It is easy to see that the algorithm always terminate if there is a finite set of non-recursive types, since in the worse case all of them a checked and none is checked more than once. To deal with the

1: Let d be the data to start from	1: Let s be the service to start from
2: Let **Found**=∅	2: Let u be a type s.t. s <u>USE</u> u
3: Let $A = associate(d)$	3: Let **Found** = ∅
4: ∀ types $t \in A$	4: ∀ types $t \in u \cup super(u) \cup sub(u)$
5: ∀ types $u, u \sqsubseteq t \vee u \sqsupseteq t \vee u \equiv t$	5: ∀ data d s.t. $type\_of(d, t)$
6: ∀ services s s.t. s <u>USE</u> u	6: if $\neg filter(d, s)$ then
7: if $\neg filter(d, s)$ then	7: **Found** = **Found**$\cup foundAssociated(d)$
8: **Found** = **Found**$\cup\{s\}$	8: return **Found**
9: return **Found**	

Fig. 2. Use-based service discovery algorithms

recursive types it is enough to keep track of the types crossed in the associate function, and make sure that no type is visited twice.

Line 5 and 7 show may prove confusing and potentially controversial aspect of the algorithm. Specifically, line 5 directs the discovery to both the sub-types as well as the super-types of the type t. Indeed, in object oriented programming given a piece of data, the search for appropriate methods proceeds upwards toward crossing all super-types up to the root of the type tree and does not analyze all the methods available to the sub-types. The point to be noticed here is that the <u>USE</u> relation is looser than the type association in object oriented programming, therefore although the a service is specified for more specific data, there may be enough information in the data d to invoke the service anyway. Instead the filters specified in line 7 have been introduced as placeholders to enforce user customization. Such filters may select only services from a given provider, or it may provide only services that are proven to be of a given quality and so on. In this paper we do not concern ourselves with the definition of such filters; we just notice that they will be indeed required.

Example 1. In the example in Fig 1, when the train ticket tkt is added to the data of the user, the above algorithm is performed and the first step is to compute the list of associated types to be stored in the variable A resulting in types *Train Ticket* (through the first condition of the definition), Money, Time, and Location (through the second condition). The next step in the algorithm is the search through the type system. In this way, the services e-wallet, Calendar, and Navigation are found, while the service Check-in is not discovered because it is to be used with a type that is not recognized as associated with tkt. Under the assumption that none of the found services is filtered, they are added to the Found and returned as discovered services.

3.2 Discovery Objects for Services

In mobile and ubiquitous computing, services may be local; therefore new services may become available when the user enters a new environment. An example of this type of services is the gate notification service at the Manchester airport that reports directly on the mobile phone gate that reports directly on the mobile phone gate information on the flights departing at the airport. Ideally, the discovery process should relate gate information to the airline ticket that is bound to be also stored in the mobile phone.

The algorithm shown in Fig. 2 (left) above fails to find to which data a new service can be applied. This problem is addressed by algorithm shown in Fig. 2 (right) that maps a service to the available data exploiting both the USE relation and the "attribute" relation between data objects in the mobile phone.

More in details, the algorithm starts from a service s (line 1) that is in USE relation with a type u (line 2) and establishes a storage variable Found (line 3). In Line 4 starts the search for each super- and sub-types of the type u and for each data instance of those types. If the data is not filtered out (line 6), a call to the procedure *findAssociated* is issued to find the data that is associated with the type (line 7). The procedure *findAssociated* is the inverse of the *associate* function defined above. The filtering mentioned in line 6 are similar to the filters proposed in the algorithm in Fig. 2 (left).

Example 2. Assume in our example that upon arriving at the train station, the mobile phone of the user detects a service, which reports track information (Track Info service), whose declared use is the type *Train Ticket*. Line 4 and 5 of the algorithm perform the search for relevant data in the type *Train Ticket* or in its super and sub types. The tkt, being an instance of *Train Ticket*, is found among that data. A search for the data that is associated with it is then triggered by invoking *findAssociated*. Finally, all data found is returned as candidate data for the specified service, and the user can be notified that a new service relevant for her train ticket has been found. In a similar way for a Calendar service that uses *Time* objects the algorithm will identify the set tkt and the calendar can be offered to store the date of the train travel. Crucially, in this case the calendar is not associated to the time instance, but directly with the train ticket. This is because the train travel is the activity that the user intends to perform and the train ticket is the object that has the greater relevance to the user. It is now an empirical question whether any service should also be reflected in the time of travel.

4   Initial Evaluation

Whereas "use-based" discovery made sense in the scenarios that we were looking while developing our work, we run the risk that the overall framework works only under the idealistic assumptions that are made while developing a new way to think of a problem. Under those idealistic assumptions we cannot predict whether "use-based" discovery will fail to provide any interesting result.

To provide an initial evaluation of our ideas, we referred to the OWLS-MX [6] toolkit, which provide the description of more than 1000 services and the ontologies that are used to interpret those service descriptions. These services have been defined completely independently of our project, and therefore potentially challenging for our framework. The goal of our experimentation was twofold.

Performance evaluation. First, we wanted to evaluate "quantitative" characteristics of the algorithm with respect to the OWLS-MX toolkit and the OWL-based reasoning. For this experiment we have chosen 81 services from different domains, relevant for our scenarios, since most of the services in OWLS-MX can hardly be deployed in pervasive settings. We have augmented the service definitions with the necessary USE descriptions, and also created a set of 37 data objects with complex structure. We conducted

Table 1. Results of the experimental evaluation

	Classes	Object properties	Data properties	Individuals	Services / Objects	Time
1	2678	527	78	895	36	16
2	2673	527	78	860	18	12
3	2674	527	78	887	36	17

a series of experiments with different ontologies and services for both discovery algorithms implemented using OWL-API 2.2.0. The results are summarized in Table 1, which shows the size of the domain, number of services/objects tested, and the total amount of time (sec.) to discover those objects and services. As it follows from the results, the algorithm is rather efficient.

Usability evaluation. Second, we aimed at evaluating the "qualitative" applicability of the use-based service discovery approach. Indeed, while the description of service inputs and outputs is uniquely defined by the service signatures, the definition of service "use" may be rather subjective, and thus the discovery results may vary when the same services are advertisements are defined by different engineers. The goal of the evaluation was, therefore, to understand the "degree" of these variations, and to see how much the presented algorithm is able to abstract them away. To perform such an evaluation, we split in two independent teams and independently used the available ontologies to describe the selected services. Finally, we performed two different evaluations: to analyze the differences between the descriptions provided by two groups, and to match the selections of one team against the other using the implementation of the above algorithms. With the first evaluation we measured the intuition about the "use" of a service. With the second evaluation we verified the ability of the algorithm to abstract the differences between different advertisements.

With the first evaluation we obtained the following results. Among the 81 service descriptions, 9 were equivalent (i.e., the same or equivalent concepts were used for specifying service use), 22 were overlapping (i.e., the concept used by one group was more general than the other), and completely disjoint in remaining cases. These numbers show that service use is very subjective, and it is possible to advertise the same services in very different ways. Consequently, one could expect that objects will be mapped to completely different (and arbitrary) set of services. However, if we consider only those services that intrinsically refer to certain context (e.g., to locations), this number becomes much better: only for 21 services the description is disjoint.

In defining these descriptions we had to deal with two important problems. First, the derivation of USE relation from the service description in many cases degrades to the input specification (e.g., relating parking service to the "car" rather than to the location). Second, the ontologies loaded by OWLS-MX that do not describe user objects and user activities, and therefore were hardly applicable in our case. Furthermore, they make a very limited use of OWL properties, restricting possible specializations. For example, one could claim that FrancemapService is restricted to locations in France, but when we specified that the service is used in conjunction with "SUMO:Geographic Region" or "SUMO:Translocation" it was impossible to restrict the scope to France only. Lack of expressivity of the ontologies contained in OWLS-MX worsened the quality of the service representation, providing an intrinsic bias against our algorithm.

As a result of the matching evaluation, we obtained the following outcome. The precision and recall of the test was 0.63/0.54, and in only 8% of cases no services were found. One of the main reasons behind this result is that we considered not only the services and object that are related to specific user activities, but also generic services and object that were unrelated and created considerable noise. In a second measurement, we restricted to objects and services that correspond to certain context and activities and the results improved: the precision and recall rose to 0.73/0.55. Furthermore, even if the service advertisements do not match, the discovery algorithm manages to smooth down the differences between the service advertisements. The reason of this abstraction is that the relation between services and objects is based on the properties of these objects, which lowers the dependence on the specific advertisement.

The experiments show that the proposed approach is indeed applicable in the scenarios and domains like those discussed in this paper: where pervasive services are explored and where objects associated to the user activities are considered. In addition it shows that the more structured objects are, the better discovery results may be obtained.

5   Discussion

Use-based discovery has a form of advertisement that is completely unrelated from the description of the service as a computation entity. In this sense, it is different from our Web services languages, such as OWL-S [7], SAWSDL [2] or WSML [10], which directly ground on computational features of the service like inputs, outputs, or message formats. On the other hand, a service can be declared to be useful for all sort of objects, even though such definition is totally meaningless.

The lack of direct relation with the service has both positive and negative effects. On the positive side, the USE relation can be used to describe not only services but any information source, such as Web pages and of RSS feeds. Ultimately, Use-based discovery contributes to bridge the gap between Web services and the rest of the Web.

On the negative side, the USE relation is too subjective; the same service may be described in very different ways. Furthermore, the USE relation does not provide any information about the invocation of services: ultimately the user may know that a service relates to a given object, but she may still have the problem of invoking the service.

In our current work, we are trying to address these two problems. Specifically, the problem of handling invocation can be addressed by defining lifting and lowering functions that given a USE description describe how to generate the service inputs and interpret service outputs. As for a more precise definition of the relation between USE and services, we are evaluating different approaches of tightening the relation between the service and the description.

Exploiting USE for Invocation. As pointed out above, the USE relation specifies what a service can be used for, but not how to invoke the service. Therefore, it does not provide any guidance in the actual use and invocation of the service. To address the problems listed above, the USE specification needs to be supplemented with invocation information. Although the service invocation is beyond the scope of this paper, the relation to the invocation information can be performed by leveraging on existing solutions such as OWL-S [7] and SAWSDL [2]. Specifically, the USE specification can be mapped to a workflow language that specifies which operations of the service should be performed to realise a given use of the service and in which order.

In addition, the specification of the USE of a service needs to relate to the inputs of the service, so that, given the use, it is possible to generate the inputs of the service. We adopt the solution of SAWSDL that proposes the use of lowering functions to map the inputs and outputs of operations to the corresponding concepts in the ontology. That is, the USE definition is enriched with lowering functions that specify how the data specified in the use can be mapped to the data needed by the operations.

Specifically, given a service s, let I be the inputs of service s and U be the specified usage for s. We define the lowering function for s, $lower_s : U_s \rightarrow I_s$, which specifies how data that is defined consistently with the use of s (U_s) can be transformed in the set of inputs of s (I_s) required by the service. Crucially, when such a lowering function is complete, then the usage specification U_s guarantees that the service client has all the information that is required to invoke the service.

In addition to a lowering function we need to specify lifting functions that show how the outputs of the service reflect in the ontology. Formally, the lifting functions is defined as $lift_s : Os \rightarrow C$, where O_s are the outputs of the service, while C is a concept in the ontology. The definition of the lifting function highlights an asymmetry with lowering functions. Whereas, lowering functions need to relate the service directly to the service specification because at the time of invocation the service client needs to map the use into the inputs of the service, the lifting functions may result in the specification of data items that may be unrelated to the specified use.

Improving USE precision. To improve the precision of the USE relation, we need to tighten the relation between a service and its specified use. To achieve this, we are following two approaches. The first one aims at weakening the original statements on which USE was based, i.e., that USE is independent of inputs and outputs of the service. In this sense, we could exploit the lifting functions described above to figure out what inputs the service requires, and then relate these inputs to objects. From an implementation point of view, this process can proceed along the "associate" links described in section 3. While this approach may prove able to describe the relation between services and objects to which they relate, its weakness is that it is not able to describe Web pages and RSS feeds that could be useful for the user.

Another approach is to relate services and objects with the concepts representing the "assets" that they manipulate. The idea of relating services with resources is described in [5]. In this approach the USE relation will then be inferred by the declarations of both services and objects, and likely it will not be as generic as it is defined here.

6   Conclusions

The service discovery mechanisms that have been proposed in literature fail to address the problem of discovering the pervasive services that are emerging in the mobile environment The reason of this failure is that discovery is thought as a goal directed process in which the service is found because it achieves (part of) the goals of the requester. As a consequence service discovery concentrates on the technical aspects of the services, such as its inputs and outputs trying to derive from them the potential use of the service.

On the opposite of this view, discovery in pervasive service provisioning is not goal directed, rather services associate to the activities of the user adding value and information to the data that the user needs to utilize. To address this different use of services, in this paper we proposed "use-based" discovery: a framework in which the expected use of a service is specified, and then this use is matched against the data that to the services. To evaluate this idea, we tried to describe 81 services from the OWLS-TC testbed and we showed that even if the service advertisement strongly depends on the requirements and objectives of the provider of the description, the proposed algorithm is able to smooth this aspect and to deliver reasonable enough set of services that correspond to the user objects and activities.

References

1. Akkiraju, R., Srivastava, B., Ivan, A., Goodwin, R., Syeda-Mahmood, T.: Semantic Matching to Achieve Web Service Discovery. In: Conference on Enterprise Computing (2006)
2. Farrell, J., Lausen, H.: Semantic Annotations for WSDL and XML Schema -W3C Recommendation (August 28, 2007) (2006), http://www.w3.org/TR/sawsdl/ (l.v. 01.10.2008)
3. Goland, Y.Y., Cai, T., Leach, P., Gu, Y., Albright, S.: Simple Service Discovery Protocol 1.0
4. Hamdy, M., Konig-Ries, B., Kuster, U.: Non-functional Parameters as First Class Citizens in Service Description and Matchmaking - An Integrated Approach. In: NFPSLA-SOC (2007)
5. Kazhamiakin, R., Bertoli, P., Paolucci, M., Pistore, M., Wagner, M.: Having Services "Your-Way!" Towards User-Centric Composition of Mobile Services. In: Future Internet Symposium (FIS), pp. 94–106 (2008)
6. Klusch, M., Fries, B., Sycara, K.: Automated Semantic Web Service Discovery with OWLS-MX. In: Conference on Autonomous Agents and Multi-Agent Systems, AAMAS (2006)
7. Martin, D.L., Burstein, M.H., McDermott, D.V., McGuinness, D.L., McIlraith, S.A., Paolucci, M., Sirin, E., Srinivasan, N., Sycara, K.: Bringing Semantics to Web Services with OWL-S. World Wide Web Journal (2006)
8. Ortiz, C.E.: An Introduction to Near-Field Communication and the Contactless Communication API, http://java.sun.com/developer/technicalArticles/javame/nfc/ (lv: 25.06.2009)
9. Paolucci, M., Kawamura, T., Payne, T.R., Sycara, K.: Semantic Matching of Web Services Capabilities. In: 1st International Semantic Web Conference (2002)
10. Roman, D., Lausen, H., Keller, U.: Web Service Modeling Ontology (WSMO), http://www.wsmo.org/TR/d2/v1.2/ (lv: 25.06.2009)
11. Sivashanmugam, K., Verma, K., Sheth, A., Miller, J.: Adding Semantics to Web Services Standards. In: International Conference on Web Services, ICWS (2003)
12. Sycara, K., Widoff, S., Klusch, M., Lu, J.: Larks: Dynamic matchmaking among heterogeneous software agents in cyberspace. Autonomous Agents and Multi-Agent Systems 5, 173–203 (2002)
13. Zhou, G., Yu, J., Chen, R., Zhang, H.: Scalable Web Service Discovery on P2P Overlay. In: Service Computing Conference, SCC (2007)

Expressive Approximations in *DL-Lite* Ontologies[*]

Elena Botoeva, Diego Calvanese, and Mariano Rodriguez-Muro

KRDB Research Centre
Free University of Bozen-Bolzano, Italy
{botoeva,calvanese,rodriguez}@inf.unibz.it

Abstract. Ontology based data access (OBDA) is concerned with providing access to typically very large data sources through a mediating conceptual layer that allows one to improve answers to user queries by taking into account domain knowledge. In the context of OBDA applications, an important issue is that of reusing existing domain ontologies. However, such ontologies are often formulated in expressive languages, which are incompatible with the requirements of efficiently accessing large amounts of data. Approximation of such ontologies by means of less expressive ones has been proposed as a possible solution to this problem. In this work we present our approach to semantic (as opposed to syntactic) approximation of OWL 2 TBoxes by means of TBoxes in *DL-Lite$_A$*. The point of interest in *DL-Lite$_A$* approximations is capturing entailments involving chains of existential role restrictions, which can play an essential role in query answering. The presence of TBox assertions involving existential chains affects query answering by enriching the number of obtained rewritings, and hence allows us to cope better with incomplete information about object and data properties. We provide an approximation algorithm and show its soundness and completeness. We also discuss the implementation of the algorithm.

1 Introduction

Ontology based data access (OBDA) [7,15,4] is concerned with providing access to typically large data sources through a conceptual layer constituted by an ontology. Such a mediating ontology on the one hand provides a high level conceptual view of the data residing at the sources, thus freeing users from the need to be aware of the precise organization of the data, and on the other hand allows for improving answers to user queries by taking into account the domain knowledge encoded in the ontology.

In the context of OBDA applications, an important issue is that of reusing existing domain ontologies. However, such ontologies are designed to be as general as possible, and hence are often formulated in expressive languages, such as the Web Ontology Language OWL 2 [6], that are incompatible with the requirements of efficiently accessing large amounts of data. Given the well known trade-off of language expressiveness vs. complexity of reasoning in ontology languages [3], in order to regain efficiency of inference in data access, it might be necessary to *approximate* an ontology formulated in an expressive language by means of an ontology formulated in a less expressive

---

[*] This work has been partially supported by the ICT Collaborative Project ACSI, funded by the EU under FP7 grant agreement n. 257593.

D. Dicheva and D. Dochev (Eds.): AIMSA 2010, LNAI 6304, pp. 21–31, 2010.

language that exhibits nice computational properties, in particular with respect to data complexity. Examples of such languages are those of the $\mathcal{EL}$ family [2,11], and those of the *DL-Lite* family [5,4], where the latter languages have been designed specifically for allowing efficient access to large amounts of data. The problem of computing approximations of ontologies in OWL 2 (or its expressive fragments) has been addressed considering languages of both families as target [13,14,17].

An approximation should be as faithful as possible, i.e., capture as much as possible of the semantics of the original ontology. A basic requirement is that the approximation is *sound*, i.e., that it does not imply additional unwanted inferences. Instead, *completeness* guarantees that all entailments of the original ontology that are also expressible in the target language are preserved in the approximated ontology. A common type of approximations are syntactic approximations, which are transformation of the original ontology that only consider the syntactic form of the axioms that are to be approximated [19]. This kind of approximation generally allows for fast and simple algorithms, however in general it does not guarantee soundness and/or completeness. More interesting are *semantic approximations*, which exploit the semantics of the original ontology to compute the approximated one [18]. Algorithms for computing this kind of approximations tend to be slower since they often involve sound and complete reasoning in the expressive language of the original ontology, e.g., to perform a complete classification of the concepts. However, they can also provide better guarantees with respect to soundness and completeness of the result. Selman and Kautz [18] introduced the term knowledge compilation for computing such approximations.

In this paper we focus on sound and complete semantic approximations of OWL 2 ontologies by means of *DL-Lite* ontologies: specifically, as target ontology language we consider *DL-Lite$_A$*, which is an expressive member of the *DL-Lite* family [15,4] known to have very nice computational properties. Moreover, polynomial reasoning techniques for this logic were developed and implemented in the system QUONTO [1,16].

For the purpose of approximation of OWL 2 in *DL-Lite$_A$*, it suffices to consider only the TBox of an ontology, which represents the intensional information about the domain of interest. Indeed, the extensional knowledge is represented by data sources that are accessed through the TBox of the ontology. The objective is to compile, in the best possible way, the knowledge expressed in the OWL 2 TBox into a *DL-Lite$_A$* TBox approximation. The latter can then be used by application designers in scenarios where they need to access large amounts of data (i.e., ABoxes), and in which reasoning over the original, expressive ontology would be practically unfeasible.

Our work represents an important extension of previous work on semantic approximation in the *DL-Lite* family [13], which proposes an algorithm that approximates OWL DL ontologies in *DL-Lite$_\mathcal{F}$*. A crucial difference between *DL-Lite$_\mathcal{F}$* and *DL-Lite$_A$* is that the former, but not the latter, rules out role hierarchies, and nested qualified existentials on the right-hand part of concept inclusion assertions [15,4]. On the one hand, this added expressive power is of importance in applications [10]. On the other hand it makes the task of computing sound and complete semantic approximations significantly more challenging. Indeed, while for *DL-Lite$_\mathcal{F}$* the number of different concepts that can be expressed with a given finite alphabet of concept (i.e., unary relation) and role (i.e., binary relation) symbols is finite, this is not the case for *DL-Lite$_A$*, due to the presence of concepts of the form $\exists R_1. \ldots \exists R_n.A$, where $R_1, \ldots, R_n$ are roles and A is a concept name. We call such concepts making use of nested qualified existentials *existential role chains*. They can be used in the rewriting step of query answering algorithms (see, e.g., [5]), and hence they play an essential role in query answering, as illustrated by the following example.

Example 1. Consider the medical OWL 2 TBox containing the following:

Assume that an ABox, built from a large database containing patient records, contains the assertions *Pharyngitis*(c) and *hasCondition*(john, c). Consider a clinical trial query asking for patients that have a treatment that might cause or be involved in acute renal failure symptoms:

$$q(x) \leftarrow hasCondition(x, y), hasTreatment(y, z), sideEffect(z, m),$$
$$symptomOf(m, n), AcuteRenalFailure(n)$$

Trying to answer the query with an OWL 2 reasoner might fail due to the amount of data in the ABox and the complexity of the query. On the other hand, approximating syntactically the ontology or by means of the algorithm in [13] will fail to give the expected answer, i.e., {john} because the entailment *Pharyngitis* $\sqsubseteq$ $\exists hasTreatment.\exists sideEffect.\exists symptomOf.AcuteRenalFailure$ is not captured. To capture such an entailment, a form of approximation taking into account existential role chains is required.   ∎

Approaching the problem is non-trivial if one wants to keep soundness and completeness of the approximation. On the one hand, entailments involving existential chains can come from complex OWL 2 concept descriptions, as we have seen in the example. On the other hand, there is no a priori bound on the length of the chains that have to be considered. In this paper we show that by suitably extending the alphabet, it is possible to capture all *DL-Lite$_A$* entailments involving existential chains (of arbitrary length). We also show that it is not possible to capture all *DL-Lite$_A$* entailments if the alphabet is not extended; however, we propose a compromise on the length of the entailed formulas that provides useful guarantees of completeness. We demonstrate the proposed approach in a Java based, open source, approximation engine that is available as a Java library, a command line OWL 2-to-*DL-Lite$_A$* approximation tool, and a Protege 4.0 plugin.

2 Preliminaries on Description Logics

Description Logics (DLs) [3] are logics specifically designed for representing structured knowledge, and they provide the formal underpinning for the standard ontology languages OWL and OWL 2 [6]. We introduce now OWL 2 and *DL-Lite$_A$*, the two DLs that we deal with in this paper.

OWL 2. The Web Ontology language[1] OWL 2 is an ontology language for the Semantic Web that has been designed by W3C to represent rich and complex knowledge and to reason about it. OWL 2 corresponds to the DL $\mathcal{SROIQ}$ [8], that we now define.[2]

In DLs, the domain of interest is modeled by means of concepts and roles, denoting respectively unary and binary predicates. The language of OWL 2 contains atomic concept names A, atomic role names P, and individual names a. *Complex concepts* and *roles*, denoted respectively by C and R, are defined as:

$$R ::= P_\top \mid P_\perp \mid \quad C ::= \top \mid \perp \mid A \mid \neg C \mid C_1 \sqcap \cdots \sqcap C_n \mid C_1 \sqcup \cdots \sqcup C_n \mid \forall R.C \mid$$
$$P \mid P^- \qquad \exists R.C \mid \exists R.\mathsf{Self} \mid \geq k\, R.C \mid \leq k\, R.C \mid \{a_1, \ldots, a_n\}$$

A concept of the form $\exists R.C$ is called a *qualified existential (restriction)*, and the simpler form $\exists R.\top$, in the following abbreviated as $\exists R$, is called an *unqualified existential*.

In DLs, the intensional knowledge about the domain is represented in a TBox, consisting of a finite set of axioms and constraints involving concepts and roles. An OWL 2 *TBox*, $\mathcal{T}$, is a finite set of:

(i)   *concept inclusion axioms* of the form $C_1 \sqsubseteq C_2$,
(ii) *role inclusion axioms* of the form $R_1 \circ \cdots \circ R_n \sqsubseteq R$, $n \geq 1$, and
(iii) *role constraints*, such as disjointness, functionality, transitivity, asymmetry, symmetry, irreflexivity, and reflexivity, expressed respectively with $\mathsf{Dis}(R_1, R_2)$, $\mathsf{Fun}(R)$, $\mathsf{Trans}(P)$, $\mathsf{Asym}(P)$, $\mathsf{Sym}(P)$, $\mathsf{Irr}(P)$, and $\mathsf{Ref}(P)$.

Note that some of the role constraints can be expressed using concept or role inclusion axioms [8]. OWL 2 TBoxes satisfy some syntactic conditions involving the role hierarchy and the appearance of roles in concepts of the form $\exists R.\mathsf{Self}$, $\geq k\, R.C$, $\leq k\, R.C$ and in the assertions $\mathsf{Irr}(P)$, $\mathsf{Dis}(R_1, R_2)$. See [8] for details. The *role depth* of $\mathcal{T}$ is the maximal nesting of constructors involving roles in $\mathcal{T}$.

In DLs, the extensional knowledge about individuals is represented in an ABox. An OWL 2 *ABox*, $\mathcal{A}$, is a finite set of *membership assertions* of the form $C(a)$, $P(a, b)$, and $\neg P(a, b)$. TBox and ABox constitute a *knowledge base* $\langle \mathcal{T}, \mathcal{A} \rangle$.

The semantics of DLs is given in terms of first-order interpretations [3], and the constructs and axioms of OWL 2 are interpreted in the standard way, see [8]. We just mention that for an interpretation $\mathcal{I}$, we have that $(\exists R.\mathsf{Self})^{\mathcal{I}} = \{x \mid (x, x) \in R^{\mathcal{I}}\}$, since this construct is not usually found in DL languages.

OWL 2 is a very expressive DL, but this expressiveness comes at a price. Indeed, reasoning over an OWL 2 ontology is 2ExpTime-hard, and the best known upper bound is 2NExpTime [9]. Also, it is open whether answering conjunctive queries is decidable.

DL-Lite$_{\mathcal{A}}$. *DL-Lite$_{\mathcal{A}}$* has been specifically designed for efficient reasoning and query answering over large amounts of data [15,4]. A *DL-Lite$_{\mathcal{A}}$* ontology is formed using atomic concept names A, atomic role names P, and individual names a. In *DL-Lite$_{\mathcal{A}}$*, we distinguish *basic concepts* B, that may appear in the lhs of concept inclusions, from arbitrary concepts L (for 'light') that may appear only in the rhs of inclusions:

---

[1] http://www.w3.org/TR/owl2-overview/

[2] For simplicity, we restrict the attention to the features of $\mathcal{SROIQ}$/OWL 2 that are relevant for our purposes.

$$B ::= \bot \mid A \mid \exists R \qquad\qquad L ::= B \mid \neg B \mid \exists R.L \qquad\qquad R ::= P \mid P^-$$

We call *chain* a sequence $S = R_1 \circ \cdots \circ R_n$ of roles, and use $\exists S$ for $\exists R_1. \ldots .\exists R_n$.

A *DL-Lite$_A$ TBox*, $\mathcal{T}$, is a finite set of: *(i) concept inclusion axioms* $B \sqsubseteq L$, *(ii) role inclusion axioms* $R_1 \sqsubseteq R_2$, and *(iii) role constraints* $\mathsf{Fun}(R)$, $\mathsf{Dis}(R_1, R_2)$, $\mathsf{Asym}(P)$, and $\mathsf{Sym}(P)$, with the syntactic condition that no role that is functional or whose inverse is functional can appear in the rhs of a role inclusion axiom or in a qualified existential restriction [15,4]. Concept inclusions of the form $B \sqsubseteq \exists R_1. \ldots .\exists R_n.B'$ are called *chain inclusions*, and n is the *length* of the chain inclusion.

3   Semantic Approximation from OWL 2 to *DL-Lite$_A$*

We start by providing the formal definition of the problem we are addressing, that is we define the notion of sound and complete approximation.

Let $\mathcal{L}$ be a description logic. An $\mathcal{L}$ *axiom* is an axiom allowed in $\mathcal{L}$, and an $\mathcal{L}$ *TBox* is a TBox that contains only $\mathcal{L}$ axioms. The *signature* Σ of an $\mathcal{L}$ TBox $\mathcal{T}$ is the alphabet of concept, role, and individual names occurring in $\mathcal{T}$. Let $\mathcal{T}$ and $\mathcal{T}'$ be two $\mathcal{L}$ TBoxes such that $\mathcal{T} \subseteq \mathcal{T}'$. Following [12], we say that $\mathcal{T}'$ is a *conservative extension* of $\mathcal{T}$, if for every axiom I over the signature Σ of $\mathcal{T}$ s.t. $\mathcal{T}' \models I$ we also have that $\mathcal{T} \models I$.

Given two DLs $\mathcal{L}$ and $\mathcal{L}'$, we say that $\mathcal{L}$ is *(syntactically) more expressive* than $\mathcal{L}'$, denoted $\mathcal{L}' \preceq \mathcal{L}$, if every $\mathcal{L}'$ TBox is also an $\mathcal{L}$ TBox.

Definition 1. *Let $\mathcal{L}'$ and $\mathcal{L}$ be two DLs with $\mathcal{L}' \preceq \mathcal{L}$, and let $\mathcal{T}$ be an $\mathcal{L}$ TBox with signature Σ. An* approximation *of $\mathcal{T}$ in $\mathcal{L}'$ is an $\mathcal{L}'$ TBox $\mathcal{T}'$ over a signature $\Sigma' = \Sigma \cup \Sigma_{new}$, where Σ_{new} is a possibly empty set of new names.*

- *$\mathcal{T}'$ is a* sound *approximation (w.r.t. TBox reasoning) if for every $\mathcal{L}'$ axiom I over Σ s.t. $\mathcal{T}' \models I$, we have that $\mathcal{T} \models I$.*
- *$\mathcal{T}'$ is a* complete *approximation (w.r.t. TBox reasoning) if for every $\mathcal{L}'$ axiom I over Σ s.t. $\mathcal{T}' \cup \{I\}$ is an $\mathcal{L}'$ TBox and $\mathcal{T} \models I$, we have that $\mathcal{T}' \models I$.*

The work of [13] allows one to capture in an approximation the basic concept hierarchy entailed by the input ontology that is formulated in an expressive DL such as OWL 2. The approach can be summarized as follows: *(i)* for each pair of basic concepts B_1, B_2 in the signature of the original ontology, check (using a DL reasoner) whether the original ontology implies $B_1 \sqsubseteq B_2$ or $B_1 \sqsubseteq \neg B_2$, *(ii)* for each direct and inverse role R check whether $\mathsf{Fun}(R)$ is implied, and *(iii)* collect all entailments (the so called entailment set) in a *DL-Lite$_\mathcal{F}$* ontology, i.e., the approximated ontology.

The algorithm in [13] is sound and complete when the target language is *DL-Lite$_\mathcal{F}$*. However, proceeding in this way when the target language is *DL-Lite$_A$* will result in incomplete approximations, specifically w.r.t. to entailments involving existential chains. The following example demonstrates this.

Example 2. Consider the TBox $\mathcal{T}_1$ constituted by the axioms: $A \sqsubseteq \exists R_1.(A_1 \sqcup A_2)$, $A_1 \sqsubseteq \exists R_2.A$, and $A_2 \sqsubseteq \exists R_2.A$. One can see that $\mathcal{T}_1$ implies the chain inclusion $A \sqsubseteq \exists R_1.\exists R_2.A$, which is not in the entailment set computed as illustrated above. ∎

In *DL-Lite$_A$*, incompleteness of the approximation is due to missed entailments involving existential chains. Assuring that the approximation entails all possible existential chains is non trivial, since in principle there can be an infinite number of these entailments. For instance, in Example 2, T_1 implies all chain inclusions of even length of the form $A \sqsubseteq \exists R_1.\exists R_2.\ldots.\exists R_1.\exists R_2.A$. Hence, it is clear that the naive approach of enumerating all possible entailments is not viable for *DL-Lite$_A$*. However, we show that we can preprocess the original ontology by introducing new concepts so that we can resort to checking the existence of chains of a limited length (determined by the role depth in the original ontology), and therefore, limit the number of entailments that we need to check.

Specifically, our algorithm computes the approximation in two steps:

1. We analyze T to understand which *given* complex concepts can give rise to existential chains. Based on such an analysis, we create an intermediate TBox T' in which we introduce new named concepts, one for each discovered complex concept. T' turns out to be a conservative extension of T. In this way, in T' we can detect all chain inclusions of the form $B \sqsubseteq \exists S.B'$, where S is a chain of limited length, and B, B' are basic concepts in T'.
2. We approximate T to a *DL-Lite$_A$* TBox T_A by checking all relevant entailments of T'. Due to the extended alphabet we are able to guarantee completeness in a finite number of entailment checks.

We now elaborate on the details of our approach. In Section 3.1, we describe the construction of T' and in Section 3.2, we provide an algorithm for constructing T_A and show its correctness.

3.1 Preparation: Introducing New Names

Let T be an OWL 2 TBox that we want to approximate. We first construct a new OWL 2 TBox T', which is a conservative extension of T, by introducing named concepts for some of the complex concepts occurring in T. The intuition behind this operation is that giving names to non-*DL-Lite$_A$* concepts that qualify existential chains in T will later allow us to use these 'names' as junctions between several chain inclusions of a restricted length. The difficulty here is to introduce just enough names so that we can guarantee that all existential chains are captured but not more. Thus, we do not give names to the concepts of the form $\exists R_1.\ldots.\exists R_n.A$ occurring on the right-hand side of concept inclusions or of the form $A_1 \sqcup A_2$ on the left-hand side of concept inclusions: these are valid *DL-Lite$_A$* expressions. However, obviously, we need to name any non *DL-Lite$_A$* concept, such as $A_1 \sqcup A_2$ in Example 2.

We define now how to construct T'.

Definition 2. *Let T be an OWL 2 TBox. Then for every axiom $I \in T$ we have that $I \in T'$. Moreover:*
  - *if a concept C of the form $\{a_1, \ldots, a_m\}$, $\forall R.C_1$, $\neg C_1$, or $C_1 \sqcup \cdots \sqcup C_m$ appears in T, then add to T' the concept inclusion axiom $A_C \equiv C$,*
  - *if a concept C of the form $C_1 \sqcap \cdots \sqcap C_m$ appears in T in an inclusion of the form $C' \sqsubseteq \exists S.C$, where S is a chain of roles, then add to T' the axiom $A_C \equiv C$,*
  - *if a concept C of the form $C_1 \sqcap \cdots \sqcap C_m$ or $\exists R_1.\ldots.\exists R_m.C'$ appears in T on the left-hand side of a concept inclusion, then add to T' the axiom $A_C \equiv C$,*
where A_C is a newly introduced concept name.

The following result is an immediate consequence of the fact that in $\mathcal{T}'$ the newly introduced concept names are asserted to be equivalent to the concepts they stand for.

Lemma 1. *Let $\mathcal{T}$ be an OWL 2 TBox and $\mathcal{T}'$ the TBox obtained from $\mathcal{T}$ according to Definition 2. Then $\mathcal{T}'$ is a conservative extension of $\mathcal{T}$.*

Now, we show how we use $\mathcal{T}'$ for computing approximations. The purpose of extending $\mathcal{T}$ with new names is to be able to restrict the attention to a limited number of entailment checks, specifically checks of chain inclusions. We can detect all chain inclusions implied by $\mathcal{T}$ by looking at chain inclusions of the form $B \sqsubseteq \exists S.B'$ with B, B' basic concepts of $\mathcal{T}'$, and S a chain of roles of length limited by the *role depth* in $\mathcal{T}$. The following result establishes a useful property of the constructed TBox $\mathcal{T}'$.

Proposition 1. *Let $\mathcal{T}$ be an OWL 2 TBox of role depth k, and $\mathcal{T}'$ the TBox obtained from $\mathcal{T}$ according to Definition 2. Let further $\mathcal{T} \models B \sqsubseteq \exists S.B'$, where B, B' are basic concepts of $\mathcal{T}$, and S is a chain of roles of $\mathcal{T}$ (of arbitrary length). Then there are chains $S_1, \ldots, S_m$ of roles of $\mathcal{T}$, all of length at most k, and basic concepts $B_0, \ldots, B_m$ of $\mathcal{T}'$ such that $B_0 = B$, $B_m = B'$, $S = S_1 \circ \cdots \circ S_m$, and $\mathcal{T}' \models B_{i-1} \sqsubseteq \exists S_i.B_i$ for $1 \leq i \leq m$.*

3.2 Constructing a *DL-Lite$_{\mathcal{A}}$* TBox

Now, using $\mathcal{T}'$, we show how to construct a *DL-Lite$_{\mathcal{A}}$* TBox $\mathcal{T}_A$ that is a sound and complete approximation of $\mathcal{T}$.

Definition 3. *Let $\mathcal{T}$ be an OWL 2 TBox of role depth k, and $\mathcal{T}'$ the TBox obtained from $\mathcal{T}$ according to Definition 2. Then, the DL-Lite$_{\mathcal{A}}$ approximation of $\mathcal{T}$ is the TBox $\mathcal{T}_A$ constructed as follows. We set $\mathcal{T}_A = \emptyset$ and execute the following sequence of steps:*

1. *for all basic concepts B_1, B_2 of $\mathcal{T}'$, if $\mathcal{T}' \models B_1 \sqsubseteq B_2$, then add $B_1 \sqsubseteq B_2$ to $\mathcal{T}_A$;*
2. *for all basic concepts B_1, B_2 of $\mathcal{T}'$, if $\mathcal{T}' \models B_1 \sqsubseteq \neg B_2$, then add $B_1 \sqsubseteq \neg B_2$ to $\mathcal{T}_A$;*
3. *for all atomic or inverse roles R_1, R_2, if $\mathcal{T}' \models R_1 \sqsubseteq R_2$, then add $R_1 \sqsubseteq R_2$ to $\mathcal{T}_A$;*
4. *for all atomic or inverse roles R_1, R_2 (atomic roles P), if $\mathcal{T}' \models \mathsf{Dis}(R_1, R_2)$ (resp., $\mathsf{Asym}(P)$, $\mathsf{Sym}(P)$), then add $\mathsf{Dis}(R_1, R_2)$ (resp., $\mathsf{Asym}(P)$, $\mathsf{Sym}(P)$) to $\mathcal{T}_A$;*
5. *for all basic concepts B_1, B_2 of $\mathcal{T}'$ and atomic or inverse roles $R_1, \ldots, R_l$, $l < k$, if $\mathcal{T}' \models B_1 \sqsubseteq \exists R_1. \ldots. \exists R_l.B_2$, then add $B_1 \sqsubseteq \exists R_1. \ldots. \exists R_l.B_2$ to $\mathcal{T}_A$;*
6. *for all atomic or inverse roles R, if $\mathcal{T}' \models \mathsf{Fun}(R)$, R does not have proper subroles in $\mathcal{T}_A$, and neither R nor R^- appear in a qualified existential of $\mathcal{T}_A$, then add $\mathsf{Fun}(R)$ to $\mathcal{T}_A$.*

Theorem 1. *Let $\mathcal{T}$ be an OWL 2 TBox and $\mathcal{T}_A$ defined according to Definition 3. Then, $\mathcal{T}_A$ is a sound and complete DL-Lite$_{\mathcal{A}}$ approximation of $\mathcal{T}$.*

The following result establishes the complexity of computing a sound and complete approximation in *DL-Lite$_{\mathcal{A}}$*.

Theorem 2. *Let T be an OWL 2 TBox of role depth k, and Σ the signature of T. Then the algorithm for constructing T_A according to Definition 3 performs a number of OWL 2 entailment checks that is exponential in k and polynomial in the number of elements of Σ.*

3.3   Cleaning the Alphabet: Removing New Named Concepts

The sound and complete approximation constructed as described above requires to extend the original alphabet. In some situations this might not be desirable, e.g., in those cases where all terms in the ontology need to be 'understandable' by the end-user. However, in order to achieve completeness of the approximation w.r.t. existential chains, such an alphabet extension is in general unavoidable. We believe that a good compromise is a limit on the length of the existential chain entailments captured by the approximation. This limit is reasonable since the presence of existential chains becomes relevant mostly in the context of query answering. We have seen that in this context one can safely assume that the length of the queries will not go beyond a certain limit. Moreover, in cases, in which the ontology is used in a running application, it is reasonable to expect that the queries that are going to be asked to the reasoner are known in advance, as is the case in applications built on top of traditional RDBMS engines. With these observations in mind we can define a limit on the chains based on this length and we will be certain that we are sound and complete in the context of our queries/application.

In order to achieve this, we need to modify the construction of the approximated ontology T_A. Let k be the role depth in T, and ℓ the maximum length of queries. Then, we replace Rule 5 in Definition 3 with the following:

5a. if $T' \models B \sqsubseteq \exists R_1.\ldots.\exists R_{l_1}.A_1$, $T' \models A_1 \sqsubseteq \exists R_{l_1+1}.\ldots.\exists R_{l_2}.A_2$, ..., $T' \models A_{m-1} \sqsubseteq \exists R_{l_{m-1}+1}.\ldots.\exists R_n.B'$, with $m \geq 1$, A_i, for $1 \leq i \leq m-1$ new names in T', $l_i < k$, and B, B' basic concepts of T, then $B \sqsubseteq \exists R_1.\ldots.\exists R_n.B'$ is in T_A;

5b. if $T' \models B \sqsubseteq \exists R_1.\ldots.\exists R_{l_1}.A_1$, $T' \models A_1 \sqsubseteq \exists R_{l_1+1}.\ldots.\exists R_{l_2}.A_2$, ..., $T' \models A_{m-1} \sqsubseteq \exists R_{l_{m-1}+1}.\ldots.\exists R_n.A_m$, with A_i, for $1 \leq i \leq m$ new names in T', $l_i < k$, $n \leq \ell$, and B a basic concept of T, then $B \sqsubseteq \exists R_1.\ldots.\exists R_n$ is in T_A.

Theorem 3. *Let T be an OWL 2 TBox of role depth k and T_A the TBox obtained by the Rules 1-4, 5a, 5b, and 6. Then T_A is a sound and complete approximation of T in the languages of DL-Lite$_A$ in which existential chains are limited to the maximum length ℓ.*

4   Implementation

We have implemented the proposed algorithm for *DL-Lite$_A$* approximations, as well as a slightly extended version of the algorithm proposed in [13]. The former is a naive, straightforward implementation of the described technique. It is neither optimized w.r.t. run-time nor w.r.t. the size of the output ontology. These implementations are available at `https://babbage.inf.unibz.it/trac/obdapublic/wiki/approx_semantic_index` in three forms: *(i)* a Java API; *(ii)* a command line application suitable for batch approximations; *(iii)* a plug-in for Protégé 4.0. The core algorithm of these modules can work in two modes:

Simple Approximations. This is the algorithm from [13], extended with the ability to capture qualified existential restrictions of length 1 on the right-hand side of concept inclusions. This mode provides sound approximations, which however are incomplete for *DL-Lite*$_A$ due to the reasons we have explained above;

Complete approximations. This is the algorithm presented in this paper. It is able to construct sound and complete *DL-Lite*$_A$ approximations with a possibly *extended alphabet*, or *DL-Lite*$_A$ approximations in the *original alphabet* that are sound and complete w.r.t. chains of maximum length ℓ. In both modes, we resort to publicly available OWL 2 (or OWL) reasoners, such as Pellet[3], to check for entailments.

Using this implementation we confirmed that indeed even relatively simple ontologies, such as the 'Pizza'[4] ontology, do entail the kind of existential chains that we are interested in. Moreover, as intended, our algorithms are able to capture these chains in practice and we can use our approximations for query answering successfully.

With respect to performance, we found that the exponential nature of the algorithm for complete *DL-Lite*$_A$ approximations does limit the scope of the usage scenarios in which the technique is applicable. Consider that the approximation of the 'Pizza' ontology took approximately 30 minutes to complete on a windows Vista machine equipped with 2Gb of RAM and an Intel Core 2 Duo processor. In usage scenarios, in which the ontology rarely changes, this performance is acceptable, for example, when the domain ontology is finished and the approximated ontology is only recalculated when there are updates on the former. In contrast, in scenarios where the ontology is dynamic, the high cost of our approximations will be problematic.

With respect to the output ontology, we found that the number of generated assertions is not adequate, especially for those scenarios in which the result should be inspected by humans. For example, in the case of the Pizza ontology we found that an approximation keeping the original alphabet and sound and complete with respect to chains of length 3 generated 130,424 axioms.

5   Conclusions

We have shown that providing sound and complete approximations of OWL 2 to *DL-Lite*$_A$ ontologies is non-trivial due to the entailment of axioms involving existential chains of unbounded length. We have also provided an algorithm that is able to compute these approximations by locating the sources of these entailments. The core idea of the algorithm is the introduction of new named concepts that allow us to only consider a limited number of chain inclusions. We have also shown that if the approximation is to be complete, in general it is necessary to extend the alphabet. However, if this extended alphabet is not desirable, then it is possible to maintain the original alphabet as long as we put a limit on the length of the entailed chains. A reasonable reference limit is naturally given by the maximum length of queries that are issued over the ontology.

We will focus on further refinements to the proposed techniques. On the one hand, we aim at devising methods to eliminate redundant axioms, generated by the current

---

[3] http://clarkparsia.com/pellet
[4] http://www.co-ode.org/ontologies/pizza/

algorithm. On the other hand we aim at developing methods for the incremental computation of the approximation in dynamic scenarios, to overcome the long processing times that we have currently observed also with 'average case' ontologies such as the ones in public repositories. We are further working on extending our technique towards obtaining (possibly sound and complete) approximations for TBox and ABox reasoning, and for query answering in OWL 2 (See [14] for a first simple solution in this direction).

References

1. Acciarri, A., Calvanese, D., De Giacomo, G., Lembo, D., Lenzerini, M., Palmieri, M., Rosati, R.: QuONTO: QUerying ONTOlogies. In: Proc. of AAAI 2005, pp. 1670–1671 (2005)
2. Baader, F., Brandt, S., Lutz, C.: Pushing the $\mathcal{EL}$ envelope. In: Proc. of IJCAI 2005, pp. 364–369 (2005)
3. Baader, F., Calvanese, D., McGuinness, D., Nardi, D., Patel-Schneider, P.F. (eds.): The Description Logic Handbook: Theory, Implementation and Applications, 2nd edn. Cambridge University Press, Cambridge (2007)
4. Calvanese, D., De Giacomo, G., Lembo, D., Lenzerini, M., Poggi, A., Rodriguez-Muro, M., Rosati, R.: Ontologies and databases: The DL-Lite approach. In: Tessaris, S., Franconi, E., Eiter, T., Gutierrez, C., Handschuh, S., Rousset, M.-C., Schmidt, R.A. (eds.) Reasoning Web. Semantic Technologies for Information Systems. LNCS, vol. 5689, pp. 255–356. Springer, Heidelberg (2009)
5. Calvanese, D., De Giacomo, G., Lembo, D., Lenzerini, M., Rosati, R.: Tractable reasoning and efficient query answering in description logics: The DL-Lite family. J. of Automated Reasoning 39(3), 385–429 (2007)
6. Cuenca Grau, B., Horrocks, I., Motik, B., Parsia, B., Patel-Schneider, P., Sattler, U.: OWL 2: The next step for OWL. J. of Web Semantics 6(4), 309–322 (2008)
7. Decker, S., Erdmann, M., Fensel, D., Studer, R.: Ontobroker: Ontology based access to distributed and semi-structured information. In: Meersman, R., Tari, Z., Stevens, S. (eds.) Database Semantic: Semantic Issues in Multimedia Systems, ch. 20, pp. 351–370. Kluwer Academic Publishers, Dordrecht (1999)
8. Horrocks, I., Kutz, O., Sattler, U.: The even more irresistible $\mathcal{SROIQ}$. In: Proc. of KR 2006, pp. 57–67 (2006)
9. Kazakov, Y.: $\mathcal{RIQ}$ and $\mathcal{SROIQ}$ are harder than $\mathcal{SHOIQ}$. In: Proc. of KR 2008, pp. 274–284 (2008)
10. Keet, C.M., Alberts, R., Gerber, A., Chimamiwa, G.: Enhancing web portals with Ontology-Based Data Access: the case study of South Africa's Accessibility Portal for people with disabilities. In: Proc. of OWLED 2008 (2008)
11. Krisnadhi, A., Lutz, C.: Data complexity in the $\mathcal{EL}$ family of description logics. In: Proc. of LPAR 2007, pp. 333–347 (2007)
12. Lutz, C., Walther, D., Wolter, F.: Conservative extensions in expressive description logics. In: Proc. of IJCAI 2007, pp. 453–458 (2007)
13. Pan, J.Z., Thomas, E.: Approximating OWL-DL ontologies. In: Proc. of AAAI 2007, pp. 1434–1439 (2007)

14. Pan, J.Z., Thomas, E., Zhao, Y.: Completeness guaranteed approximations for OWL-DL query answering. In: Proc. of DL 2009. CEUR, vol. 477 (2009), ceur-ws.org
15. Poggi, A., Lembo, D., Calvanese, D., De Giacomo, G., Lenzerini, M., Rosati, R.: Linking data to ontologies. J. on Data Semantics X, 133–173 (2008)
16. Poggi, A., Rodríguez-Muro, M., Ruzzi, M.: Ontology-based database access with DIG-Mastro and the OBDA Plugin for Protégé. In: Clark, K., Patel-Schneider, P.F. (eds.) Proc. of OWLED 2008 DC (2008)
17. Ren, Y., Pan, J.Z., Zhao, Y.: Soundness preserving approximation for TBox reasoning. In: Proc. of AAAI 2010 (2010)
18. Selman, B., Kautz, H.: Knowledge compilation and theory approximation. J. of the ACM 43(2), 193–224 (1996)
19. Tserendorj, T., Rudolph, S., Krötzsch, M., Hitzler, P.: Approximate OWL-reasoning with Screech. In: Calvanese, D., Lausen, G. (eds.) RR 2008. LNCS, vol. 5341, pp. 165–180. Springer, Heidelberg (2008)

Reasoning Mechanism for Cardinal Direction Relations

Ah-Lian Kor[1] and Brandon Bennett[2]

[1] Arts Environment and Technology Faculty, Leeds Metropolitan University, Headingley Campus, Leeds LS6 3QS, UK
[2] School of Computing, Leeds University, Leeds LS2 9JT, UK
A.Kor@leedsmet.ac.uk, Brandon@Comp.leeds.ac.uk

Abstract. In the classical Projection-based Model for cardinal directions [6], a two-dimensional Euclidean space relative to an arbitrary single-piece region, a, is partitioned into the following nine tiles: North-West, NW(a); North, N(a); North-East, NE(a); West, W(a); Neutral Zone, O(a);East, E(a); South-West, SW(a); South, S(a); and South-East,SE(a). In our Horizontal and Vertical Constraints Model [9], [10] these cardinal directions are decomposed into sets corresponding to horizontal and vertical constraints. Composition is computed for these sets instead of the typical individual cardinal directions. In this paper, we define several whole and part direction relations followed by showing how to compose such relations using a formula introduced in our previous paper [10]. In order to develop a more versatile reasoning system for direction relations, we shall integrate mereology, topology, cardinal directions and include their negations as well.

Keywords: Cardinal directions, composition table, mereology, topology, qualitative spatial reasoning, vertical and horizontal constraints model.

1   Introduction

Cardinal directions are generally used to describe relative positions of objects in large-scale spaces. The two classical models for reasoning about cardinal direction relations are the cone-shaped and projection-based models [6] where the latter forms the basis of our Horizontal and Vertical Constraints Model.

Composition tables are typically used to make inferences about spatial relations between objects. Work has been done on the composition of cardinal direction relations of points [6], [7], [13] which is more suitable for describing positions of point-like objects in a map. Goyal et. al [8] used the direction-relation matrix to compose cardinal direction relations for points, lines as well as extended objects. Skiadopoulos et. al [15] highlighted some of the flaws in their reasoning system and thus developed a method for correctly computing cardinal direction relations. However, the set of basic cardinal relations in their model consists of 218 elements which is the set of all disjunctions of the nine cardinal directions. In our Horizontal and Vertical Constraints Model, the nine cardinal directions are partitioned into sets based on horizontal and vertical constraints. Composition is computed for these sets instead of the individual cardinal directions, thus helping collapse the typical disjunctive relations into smaller sets. We employed the constraint network of binary direction relations to evaluate the consistency of the composed set relations. Ligozat

D. Dicheva and D. Dochev (Eds.): AIMSA 2010, LNAI 6304, pp. 32–41, 2010.

[11] has worked on constraint networks for the individual tiles but not on their corresponding vertical and horizontal sets. Some work relating to hybrid cardinal direction models has been done. Escrig et.al [5] and Clementini et.al [2] combined qualitative orientation combined with distance, while Sharma et. al [14] integrated topological and cardinal direction relations. In order to come up with a more expressive model for direction relations, have extended existing spatial language for directions by integrating mereology, topology, and cardinal direction relations. Additionally, to develop a more versatile reasoning system for such relations, we have included their negations as well.

2 Cardinal Directions Reasoning Model

2.1 Projection-Based Model

In the Projection-based Model for cardinal directions [6], a two-dimensional Euclidean space of an arbitrary single-piece region, a, is partitioned into nine tiles. They are North-West, NW(a); North, N(a); North-East, NE(a); West, W(a); Neutral Zone, O(a); East, E(a); South-West, SW(a); South, S(a); and South-East, SE(a). In this paper, we only address finite regions which are bounded. Thus every region will have a minimal bounding box with specific minimum and maximum x (and y) values (in Table 1). The boundaries of the minimal bounding box of a region a is illustrated in Figure 1. The definition of the nine tiles in terms of the boundaries of the minimal bounding box is listed below. Note that all the tiles are regarded as closed regions. Thus neighboring tiles share common boundaries but their interior will remain disjoint.

Table 1. Definition of Tiles

Definition of tiles	
$N(a) \equiv \{\langle x,y \rangle \mid Xmin(a) \leq x \leq Xmax(a) \wedge y \geq Ymax(a)\}$	$SW(a) \equiv \{\langle x,y \rangle \mid x \leq Xmin(a) \wedge y \leq Ymin(a)\}$
$NE(a) \equiv \{\langle x,y \rangle \mid x \geq Xmax(a) \wedge y \geq Ymax(a)\}$	$E(a) \equiv \{\langle x,y \rangle \mid x \geq Xmax(a) \wedge Ymin(a) \leq y \leq Ymax(a)\}$
$NW(a) \equiv \{\langle x,y \rangle \mid x \leq Xmin(a) \wedge y \geq Ymax(a)\}$	$W(a) \equiv \{\langle x,y \rangle \mid x \leq Xmin(a) \wedge Ymin(a) \leq y \leq Ymax(a)\}$
$S(a) \equiv \{\langle x,y \rangle \mid Xmin(a) \leq x \leq Xmax(a) \wedge y \leq Ymin(a)\}$	$O(a) \equiv \{\langle x,y \rangle \mid Xmin(a) \leq x \leq Xmax(a) \wedge Ymin(a) \leq y \leq Ymax(a)\}$
$SE(a) \equiv \{\langle x,y \rangle \mid x \geq Xmax(a) \wedge y \leq Ymin(a)\}$	

Table 2. Definitions for the Horizontal and Vertical Constraints Model

Definitions for the Horizontal and Vertical Constraints Model	
WeakNorth(a) is the region that covers the *tiles* NW(a), N(a), and NE(a); WeakNorth(a) ≡ NW(a) ∪ N(a) ∪ NE(a).	WeakWest(a) is the region that covers the *tiles* SW(a), W(a), and NW(a); WeakWest(a) ≡ SW(a) ∪ W(a) ∪ NW(a).
Horizontal(a) is the region that covers the *tiles* W(a), O(a), and E(a); Horizontal(a) ≡ W(a), O(a), and E(a).	Vertical(a) is the region that covers the *tiles* S(a), O(a), and N(a); Vertical(a) ≡ S(a) ∪ O(a) ∪ N(a).
WeakSouth(a) is the region that covers the *tiles* SW(a), S(a), and SE(a); WeakSouth(a) ≡ SW(a) ∪ S(a) ∪ SE(a).	WeakEast(a) is the region that covers the *tiles* NE(a), E(a), and SE(a); WeakEast(a) ≡ NE(a) ∪ E(a) ∪ SE(a).

2.2 Horizontal and Vertical Constraints Model

In the Horizontal and Vertical Constraints Model [9, 10], the nine tiles are collapsed into six sets based on horizontal and vertical constraints as shown in Figure 1. The definitions of the partitioned regions are shown in Table 2 and the nine cardinal direction *tiles* can be defined in terms of horizontal and vertical sets (see Table 3).

Table 3. Definition of the tiles in terms of Horizontal and Vertical Constraints Sets

NW(a)≡ WeakNorth(a) ∩ WeakWest(a) N(a)≡ WeakNorth(a) ∩ Vertical(a) NE(a)≡ WeakNorth(a) ∩ WeakEast(a)	W(a)≡ Horizontal(a) ∩ WeakWest(a) O(a)≡ Horizontal(a) ∩ Vertical(a) E(a)≡ Horizontal(a) ∩ WeakEast(a)	SW(a)≡ WeakSouth(a) ∩ WeakWest(a) S(a)≡ WeakSouth(a) ∩ Vertical(a) SE(a)≡ WeakSouth(a) ∩ WeakEast(a)

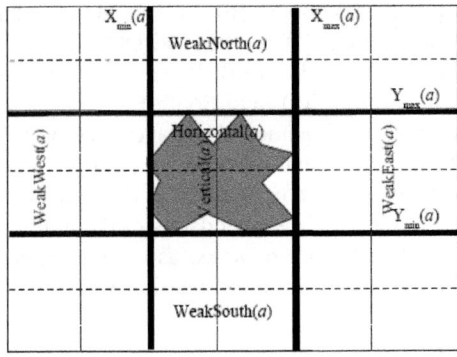

Fig. 1. Horizontal and Vertical Sets of *Tiles*

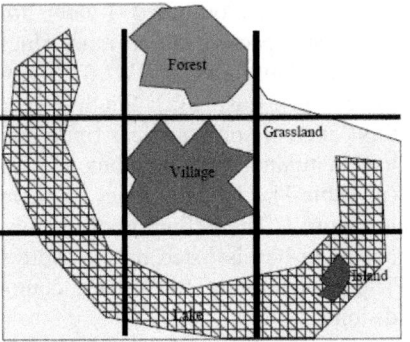

Fig. 2. Spatial Relationships between regions

2.3   RCC Binary Relations

In this paper, we shall use the RCC-5 [3] JPED binary topological relations for regions. They are: PP(x, y) which means 'x is a proper part of y'; PPi(x, y) which means 'y is a proper part of x'; EQ(x, y) which means 'x is identical with y'; PO(x, y) which means 'x partially overlaps y'; DR(x,y) which means 'x is discrete from y'. The relations EQ, PO, and DR are symmetric while the rest are not. PPi is also regarded as the inverse of PP. However, in this paper, the relationship PPi will not be considered because all tiles (except for tile O) are unbounded.

2.4   *Whole* or *Part* Cardinal Direction Relations

In our previous paper [8], we created an expressive hybrid mereological, topological and cardinal direction relation model. Here we shall improve the definitions of $A_R(b, a)$ which means that the *whole* destination region, b, is in the tile $R(a)$ while $P_R(b, a)$ means that *part* of b is in tile $R(a)$.

Cardinal direction relations defined in terms of *tiles*

In this section, we shall introduce several terms to extend the existing spatial language for cardinal directions to facilitate a more versatile reasoning about their relations. We shall use RCC-5 relations to define three categories of direction relations: *whole*, *part*, and *no part*. $A_N(b, a)$ means *whole* of *b* is in the North *tile* of *a*: $A_N(b, a)$ $\equiv PP(b, N(a)) \vee EQ(b, N(a))$

Here we adopt the natural language meaning for the word *part* which is 'some but not all'. $P_N(b, a)$ represents *part* of *b* is in the North tile of *a*. When *part* of *b* is in the North *tile* of *a*, this means that *part* of *b* covers the North *tile* and possibly one or more of the complementary *tiles* of North.

$$P_N(b, a) \equiv PO(b, N(a))$$

We shall use the Skiadopoulos et. al [2004] definition of multi-tile cardinal direction relations. As an example, if *part* of *b* is in the North *tile* and the remaining *part* of *b* is in the NorthWest *tile* of *a* (or in other words, *part* of *b* is only in the North and NorthWest *tiles* of *a*) and vice versa, then its representation is

$$P_{N:NW}(b, a) \equiv PO(b, N(a)) \wedge PO(b, NW(a)) \wedge DR(b, NE(a)) \wedge DR(b, W(a)) \wedge DR(b, O(a))$$
$$\wedge DR(b, E(a)) \wedge DR(b, SE(a)) \wedge DR(b, S(a)) \wedge DR(b, SW(a))$$
$$\text{or } P_{N:NW}(b, a) \equiv A_N(b1, a) \wedge A_{NW}(b2, a) \text{ where } b = b1 \cup b2.$$

$\Phi_N(b, a)$ means *no part* of *b* is in the North *tile* of *a*. When b has no part in the North tile of a, this means that b could be in one or more the complementary tiles of North so

$$\Phi_N(b, a) \equiv DR(b, N(a))$$

If *no part* of *b* is in North and Northwest tiles (or in other words, *b* could only be in one or more of the complementary *tiles* of North and Northwest), then the representation is

$$\Phi_{N:NW}(b, a) \equiv DR(b, N(a)) \wedge DR(b, NW(a))$$

Assume U = {N, NW, NE, O, W, E, S, SW, SE}. The general definition of the following direction relations are in Table 4:

Table 4. Definition of direction relations

D1. $A_R(b, a) \equiv PP(b, R(a)) \vee EQ(b, R(a))$ where $R \in U$	**D4.** $\Phi_R(b, a) \equiv DR(b, R(a))$ where $R \in U$
D2. $P_R(b, a) \equiv PO(b, R(a))$ where $R \in U$	**D5.** $\Phi_{R1:...:Rn}(b, a) \equiv DR(b, R1(a))$ $\wedge ... \wedge DR(b, Rn(a))$ where R,...,Rn∈ U and 1 $\leq n \leq 9$.
D3.1. $P_{R1:...:Rn}(b,a) \equiv PO(b, R1(a))$ $\wedge ... \wedge PO(b, Rn(a)) \wedge DR(b, R'(a))$ where R1,...,Rn∈ U, $1 \leq n \leq 9$ and R'∈ U - {R1,...,Rn}	**D6.** $\neg A_R(b, a) \equiv \Phi_R(b, a) \vee P_R(b, a)$ where R ∈ U.
D3.1. $P_{R1:...:Rn}(b,a) \equiv A_{R1}(b1, a) \wedge ... \wedge A_{Rn}(bn, a)$ where $b=b1 \cup ... \cup bn$, where R1,...,Rn∈ U and $1 \leq n \leq 9$	**D7.** $\neg P_R(b,a) \equiv A_R(b,a) \vee \Phi_R(b,a)$ where R ∈ U.
	D8. $\neg \Phi_R(b,a) \equiv A_R(b,a) \vee P_R(b,a)$ where R∈ U.

Negated cardinal direction relations defined in terms of *tiles*

In this section, we shall define three categories of negated cardinal direction relations: not whole, not part, and not no part. Negated direction relations could be used when reasoning with incomplete knowledge. Assume B is a set of the relations, {PP, EQ, PO, DR}. $\neg A_N(b, a)$ means that b is not wholly in North tile of a. It is represented by:

$$\neg A_N (b,\ a) \equiv \neg[PP(b,\ N(a)) \vee EQ(b,\ N(a))]$$

Use De Morgan's Law and we have $\neg A_N(b,\ a) \equiv \neg PP(b,\ N(a)) \wedge \neg EQ(b,\ N(a))$

The complement of PP and EQ is {PO, DR} so the following holds:

$$\neg A_N(b,\ a) \equiv [PO(b,\ N(a))] \vee DR(b,\ N(a))$$

Use **D2** and **D4** and we have *part* of b is **not** or *no part* of b is in North *tile* of a so

$$\neg A_N(b,\ a) \equiv \Phi_N(b,\ a) \vee P_N(b,\ a)$$

$\neg P_N(b,\ a)$ means b is not partly in North tile of a so $\neg P_N(b,\ a) \equiv \neg PO(b,\ N(a))$

The complement of PO is {PP, EQ, DR} so the following holds:

$$\neg P_N(b,\ a) \equiv [PP(b,\ N(a)) \vee EQ(b,\ N(a))] \vee DR(b,\ N(a))$$

Use **D1**, **D4**, we have $\neg P_N(b,\ a) \equiv A_N(b,\ a) \vee \Phi_N(b,\ a)$

$\neg \Phi_N(b,\ a)$ means not no part of b is in the North tile of a. Thus

$\neg \Phi_N(b,\ a) \equiv \neg DR(b,\ N(a))$ or $\neg \Phi_N(b,\ a) \equiv [PP(b,\ N(a)) \vee EQ(b,\ N(a))] \vee PO(b,\ N(a))$

Use **D1**, **D2** and **D4**, we have the following: $\neg \Phi_N(b,\ a) \equiv A_N(b,a) \vee P_N(b,\ a)$

Assume U = {N, NW, NE, O, W, E, S, SW, SE}. The general definition of the *negated* direction relations are in Table 4. Here we shall give an example to show how some of the aforementioned *whole-part* relations could be employed to describe the spatial relationships between regions. In Figure 2, we shall take the village as the referent region while the rest will be destination regions. The following is a list of possible direction relations between the village and the other regions in the scene:

- A_N(forest,village): The whole forest is in the North tile of the village and A_{SE}(island,village): the whole island is in the SouthEast tile of the village.
- $P_{NW:W:SW:S:SE:E}$(lake,village): Part of the lake is in the NorthWest, West, SouthWest, South, SouthEast and East tiles of the village.
- $\Phi_{O:N:NE}$(lake,village): This is another way to represent the direction relationship between the lake and village. t means no part of the lake is in the Neutral, North and NorthEast tiles of the village.
- $P_{O:N:NE:NW:W:SW:S:SE:E}$(grassland,village): Part of the grassland is in all the tiles of the village.

Next we shall show how negated direction relations could be used to represent incomplete knowledge about the direction relations between two regions. Assume that we have a situation where the hills are not wholly in the North tile of the village. We can interpret such incomplete knowledge using D6, part or no part direction relations: P_N(hills, village)$\vee \Phi_N$(hills,village). In other words, either there is no hilly region is in the North tile of the village or part of the hilly region covers the North tile of the village. If we are given this piece of information 'it is not true that no part of the lake lies in the North tile of the village', we shall use D8 to interpret it. Thus we have the following possible relations: A_N(lake,village)$\vee P_N$(lake,village). This means that the whole or only part of the lake is in the North tile of the village.

2.5 Cardinal Direction Relations Defined in Terms of Horizontal or Vertical Constraints

The definitions of cardinal direction relations expressed in terms of horizontal and vertical constraints are similar to those shown in the previous section (**D1** to **D8**). The only difference is that the universal set, U is {WeakNorth (WN), Horizontal (H), WeakSouth (WS), WeakEast (WE), Vertical (V), WeakWest (WW)}.

Whole and **part** cardinal direction relations defined in terms of horizontal and vertical constraints

In this section, we use examples to show how *whole* and *part* cardinal direction relations could be represented in terms of horizontal and vertical constraints. We shall exclude the inverse and negated relations for reasons that will be given in the later part of this paper. We shall use abbreviations {*WN, H, WS*} for {*WeakNorth, Horizontal, WeakSouth*} and {*WE, V, WW*} for {*WeakEast, Vertical, WeakWest*} respectively.

D9. $A_N(b, a) \equiv A_{WN}(b, a) \wedge A_V(b, a)$
D10. $P_N(b, a) \equiv P_{WN}(b, a) \wedge P_V(b, a)$
D11. $P_{N:NW}(b, a) \equiv A_N(b_1, a) \wedge A_{NW}(b_2, a) \equiv [A_{WN}(b_1, a) \wedge A_V(b_1, a)] \wedge [A_{WN}(b_2, a) \wedge A_{WW}(b_2, a)]$ where $b = b_1 \cup b_2$
D12. $\Phi_M(b, a) \equiv \Phi_{WM}(b, a) \wedge \Phi_V(b, a)$
D13. $\Phi_{N:NW}(b,a) \equiv \Phi_N(b,a) \wedge \Phi_N(b,a) \equiv [\Phi_{WN}(b,a) \wedge \Phi_V(b,a)] \wedge [\Phi_{WN}(b, a) \wedge \Phi_{WW}(b, a)]$

Next we shall use the *part* relation as a primitive for the definitions of the *whole* and *no part* relations. Once again assume U = {N, NW, NE, O, W, E, S, SW, SE}.
D14.1. $A_R(b, a) \equiv P_R(b, a) \wedge [\neg P_{R1}(b, a) \wedge \neg P_{R2}(b, a) \wedge ... \wedge \neg P_{Rm}(b, a)]$ where R∈ U, Rm ∈ U – {R} (which is the complement of R), and $1 \le m \le 8$. As an example,
$A_N(b,a) \equiv P_N(b,a) \wedge [\neg P_{NE}(b,a) \wedge \neg P_{NW}(b,a) \wedge \neg P_W(b,a) \wedge \neg P_O(b,a) \wedge \neg P_E(b,a) \wedge \neg P_{SW}(b,a) \wedge \neg P_S(b,a) \wedge \neg P_{SE}(b,a)]$
D14.2. $A_{HR}(b, a) \equiv P_{HR}(b, a) \wedge [\neg P_{HR1}(b, a) \wedge ... \wedge \neg P_{HRn}(b, a)]$ where HR ∈ {WN, H, WS}, HRm is the complement of HR, and $1 \le n \le 3$. As an example, $A_{WN}(b, a) \equiv P_{WN}(b, a) \wedge [\neg P_H(b, a) \wedge \neg P_{WS}(b, a)]$
D14.3. $A_{VR}(b, a) \equiv P_{VR}(b, a) \wedge [\neg P_{VR1}(b, a) \wedge ... \wedge \neg P_{VRn}(b, a)]$ where VR∈ {WW, V, WE}, VRm is the complement of VR), and $1 \le n \le 3$. As an example, $A_{WW}(b, a) \equiv P_{WW}(b, a) \wedge [\neg P_V(b, a) \wedge \neg P_{WE}(b, a)]$
D15.1. $\Phi_R(b, a) \equiv \neg P_R(b, a) \wedge [P_{R1}(b, a) \vee P_{R2}(b, a) \vee ... \vee P_{Rm}(b, a)]$ where R∈ U, Rm ∈ U – {R}, and $1 \le m \le 8$. As an example,
$\Phi_N(b,a) \equiv \neg P_N(b,a) \wedge [P_{NE}(b, a) \vee P_{NW}(b,a) \vee P_W(b,a) \vee P_O(b,a) \vee P_E(b,a) \vee P_{SW}(b,a) \vee P_S(b,a) \vee P_{SE}(b,a)]$
D15.2. $\Phi_{HR}(b, a) \equiv \neg P_{HR}(b, a) \wedge [P_{HR1}(b, a) \vee P_{HR2}(b, a)]$ where HR∈ {WN,H, WS}, while HR1 and HR2 constitute its complement. As an example, $\Phi_{WN}(b, a) \equiv \neg P_{WN}(b, a) \wedge [P_H(b, a) \vee P_{WS}(b, a)]$.
D15.3. $\Phi_{VR}(b, a) \equiv \neg P_{VR}(b, a) \wedge [P_{VR1}(b, a) \vee P_{VR2}(b, a)]$ where VR∈ {WW,V, WE}, while VR1 and VR2 constitute its complement. As an example, $\Phi_{WW}(b, a) \equiv \neg P_{WW}(b, a) \wedge [P_V(b, a) \vee P_{WE}(b, a)]$.

3 Composition Table for Cardinal Directions

Ligozat (1988) obtained the outcome of the composition of all the nine tiles in a *Projection Based Model* for point objects by composing the constraints {<, =, >}. However, our composition tables (Tables 5 and 6) are computed using the vertical and horizontal constraints of the sets of direction relations. We shall abstract several composition rules in Table 5. Similar rules apply to Table 6. Assume U is { A_{WE}, A_V, A_{WW} }. *WeakEast(WE)* is considered the converse of *WeakWest (WW)* and vice versa.
Rule 1 (Identity Rule): $R \wedge R = R$ where $R \in U$.
Rule 2 (Converse Rule): $S \wedge S' = U$, $A_V \wedge S = P_V \vee P_S$ where $S \in \{A_{WE}, A_{WW}\}$ and S' is its converse.

Here we shall introduce several axioms that are necessary for the direction reasoning mechanism. In the next section we shall show how to apply these axioms and some logic rules for making inferences about direction relations.

Axiom 1. $A_R(b_1,a) \wedge A_R(b_2,a) \wedge \ldots \wedge A_R(b_k,a) \rightarrow A_R(b,a)$ where $R \in U$, $1 \leq k \leq 9$ and
$b_1 \cup b_2 \cup \ldots \cup b_k = b$

Axiom 2. $A_{R1}(b_1,a) \wedge A_{R2}(b_2,a) \wedge \ldots \wedge A_{Rn}(b_k,a) \rightarrow P_{R1:R2:\ldots:Rn}(b,a)$ where $Rn \in U$,
$1 \leq k \leq 9$ and $b_1 \cup b_2 \cup \ldots \cup b_k = b$

Axiom 3. $P_R(c_k,a) \wedge PP(c_k,c) \rightarrow P_R(c,a)$ where $R \in U$, and $1 \leq k \leq 9$

Axiom 4. $[P_{R1}(c_1,a) \wedge PP(c_1,c)] \wedge [P_{R2}(c_2,a) \wedge PP(c_2,c)] \wedge \ldots \wedge$
$[P_{Rk}(c_k,a) \wedge PP(c_k,c)] \rightarrow P_{R1:R2:\ldots:Rk}(c,a)$ where $1 \leq k \leq 9$, and $Rk \in U$.

Axiom 5. $A_R(c_k,a) \wedge PP(c_k,c) \rightarrow P_R(c,a)$ where $R \in U$, and $1 \leq k \leq 9$

Axiom 6. $\neg\{[P_{WW}(c_1,a) \wedge PP(c_1,c)] \wedge [P_{WE}(c_2,a) \wedge PP(c_2,c)]\}$ where $c_1 \cup c_2 = c$
(because c is a single connected piece)

Axiom 7. $\neg\{[P_{WN}(c_1,a) \wedge PP(c_1,c)] \wedge [P_{WS}(c_2,a) \wedge PP(c_2,c)]\}$ where $c_1 \cup c_2 = c$
(because c is a single connected piece)

3.1 Formula for Computation of Composition

In our previous paper [10], we introduced a formula (obtained through case analyses) for computing the composition of cardinal direction relations. Here we shall modify the notations used for easy comprehension. Skiadopoulos et. al [15] introduced additional concepts such as rectangular versus nonrectangular direction relations, bounding rectangle, westernmost (etc...) to facilitate the composition of relations. They have separate formulae for the composition of rectangular and non-rectangular regions. However, in this paper we shall apply one formula for the composition of all types of direction relations. The basis of the formula is to first consider the direction relation between a and each individual part of b followed by the direction relation between each individual part of b and c. Assume that the region b covers one or more tiles of region a while region c covers one or more tiles of b. The direction relation between a and b is $R(b,a)$ while the direction relation between b and c is $S(c,b)$. The composition of direction relations could be written as follows:

$$R(b,a) \wedge S(c,b)$$

Firstly, establish the direction relation between a and each individual part of b.

$$R(b,a) \wedge S(c,b) \equiv [R_1(b_1,a) \wedge R_2(b_2,a) \ldots \wedge R_k(b_k,a)] \wedge [S(c,b)] \ldots\ldots\ldots \text{where } 1 \leq k \leq 9 \ldots\ldots\ldots\ldots\ldots\ldots(1)$$

Consider the direction relation of each individual part of b and c. Equation (1) becomes: $[R_1(b_1,a) \wedge S_1(c,b_1)] \wedge [R_2(b_2,a) \wedge S_2(c,b_2)] \wedge \ldots \wedge [R_k(b_k,a) \wedge S_k(c,b_k)] \ldots$ where $1 \leq k \leq 9 \ldots\ldots(2)$

3.2 Composition of Cardinal Direction Relations

Previously we have grouped the direction relations into three categories namely: whole, part, and no part. If we include their respective inverses and negations, there will be a total of 9 types of direction relations. However, we do not intend to delve into the composition of inverse and negated relations due to the high level of uncertainty involved. Typically, the inferences drawn would consist of the universal set of tiles, which is not beneficial. In this paper, we shall demonstrate several types of composition. The type of composition shown in this part of the paper involves the composition of vertical and horizontal sets which is different from Skiadopoulos et. al's work [15] involving the composition of individual tiles. Use Tables 5 and 6 to obtain the outcome of each composition. The meaning of the two following notations $U_V(c,a)$ and $U_H(c,a)$ are in Tables 5 and 6.

Table 5. Composition of Vertical Set Relations

		WeakEast $A_{WE}(c,b)$	Vertical $A_V(c,b)$	WeakWest $A_{WW}(c,b)$
WeakEast	$A_{WE}(b,a)$	$A_{WE}(c,a)$	$A_{WE}(c,a)$	$U_V(c,a)$
Vertical	$A_V(b,a)$	$P_{WE}\vee P_V(c,a)$	$A_V(c,a)$	$P_{WW}\vee P_V(c,a)$
WeakWest	$A_{WW}(b,a)$	$U_V(c,a)$	$A_{WW}(c,a)$	$A_{WW}(c,a)$

Note: $U_V(c,a)=[P_{WE}(c,a)\vee P_V(c,a)\vee P_{WW}(c,a)]$. Therefore the possible set of relations is $\{[A_{WE}(c,a), A_V(c,a), A_{WW}(c,a), P_{WE:V:WW}(c,a), P_{WE:V}(c,a), P_{WW:V}(c,a)]\}$.

Table 6. Composition of Horizontal Set Relations

		WeakNorth $A_{WN}(c,b)$	Horizontal $A_H(c,b)$	WeakSouth $A_{WS}(c,b)$
WeakNorth	$A_{WN}(b,a)$	$A_{WN}(c,a)$	$A_{WN}(c,a)$	$U_H(c,a)$
Horizontal	$A_H(b,a)$	$P_{WN}\vee P_H(c,a)$	$A_H(c,a)$	$P_{WS}\vee P_H(c,a)$
WeakSouth	$A_{WS}(b,a)$	$U_H(c,a)$	$A_{WS}(c,a)$	$A_{WS}(c,a)$

Note: $U_H(c,a)=[P_{WN}(c,a)\vee P_H(c,a)\vee P_{WS}(c,a)]$. Therefore the possible set of relations is $\{A_{WN}(c,a), A_H(c,a), A_{WS}(c,a), P_{WN:H:WS}(c,a), P_{WN:H}(c,a), P_{WS:H}(c,a)]\}$.

Example 1

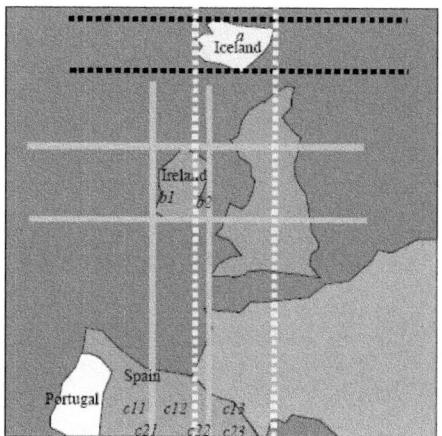

In Figure 3, *part* of Ireland (*b*) is only in the South and SouthWest *tiles* of Iceland (*a*) while the *part* of Spain (*c*) is in the SouthWest, South and SouthEast *tiles* of Ireland. We have to make an inference about the direction relation between Iceland and Spain. We shall represent the information as:

$P_{SW:S}(\text{Ireland},\text{Iceland})\wedge P_{SW:S:SE}(\text{Spain},\text{Ireland})$

Use the abbreviations *a*, *b*, *c* to represent Iceland, Ireland, and Spain respectively. The above expression is written as:

Fig. 3. Spatial relationships among regions in Europe

$P_{SW:S}(b, a)\wedge P_{SW:S:SE}(c, b)\ldots\ldots\ldots\ldots(3a)$

Firstly, establish the direction relation between *a* and each individual part of *b*. Use **D3** and expression in (3a) becomes

$[A_{SW}(b_1,a) \wedge A_S(b_2,a)]\wedge [P_{SW:S:SE}(c, b)]\ldots\ldots\ldots(3b)$

Use the extended boundaries of part region b_1 to partition *c*. As depicted in Figure 3, *c* is divided into 3 subregions (c_{11}, c_{12}, and c_{13}). Establish direction relations between these regions and b_1. We have $A_{SW}(c_{11},b_1), A_S(c_{12},b_1)$, and $A_{SE}(c_{13},b_1)$. Repeat the same procedure for b_2 and we have the following direction relations between b_2 and its corresponding subregions:

$A_{SW}(c_{21},b_2), A_S(c_{22},b_2)$ and $A_{SE}(c_{23},b_2)$

Expression (3b) becomes:

$[A_{SW}(b_1,a) \wedge A_S(b_2,a)] \wedge \{[A_{SW}(c_{11},b_1) \wedge A_S(c_{12},b_1) \wedge A_{SE}(c_{13},b_1)] \wedge [A_{SW}(c_{21},b_2) \wedge A_S(c_{22},b_2) \wedge A_{SE}(c_{23},b_2)]\}...(3c)$

Apply formula (2) into expression (3c) and we have

$\{[A_{SW}(b_1,a) \wedge A_{SW}(c_{11},b_1)] \wedge [A_{SW}(b_1,a) \wedge A_S(c_{12},b_1)] \wedge [A_{SW}(b_1,a) \wedge A_{SE}(c_{13},b_1)]\} \wedge$
$\{[A_S(b_2,a) \wedge A_{SW}(c_{21},b_2)] \wedge [A_S(b_2,a) \wedge A_S(c_{22},b_2)] \wedge [A_S(b_2,a) \wedge A_{SE}(c_{23},b_2)]\}...........(3d)$

We shall compute the vertical and horizontal constraints separately and apply formulae similar to **D9**.

Composition of Horizontal Constraints

$[[A_{WS}(b_1,a) \wedge A_{WS}(c_{11},b_1)] \wedge [A_{WS}(b_1,a) \wedge A_{WS}(c_{12},b_1)] \wedge [A_{WS}(b_1,a) \, A_{WS}(c_{13},b_1)]] \wedge$
$[[A_{WS}(b_2,a) \wedge A_{WS}(c_{21},b_2)] \wedge [A_{WS}(b_2,a) \wedge A_{WS}(c_{22},b_2)] \wedge [A_{WS}(b_2,a) \, A_{WS}(c_{23},b_2)]]$

Use Table 6 and we have

$[A_{WS}(c_{11},a) \wedge A_{WS}(c_{12},a) \wedge A_{WS}(c_{13},a)] \wedge [A_{WS}(c_{21},a) \wedge A_{WS}(c_{22},,a) \wedge A_{WS}(c_{23},a]$

However, as shown earlier, $c_{11} \cup c_{12} \cup c_{13} = c$ and $c_{21} \cup c_{22} \cup c_{23} = c$. Use Axiom 1 and the modus ponens inference rule $(P \rightarrow Q; P,$ therefore $Q)$ and the above expression becomes $A_{WS}(c,a) \wedge A_{WS}(c,a)$ which equals $A_{WS}(c,a)$.

Composition of Vertical Constraints

$[[A_{WW}(b_1,a) \wedge A_{WW}(c_{11},b_1)] \wedge [A_{WW}(b_1,a) \wedge A_V(c_{12},b_1)] \wedge [A_{WW}(b_1,a) \wedge A_{WE}(c_{13},b_1)]] \wedge$
$[[A_V(b_2,a) \wedge A_{WW}(c_{21},b_2)] \wedge [A_V(b_2,a) \wedge A_V(c_{22},b_2)] \wedge [A_V(b_2,a) \wedge A_{WE}(c_{23},b_2)]]$

Use Table 5 and we have

$[A_{WW}(c_{11},a) \wedge A_{WW}(c_{12},a) \wedge U_V(c_{13},a)] \wedge [(P_{WW} \vee P_V)(c_{21},a) \wedge A_V(c_{22},,a) \wedge (P_{WE} \vee P_V)(c_{23},a)]$

Use **Axiom 5**, **D15.3**, and the expression becomes:

$\{P_{WW}(c, a) \wedge P_{WW}(c, a) \wedge [(P_{WW} \vee P_V \vee P_{WE})(c, a)]\} \wedge \{(P_{WW} \vee P_V)(c, a)] \wedge P_V(c, a) \wedge [(P_{WE} \vee P_V)(c, a)]\}$

Use **Axiom 6**, distributivity, idempotent, and absorption rules to compute the first part of the expression

$\{P_{WW}(c, a) \wedge P_{WW}(c, a) \wedge [(P_{WW} \vee P_V \vee P_{WE})(c, a)]\}$
$= \{P_{WW}(c, a) \wedge [(P_{WW} \vee P_V \vee P_{WE})(c, a)]\}$
$= [P_{WW}(c, a) \wedge P_{WW}(c, a)] \vee [P_{WW}(c, a) \wedge P_V(c, a)] \vee [P_{WW}(c, a) \wedge P_{WE}(c, a)]$
$= [P_{WW}(c, a)] \vee [P_{WW}(c, a) \wedge P_V(c, a)]$
$= P_{WW}(c, a)...(4a)$

Use absorption rule to compute the second part of the expression

$\{[(P_{WW} \vee P_V)(c, a)] \wedge P_V(c, a) \wedge [(P_{WE} \vee P_V)(c, a)]\}$
$= P_V(c, a) \wedge [(P_{WE}(c, a) \vee P_V(c, a)]...............(4b)$

Combine the computed expressions in (4a) and (4b) and apply distributivity rule:

$P_{WW}(c, a) \wedge P_V(c, a) \wedge [(P_{WE}(c, a) \vee P_V(c, a)]$
$= [P_{WW}(c, a) \wedge P_V(c, a) \wedge (P_{WE}(c, a)] \vee [P_{WW}(c, a) \wedge P_V(c, a)]$

The outcome of the composition could be written as

$$A_{WS}(c,a) \wedge [P_{WW:V:WE}(c, a) \vee P_{WW:V}(c, a)]$$

which means c covers the SouthWest, South and SouthEast or SouthWest and South *tiles* of a. And this is confirmed by the direction relation between Iceland and Spain depicted in Figure 3.

4   Conclusion

In this paper, we have developed and formalised *whole part* cardinal direction relations to facilitate more expressive scene descriptions. We have also introduced a refined formula for computing the composition of such type of binary direction relations. Additionally, we have shown how to represent constraint networks in terms of weak cardinal direction relations. We demonstrated how to employ them for evaluating the consistency of composed weak direction relations.

References

1. Cicerone, S., Di Felice, P.: Cardinal Directions between Spatial Objects: The Pairwise consistency Problem. Information Sciences – Informatics and Computer Science: An International Journal 164(1-4), 165–188 (2004)
2. Clementini, E., Di Felice, P., Hernandez: Qualitative Representation and Positional Information. Artificial Intelligence 95, 315–356 (1997)
3. Cohn, A.G., Bennett, B., Gooday, J., Gotts, N.M.: Qualitative Spatial Representation and Reasoning with the Region Connection Calculus (1997)
4. Egenhofer, M.J., Sharma, J.: Assessing the Consistency of Complete and Incomplete Topological Information. Geographical Systems 1(1), 47–68 (1993)
5. Escrig, M.T., Toledo, F.: A framework based on CLP extended with CHRS for reasoning with qualitative orientation and positional information. JVLC 9, 81–101 (1998)
6. Frank, A.: Qualitative Spatial Reasoning with Cardinal Directions. JVLC (3), 343–371 (1992)
7. Freksa, C.: Using orientation information for qualitative spatial reasoning. In: Proceedings of International Conference GIS – From Space to Territory, Theories and Methods of Spatio-Temporal Reasoning in Geographic Space, pp. 162–178 (1992)
8. Goyal, R., Egenhofer, M.: Consistent Queries over Cardinal Directions across Different Levels of Detail. In: 11th International Workshop on Database and Expert Systems Applications, Greenwich, UK (2000)s
9. Kor, A.L., Bennett, B.: Composition for cardinal directions by decomposing horizontal and vertical constraints. In: Proceedings of AAAI 2003 Spring Symposium on Spatial and Temporal Reasoning (2003a)
10. Kor, A.L., Bennett, B.: An expressive hybrid Model for the composition of cardinal directions. In: Proceedings of IJCAI 2003 Workshop on Spatial and Temporal Reasoning, Acapulco, Mexico, August 8-15 (2003b)
11. Ligozat, G.: Reasoning about Cardinal Directions. Journal of Visual Languages and Computing 9, 23–44 (1988)
12. Mackworth, A.: Consistency in Networks of Relations. Artificial Intelligence 8, 99–118 (1977)
13. Papadias, D., Theodoridis, Y.: Spatial relations, minimum bounding rectangles, and spatial data structures. Technical Report KDBSLAB-TR-94-04 (1997)
14. Sharma, J., Flewelling, D.: Inferences from combined knowledge about topology and directions. In: Advances in Spatial Databases, 4th International Symposium, Portland, Maine, pp. 271–291 (1995)
15. Skiadopoulos, S., Koubarakis, M.: Composing Cardinal Direction Relations. Artificial Intelligence 152(2), 143–171 (2004)
16. Varzi, A.C.: Parts, Wholes, and Part-Whole Relations: The prospects of Mereotopology. Data and Knowledge Engineering 20, 259–286 (1996)

A Framework for Time-Series Analysis*

Vladimir Kurbalija[1], Miloš Radovanović[1], Zoltan Geler[2], and Mirjana Ivanović[1]

[1] Department of Mathematics and Informatics, Faculty of Science, University of Novi Sad,
Trg D. Obradovica 4, 21000 Novi Sad, Serbia
[2] Faculty of Philosophy, University of Novi Sad, Dr Zorana Đinđića 2,
21000 Novi Sad, Serbia
{kurba,radacha,zoltan.geler,mira}@dmi.uns.ac.rs

Abstract. The popularity of time-series databases in many applications has cre-ated an increasing demand for performing data-mining tasks (classification, clustering, outlier detection, etc.) on time-series data. Currently, however, no single system or library exists that specializes on providing efficient implemen-tations of data-mining techniques for time-series data, supports the necessary concepts of representations, similarity measures and preprocessing tasks, and is at the same time freely available. For these reasons we have designed a multi-purpose, multifunctional, extendable system FAP – Framework for Analysis and Prediction, which supports the aforementioned concepts and techniques for mining time-series data. This paper describes the architecture of FAP and the current version of its Java implementation which focuses on time-series similar-ity measures and nearest-neighbor classification. The correctness of the imple-mentation is verified through a battery of experiments which involve diverse time-series data sets from the UCR repository.

Keywords: time-series analysis, time-series mining, similarity measures.

1 Introduction

In different scientific fields, a time-series consists of sequence of values or events obtained over repeated measurements of time [2]. Time-series analysis comprises methods that attempt to understand such time series, often either to understand the underlying context of the data points, or to make forecasts.

Time-series databases are popular in many applications, such as stock market anal-ysis, economic and sales forecasting, budgetary analysis, process control, observation of natural phenomena, scientific and engineering experiments, medical treatments etc. As a consequence, in the last decade there has occurred an increasing amount of inter-est in querying and mining such data which resulted in a large amount of work intro-ducing new methodologies for different task types including: indexing, classification, clustering, prediction, segmentation, anomaly detection, etc. [1, 2, 3, 4].

There are several important concepts which have to be considered when dealing with time series: pre-processing transformation, time-series representation and simi-larity/distance measure.

---

* This work was supported by project *Abstract Methods and Applications in Computer Science* (no. 144017A), of the Serbian Ministry of Science and Environmental Protection.

D. Dicheva and D. Dochev (Eds.): AIMSA 2010, LNAI 6304, pp. 42–51, 2010.

Pre-processing transformation: "Raw" time series usually contain some distortions, which could be consequences of bad measurements or just a property of underlying process which generated time series. The presence of distortions can seriously deteriorate the indexing problem because the distance between two "raw" time series could be very large although their overall shape is very similar. The task of the pre-processing transformations is to remove different kinds of distortions. Some of the most common pre-processing tasks are: offset translation, amplitude scaling, removing linear trend, removing noise etc. [22] Pre-processing transformations can greatly improve the performance of time-series applications by removing different kinds of distortions.

Time-series representation: Time series are generally high-dimensional data [2] and a direct dealing with such data in its raw format is very time and memory consuming. Therefore, it is highly desirable to develop representation techniques that can reduce the dimensionality of time series. Many techniques have been proposed for representing time series with reduced dimensionality: *Discrete Fourier Transformation* (DFT) [1], *Singular Value Decomposition* (SVD) [1], *Discrete Wavelet Transf.* (DWT) [5], *Piecewise Aggregate Approximation* (PAA) [6], *Adaptive Piecewise Constant Approx.* (APCA) [7], *Symbolic Aggregate approX.* (SAX) [8], *Indexable Piecewise Linear Approx.* (IPLA) [9], *Spline Representation* [10], etc.

Similarity/distance measure: Similarity-based retrieval is used in all a fore mentioned task types of time-series analysis. However, the distance between time series needs to be carefully defined in order to reflect the underlying (dis)similarity of these specific data which is usually based on shapes and patterns. There is a number of distance measures for similarity of time-series data: L_p *distance* (L_p) [1], *Dynamic Time Warping* (DTW) [11], distance based on *Longest Common Subsequence* (LCS) [12], *Edit Distance with Real Penalty* (ERP) [13], *Edit Distance on Real sequence* (EDR) [14], *Sequence Weighted Alignment model* (Swale) [31], etc.

All these concepts, when introduced, are usually separately implemented and described in different publications. Every newly-introduced representation method or a distance measure has claimed a particular superiority [4]. However, this was usually based on comparison with only a few counterparts of the proposed concept. On the other hand, by the best of our knowledge there is no freely available system for time-series analysis and mining which supports all mentioned concepts, with the exception of the work proposed in [4].

Being motivated by these observations, we have designed a multipurpose, multifunctional system FAP - Framework for Analysis and Prediction. FAP supports all mentioned concepts: representations, similarity measures and pre-processing tasks; with the possibility to easily change some existing or to add new concrete implementation of any concept. We have implemented all major similarity measures [4], and conducted a set of experiments to validate their correctness.

The rest of the paper is organized as follows. Section 2 gives an overview and comparison of existing systems for time-series analysis. The architecture of FAP system and implemented similarity measures are described in Section 3. Section 4 presents the evaluation methodology based on [4], and gives the results of experiments. Future work is discussed in Section 5. Section 6 concludes the paper.

2   Related Work

As a consequence of importance and usefulness of time series, there is a large number of applications which deal with time series, based on different approaches.

The most popular collection of data mining algorithms is implemented within *WEKA* (Waikato Environment for Knowledge Analysis) tool [15]. *WEKA* supports a great number of data-mining and machine-learning techniques, including data pre-processing, classification, regression and visualization. However, *WEKA* is a general-purpose data-mining library, and it is not specialised for time series. As a result, the time-series support within WEKA is based on external implementations contributed by some users.

A similar system, *RapidMiner*, is a product of company *Rapid-I* [16]. It is also an open source (Community Edition) collection of data-mining and machine-learning techniques. *RapidMiner* has a very sophisticated graphical user interface, and it is also extensible with the user's implementations. It supports some aspects of statistical time-series analysis, prediction and visualisation.

Also, there are several tools specialised for summarisation and visualisation of time series: *TimeSearcher* [23], *Calendar-Based Visualisation* [24], *Spiral* [25] and *Viz-Tree* [26], but they are not specialised for real-world time-series analysis.

The above systems, which partially support time-series analysis, are mainly based on data mining methods. On the other hand, there is a large number of systems which are based on statistical and econometric modelling. Probably the most famous business system is *SAS* [17]. Among many business solutions including: Business Analytics, Business Intelligence, Data Integration, Fraud Prevention & Detection, Risk Management etc., SAS has an integrated subsystem for time series. This subsystem provides modelling of trend, cycles and seasonality of time series as well as time series forecasting, financial and econometric analysis. However, *SAS* is a complex commercial system which is not freely available.

GRETL (GNU Regression, Econometrics and Time-series Library) is open source, platform-independent library for econometric analysis [18]. It supports several least-square based statistical estimators, time-series models and several maximum-likelihood methods. *GRETL* also encloses a graphical user interface for the *X-12-ARIMA* environment. *X-12-ARIMA* is the Census Bureau's new seasonal adjustment program [27]. It supports several interesting concepts such as: alternative seasonal, trading-day and holiday effect adjustment; an alternative seasonal-trend-irregular decomposition; extensive time-series modeling and model-selection capabilities for linear regression models with ARIMA errors.

In addition, some of the most common time-series analysis systems based on statistical modelling are: *TRAMO* [28], *SEATS* [28], *SSA-MTM* [29], *JMulTi* [30], etc.

Clearly two kinds of applications can be distinguished: general purpose data-mining applications which in some extent support time series, and applications specialised for time series based on statistical and econometric models. So, it is evident that there is a huge gap between these two types of applications. There is no available system specialised for time-series mining (time-series analysis based on data-mining techniques) with the exception of the library presented in [4] which can be obtained on demand. This was our main motive for designing a FAP system. FAP will contain all main features and functionalities needed for time-series analysis (pre-processing tasks, similarity measures, time-series representations) and necessary for different data-mining tasks (indexing, classification, prediction, etc). We believe that such a system will significantly help researchers in comparing newly introduced and proposed concepts with the existing ones.

3   Architecture of the Framework for Analysis and Prediction

FAP – Framework for Analysis and Prediction is designed to incorporate all main aspects of time-series mining in one place and to combine easily some or all of them. In the moment when the system will be completed, it will contain all important realisations of proposed concepts up to then, with the possibility to add easily newly proposed realisations. Currently, all important similarity measures, mentioned in introduction are implemented and the results of their evaluation are in more details explained in the next section.

The system is implemented in Java which will make FAP to be easier for maintenance and for future upgrades.

The package diagram of the system with the main dependencies between packages is given in Fig. 1. All packages are connected with package fap.core, but the dependencies are not shown because they would burden the image.

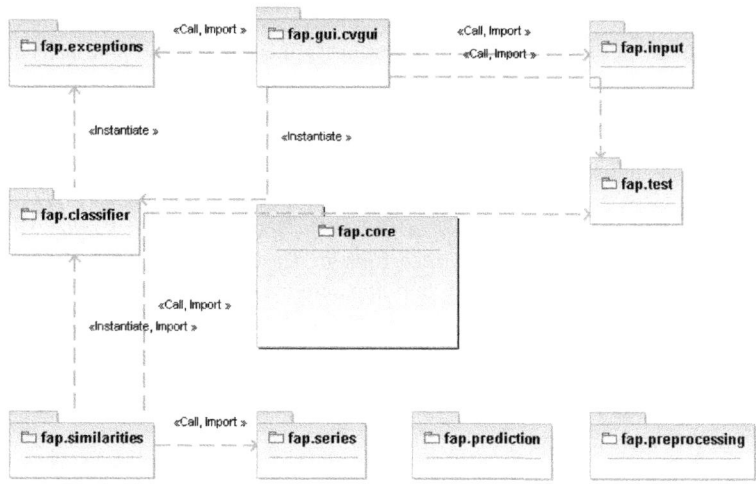

Fig. 1. Package diagram of the FAP system

The main part of the FAP system is package fap.core. It encloses the following subpackages: fap.core.math, fap.core.exceptions, fap.core.data, fap.core.series, fap.core.input, fap.core.similarities, fap.core.test, fap.core.prediction, fap.core.classifier and fap.core.preprocessing. These sub-packages contain basic classes and interfaces, which describe the functionality of the system. Almost every package from the fap.core package has its outside counterpart, where the implementations of desired concepts are stored. In the rest of this section, the contents and functionalities of the presented packages will be described.

The auxiliary packages are fap.core.math, fap.core.exceptions, fap.exceptions and fap.gui.cvgui. The first package contains implementations of additional mathematical concepts needed in the FAP (for example the implementation of polynomials, needed for the Spline representation). The fap.core.exceptions package contains the interface which describes the exceptions thrown by the FAP, while the concrete exceptions are implemented in fap.exceptions. Package fap.gui was intended to contain the implementations of various GUIs needed for different applications, while package fap.gui.cvgui contains our implementation of a GUI needed for cross-validation experiments concerning similarity measures (see the next section).

Package fap.core.data contains the basic implementations of a data point, serie and serie list. Package fap.core.series contains the raw implementation of a time series as a list of its points. It also contains an interface which describes what properties other representations should satisfy. The implementations of other representations are stored in package fap.series. Currently, the spline representation is realised as an additional one [10]. It is possible that one time serie has several representations.

The fap.core.input package is responsible for different kinds of inputs for time series. It contains interfaces which describe what classes responsible for the input should satisfy. The input can be from a file, network or any other stream. Classes which implement the basic functionality of CSV (Comma Separated Values) parser for an arbitrary separator are also stored here. The particular implementation of the CSV file reader is implemented in package fap.input.

Package fap.core.similarities contains an interface which describes the functionality of a class that computes one particular similarity measure. Some similarity measures have one or more parameters which need to be tuned based on the training part of a data set. The interface for tuning classes is also stored in this package. The actual implementations of similarity measures and their corresponding tuners are stored in package fap.similarities. The major similarity measures are implemented: L_p ($L_{1/2}$, L_1, L_2 and L_∞ distances are implemented separately for efficiency reasons), DTW, CDTW (*Constraint DTW*), LCS, CLCS (*Constraint LCS*), ERP, EDR and *Swale*. The results of experiments with these similarity measures are shown in the next section.

Similarly, packages fap.core.prediction, fap.core.classifier and fap.core.preprocessing contain the interfaces which describe prediction algorithms, classification algorithms and pre-processing tasks, respectively. The actual implementations of particular prediction algorithms, classification algorithms (e.g. kNN or 1NN classifier) and pre-processing tasks are stored in packages fap.prediction, fap.classifier and fap.preprocessing, respectively.

Package fap.core.test contains one interface which needs to be implemented when creating any statistical test in FAP. The particular implementations of these statistical tests are stored in package fap.test. We have implemented the following statistical tests: cross-validation with random splitting, cross-validation with stratified splitting and leave-one-out cross-validation.

4   Evaluation of the System

In this stage of FAP development we focused on time-series similarity measures and their complete implementation. Up to now we have implemented the following currently most important similarity measures: L_p, DTW, CDTW, LCS, CLCS, ERP, EDR and *Swale*.

All measures are very carefully implemented concerning efficiency and memory consumption. Furthermore, the measures based on dynamic programming are implemented using the Sakoe-Chiba Band [20] which limits the warping path to a band enclosed by two straight lines that are parallel to the diagonal of the warping matrix. This implementation has shown very good results especially concerning memory consumption.

Table 1. Error rates of similarity measures

Data Set	Num. of Crosses	L_1	L_2	L_∞	$L_{1/2}$	DTW	CDTW	ERP	EDR	LCS	CLCS	Swale
50words	5	0,387	0,421	0,558	0,377	0,367	0,302	0,399	0,260	0,270	0,286	0,270
Adiac	5	0,486	0,462	0,429	0,510	0,457	0,438	0,439	0,437	0,435	0,431	0,435
Beef	2	0,550	0,517	0,517	0,533	0,550	0,517	0,550	0,600	0,500	0,600	0,500
CBF	16	0,046	0,070	0,535	0,046	0,002	0,005	0,003	0,032	0,030	0,034	0,030
Coffee	2	0,161	0,161	0,107	0,250	0,125	0,143	0,161	0,196	0,304	0,286	0,304
ECG200	5	0,154	0,151	0,184	0,151	0,221	0,175	0,214	0,173	0,206	0,166	0,206
FaceAll	11	0,185	0,221	0,399	0,177	0,090	0,071	0,078	0,032	0,037	0,040	0,037
FaceFour	5	0,116	0,190	0,448	0,083	0,145	0,109	0,060	0,027	0,054	0,056	0,054
fish	5	0,302	0,283	0,304	0,328	0,309	0,270	0,190	0,155	0,169	0,184	0,169
Gun_Point	5	0,089	0,116	0,173	0,095	0,129	0,041	0,057	0,049	0,056	0,067	0,056
Lighting2	5	0,217	0,300	0,388	0,219	0,180	0,199	0,360	0,260	0,294	0,267	0,294
Lighting7	2	0,412	0,406	0,595	0,413	0,287	0,308	0,776	0,447	0,405	0,441	0,405
OliveOil	2	0,150	0,117	0,183	0,200	0,133	0,117	0,150	0,167	0,183	0,200	0,183
OSULeaf	5	0,452	0,464	0,518	0,456	0,427	0,422	0,391	0,260	0,242	0,252	0,242
SwedishLeaf	5	0,289	0,296	0,358	0,282	0,249	0,205	0,164	0,136	0,151	0,146	0,151
synthet-	5	0,145	0,132	0,223	0,166	0,012	0,015	0,035	0,058	0,044	0,048	0,044
Trace	5	0,286	0,341	0,459	0,249	0,014	0,018	0,160	0,133	0,049	0,090	0,049
Two_Patterns	5	0,038	0,102	0,796	0,049	0,000	0,000	0,000	0,001	0,001	0,001	0,001
wafer	7	0,004	0,005	0,020	0,005	0,016	0,005	0,008	0,004	0,005	0,005	0,005
yoga	11	0,162	0,159	0,181	0,167	0,151	0,140	0,129	0,113	0,124	0,119	0,124

In order to validate the correctness of our implementation, we have compared our results with the results presented in [4]. The objective evaluation method, proposed in [4], is used for comparison of different time series similarity measures: one nearest neighbour (1NN) classifier is used on labeled data to evaluate the efficiency of the similarity/distance measures. Each time series in the database has a correct class label. The classifier tries to predict the label of a time series as a label of its first nearest neighbor in the training set. There are several advantages with this approach:

- The underlying distance metric is crucial to the performance of the 1NN classifier; therefore, the accuracy of the 1NN classifier directly reflects the effectiveness of the similarity measure.
- The 1NN classifier is straightforward to be implemented and is parameter free, which reduces the possibility of appearance of errors in implementation.
- Among many other classification techniques, such as decision trees, neural networks, Bayesian networks, support vector machines, etc., some of the best results in time-series classification come from simple nearest neighbor methods [20].

In order to evaluate the effectiveness of each similarity measure, the following cross-validation algorithm has been applied. Firstly, a stratified random split has been used to divide the input data set into k subsets. However, we presented the results without randomization in order for them to be reproducible. The number of crosses in a cross-validation algorithm, k, is shown in Tab. 1. Secondly, the cross validation algorithm is applied, by using one subset at a time for the training set of the 1NN classifier, and the rest $k-1$ subsets as the test set. If the similarity measure requires parameter tuning, the training set is divided into two equal-sized stratified subsets where one is used for parameter tuning. Finally, the average error rate of 1NN classification over the k cross-validation fold is reported in Tab. 1.

For the purpose of testing similarity measures the FAP Graphical User Interface (GUI) has been implemented. The input for this FAP GUI application is the FAP experiment specification. On the basis of this specification the FAP will perform the experiments. The structure of the FAP experiment specification is very simple: the name of the specification is given in the first line while every subsequent line is reserved for one dataset. Every line consists of 9 attributes with the following meaning:

- **Name of the dataset.** FAP will look for the data sets in the subfolder with this name.
- **Number of splits (k) for cross-validation.** The default values for some or all datasets can be given in file "Datasets.csv" and can be omitted here. The values given in Tab. 1 are default values taken from [4].
- **Start fold.** The default value is 0.
- **End fold.** The default value is k-1.
- **Seed for the stratified random split.** Can be useful for reproduction of previous results. If it is not specified, the stratified split is performed without randomization.
- **Name of the similarity measure class.** The mappings between full class name and short names can be given in file "Distances.csv".
- **Name of the tuner class** if the corresponding similarity measure requires tuning.
- **Frequency of tuning serialization.** Number which tells how frequently the main FAP object will be serialized during the tuning process. Default value is 20.
- **Frequency of testing serialization.** Number which tells how frequently the main FAP object will be serialized during the testing process. Default value is 20. This and the previous attribute introduce the possibility of serializing objects at runtime. This can be very useful when long computation must be terminated for some reason. In this case the computation can be continued later.

As an output, FAP application generates four files: "*SpecName*.txt" where the code of the FAP specification is stored, "*SpecName*.csv" where only the average error rates of data sets are stored, "*SpecName*.log" where the full logging information is stored and "*SpecName*.scvp" where the object serialized during runtime is stored. The last file also contains already computed results and can be migrated to another computer where the computation will be continued.

The experiments were conducted using the described methodology, on 20 diverse time series data sets. The data is provided by the UCR Time Series Repository [21], which includes the majority of all publicly available, labeled time series data sets in the world. The average error rates of the similarity measures on each data set are shown in Tab. 1. We have compared our results with the results presented in [4] in order to verify the correctness of our implementation. The only differences appear at the second or third decimal place which is the consequence of randomization in stratified random split of the cross-validation algorithm. These facts strongly support the correctness of our implementation, which has been our main goal.

5 Usefulness of FAP and Future Work

Time series are very useful and widely applicable in different domains. In the same time, working with time series is relatively complicated, mainly because of their high dimensions. So, it is very important to correctly choose appropriate methods and algorithms when working with specified time series. Motivated by these observations, we decided to create system FAP which will incorporate all important concepts of time-series analysis and mining. FAP is mainly intended for time-series researchers for testing and comparison of existing and newly introduced concepts. However, FAP system can also be useful for non-professionals from different domains as an assistance tool for choosing appropriate methods for their own data sets.

At this stage, FAP can be successfully used for testing and comparing existing similarity measures as well as for comparison of newly proposed measures with the existing ones. However, in order to be fully functional for time series mining, several algorithms and concepts need to be implemented.

The first step in the future development of FAP system is the implementation of wide range of existing techniques for time series representation: *DFT, SVD, DWT, PAA, APCA, SAX, IPLA*, while the spline representation is already implemented.

The next step of development will be the implementation of some pre-processing tasks. These tasks (*offset translation, amplitude scaling, removing linear trend, removing noise* etc.) can remove some inappropriate distortions like noise, trend, or seasonality. These distortions can greatly influence indexing of time series and spoil the results of similarity measures.

Although, it is reported that the 1NN technique gives among the best results for time series [20] it would be valuable to test other classification techniques like: decision trees, neural networks, Bayesian networks, support vector machines, etc.

In addition, the functionalities of FAP are not limited to classification only. We intend to extend it with other time-series mining task types like: clustering, forecasting, summarization, segmentation, etc. Also, it would be very useful to have some component for visualization.

6 Conclusion

Time-series analysis and mining has been a very popular research area in the past decade. This resulted in a huge amount of proposed techniques and algorithms. A great majority of techniques and algorithms were sporadically introduced and sometimes not correctly compared with their counterparts. This is the consequence of a lack of a quality open-source system which supports different aspects of time-series mining. For all these reasons, we created a universal framework (FAP) where all main concepts, like similarity measures, representation and pre-processing, will be incorporated. Such a framework would greatly help researchers in testing and comparing newly introduced concepts with the existing ones.

In this stage all main similarity measures are implemented. Currently, the modeling and implementation of representation techniques is in progress. The current version of FAP is available online as open-source software on the address: http://perun.pmf.uns.ac.rs/fap/.

We believe that the FAP system could be intensively used in research due to its numerous advantages. First, all important up to date concepts needed for time-series mining are integrated in one place. Second, modifications of existing concepts, as well as additions of newly proposed concepts could be obtained very easily (FAP is written in Java). Finally, FAP will be open source, and everyone will be invited to contribute with newly proposed concepts. This will insure that the system is always up to date and that all major techniques in time-series mining are supported.

References

1. Faloutsos, C., Ranganathan, M., Manolopoulos, Y.: Fast Subsequence Matching in Time-Series Databases. In: SIGMOD Conference, pp. 419–429 (1994)
2. Han, J., Kamber, M.: Data Mining: Concepts and Techniques. Morgan Kaufmann Publishers, San Francisco (2005)
3. Keogh, E.J.: A Decade of Progress in Indexing and Mining Large Time Series Databases. In: VLDB, p. 1268 (2006)
4. Ding, H., Trajcevski, G., Scheuermann, P., Wang, X., Keogh, E.: Querying and Mining of Time Series Data: Experimental Comparison of Representations and Distance Measures. In: VLDB 2008, Auckland, New Zealand, pp. 1542–1552 (2008)
5. pong Chan, K., Fu, A.W.-C.: Efficient Time Series Matching by Wavelets. In: ICDE, pp. 126–133 (1999)
6. Keogh, E.J., Chakrabarti, K., Pazzani, M.J., Mehrotra, S.: Dimensionality Reduction for Fast Similarity Search in Large Time Series Databases. Knowl. Inf. Syst. 3(3), 263–286 (2001)
7. Keogh, E.J., Chakrabarti, K., Mehrotra, S., Pazzani, M.J.: Locally Adaptive Dimensionality Reduction for Indexing Large Time Series Databases. In: SIGMOD Conference, pp. 151–162 (2001)
8. Lin, J., Keogh, E.J., Wei, L., Lonardi, S.: Experiencing SAX: a novel symbolic representation of time series. Data Min. Knowl. Discov. 15(2), 107–144 (2007)
9. Chen, Q., Chen, L., Lian, X., Liu, Y., Yu, J.X.: Indexable PLA for Efficient Similarity Search. In: VLDB, pp. 435–446 (2007)
10. Kurbalija, V., Ivanović, M., Budimac, Z.: Case-Based Curve Behaviour Prediction. Software: Practice and Experience 39(1), 81–103 (2009)
11. Keogh, E.J., Ratanamahatana, C.A.: Exact indexing of dynamic time warping. Knowl. Inf. Syst. 7(3), 358–386 (2005)
12. Vlachos, M., Gunopulos, D., Kollios, G.: Discovering similar multidimensional trajectories. In: ICDE, pp. 673–684 (2002)
13. Chen, L., Ng, R.T.: On the marriage of lp-norms and edit distance. In: VLDB, pp. 792–803 (2004)

14. Chen, L., Özsu, M.T., Oria, V.: Robust and fast similarity search for moving object trajectories. In: SIGMOD Conference, pp. 491–502 (2005)
15. Hall, M., Frank, E., Holmes, G., Pfahringer, B., Reutemann, P., Witten, I.H.: The WEKA Data Mining Software: An Update. SIGKDD Explorations 11(1), 10–18 (2009)
16. http://rapid-i.com/ (January 2010)
17. http://www.sas.com (January 2010)
18. Baiocchi, G., Distaso, W.: GRETL: Econometric software for the GNU generation. Journal of Applied Econometrics 18(1), 105–110
19. Sakoe, H., Chiba, S.: Dynamic programming algorithm optimization for spoken word recognition. IEEE Transactions on Acoustics, Speech, and Signal Processing 26(1), 43–49 (1978)
20. Xi, X., Keogh, E.J., Shelton, C.R., Wei, L., Ratanamahatana, C.A.: Fast time series classification using numerosity reduction. In: ICML, pp. 1033–1040 (2006)
21. Keogh, E., Xi, X., Wei, L., Ratanamahatana, C.: The UCR Time Series dataset (2006), http://www.cs.ucr.edu/~eamonn/time_series_data/
22. Keogh, E., Pazzani, M.: Relevance Feedback Retrieval of Time Series Data. In: The Twenty-Second Annual International ACM-SIGIR Conference on Research and Development in Information Retrieval, pp. 183–190 (1999)
23. Hochheiser, H., Shneiderman, B.: Interactive Exploration of Time-Series Data. In: Proc. of the 4th Int'l Conference on Discovery Science, Washington D.C., pp. 441–446 (2001)
24. van Wijk, J.J., van Selow, E.R.: Cluster and Calendar based Visualization of Time Series Data. In: Proceedings of IEEE Symposium on Information Visualization, pp. 4–9 (1999)
25. Weber, M., Alexa, M., Müller, W.: Visualizing Time-Series on Spirals. In: Proceedings of the IEEE Symposium on Information Visualization, pp. 7–14 (2001)
26. Lin, J., Keogh, E., Lonardi, S., Lankford, J.P., Nystrom, D.M.: Visually Mining and Monitoring Massive Time Series. In: Proc. of the 10th ACM SIGKDD International Conference on Knowledge Discovery and Data Mining. Seattle, WA, pp. 460–469 (2004)
27. Findley, D.F., et al.: New Capabilities and Methods of the X-12-ARIMA Seasonal-Adjustment Program. Journal of Business & Economic Statistics, American Statistical Association 16(2), 127–152 (1998)
28. Gómez, V., Maravall, A.: Guide for using the program TRAMO and SEATS. In: Working Paper 9805, Research Department, Banco de España (1998)
29. Ghil, M., Allen, R.M., Dettinger, M.D., Ide, K., Kondrashov, D., Mann, M.E., Robertson, A., Saunders, A., Tian, Y., Varadi, F., Yiou, P.: Advanced spectral methods for climatic time series. Rev. Geophys. 40(1), 3.1–3.41 (2002)
30. Lütkepohl, H., Krätzig, M.: Applied Time Series Econometrics. Cambridge University Press, Cambridge (2004)
31. Morse, M.D., Patel, J.M.: An efficient and accurate method for evaluating time series similarity. In: SIGMOD Conference, pp. 569–580 (2007)

Cross-Language Personalization through a Semantic Content-Based Recommender System

Pasquale Lops, Cataldo Musto, Fedelucio Narducci,
Marco de Gemmis, Pierpaolo Basile, and Giovanni Semeraro

Department of Computer Science,
University of Bari "Aldo Moro", Italy
{lops,musto,narducci,degemmis,basile,semeraro}@di.uniba.it
http://www.di.uniba.it/

Abstract. The exponential growth of the Web is the most influential factor that contributes to the increasing importance of cross-lingual text retrieval and filtering systems. Indeed, relevant information exists in different languages, thus users need to find documents in languages different from the one the query is formulated in. In this context, an emerging requirement is to sift through the increasing flood of multilingual text: this poses a renewed challenge for designing effective multilingual Information Filtering systems. Content-based filtering systems adapt their behavior to individual users by learning their preferences from documents that were already deemed relevant. The learning process aims to construct a profile of the user that can be later exploited in selecting/recommending relevant items. User profiles are generally represented using keywords in a specific language. For example, if a user likes movies whose plots are written in Italian, content-based filtering algorithms will learn a profile for that user which contains Italian words, thus movies whose plots are written in English will be not recommended, although they might be definitely interesting. In this paper, we propose a language-independent content-based recommender system, called MARS (MultilAnguage Recommender System), that builds cross-language user profiles, by shifting the traditional text representation based on keywords, to a more advanced language-independent representation based on word meanings. The proposed strategy relies on a knowledge-based word sense disambiguation technique that exploits MultiWordNet as sense inventory. As a consequence, content-based user profiles become language-independent and can be exploited for recommending items represented in a language different from the one used in the content-based user profile. Experiments conducted in a movie recommendation scenario show the effectiveness of the approach.

Keywords: Cross-language Recommender System, Content-based Recommender System, Word Sense Disambiguation, MultiWordNet.

1 Introduction

Information Filtering (IF) systems are rapidly emerging as tools for overcoming information overload in the "digital era".

D. Dicheva and D. Dochev (Eds.): AIMSA 2010, LNAI 6304, pp. 52–60, 2010.
© Springer-Verlag Berlin Heidelberg 2010

Specifically, content-based filtering systems [1] analyze a set of documents (mainly textual descriptions of items previously rated as relevant by an individual user) and build a model or profile of user interests based on the features (generally keywords) that describe the target objects. The profile is compared to the item descriptions to select relevant items.

Traditional keyword-based profiles are unable to capture the semantics of user interests because they suffer from problems of *polysemy*, the presence of multiple meanings for one word, and *synonymy*, multiple words with the same meaning. Another relevant problem related to keywords is the strict connection with the user language: an English user, for example, frequently interacts with information written in English, so her profile of interests mainly contains English terms. In order to receive suggestions of items whose textual description is available in a different language, she must explicitly give her preferences on items in that specific language, as well.

The main idea presented in this paper is the adoption of MultiWordNet [2] as a bridge between different languages. MultiWordNet associates a unique identifier to each possible sense (meaning) of a word, regardless the original language. In this way we can build user profiles based on MultiWordNet senses and we can exploit them in order to provide cross-language recommendations. The paper is organized as follows. Section 2 presents the architecture of the systems proposed in the paper, while Sections 3 and 4 describe the process of building language-independent documents and profiles. Experiments carried out in a movie recommendation scenario are described in Section 5. Section 6 analyzes related works in the area of cross-language filtering and retrieval, while conclusions and future work are drawn in the last section.

2   General Architecture of MARS

MARS (MultilAnguage Recommender System) is a system capable of generating recommendations, provided that descriptions of items are available in textual form. Item properties are represented in the form of *textual slots*. For example, a movie can be described by slots *title, genre, actors, summary*. Figure 1 depicts the main components of the MARS general architecture: the *Content Analyzer*, the *Profile Learner*, and the *Recommender*.

In this work, the *Content Analyzer* is the main module involved in designing a language-independent content-based recommender system. It allows introducing semantics in the recommendation process by analyzing documents in order to identify relevant concepts representing the content. This process selects, among all the possible meanings (senses) of each (polysemous) word, the correct one according to the context in which the word occurs. The final outcome of the pre-processing step is a repository of disambiguated documents. This semantic indexing is strongly based on natural language processing techniques, such as Word Sense Disambiguation [8], and heavily relies on linguistic knowledge stored in lexical ontologies. In this work, the *Content Analyzer* relies on the *MultiWordNet* lexical ontology [2]. The generation of the cross-language user

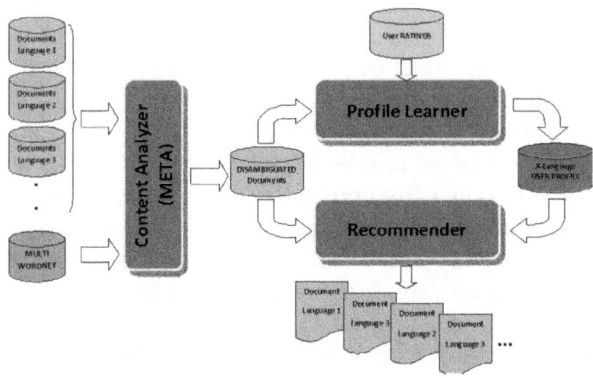

Fig. 1. General architecture of MARS

profile is performed by the *Profile Learner*, which infers the profile as a binary text classifier. Finally the *Recommender* exploits the cross-language user profile to suggest relevant items by matching concepts contained in the semantic profile against those contained in documents to be recommended (previously disambiguated).

3   Building Language-Independent Documents

Semantic indexing of documents is performed by the Content Analyzer, which relies on META (Multi Language Text Analyzer) [9], a tool able to deal with documents in English and Italian. The goal of the semantic indexing step is to obtain a concept-based document representation. To this purpose, the text is first tokenized, then for each word possible lemmas (as well as their morpho-syntactic features) are collected. Part of speech (POS) ambiguities are solved before assigning the proper sense (concept) to each word. In this work, the semantic indexing module exploits *MultiWordNet*[2] as sense-repository. MultiWordNet is a multilingual lexical database that supports the following languages: English, Italian, Spanish, Portuguese, Hebrew, Romanian and Latin.

The word-concept association is perform by META, which implements a Word Sense Disambiguation (WSD) algorithm, called JIGSAW.

The goal of a WSD algorithm is to associate a word w occurring in a document d with its appropriate meaning or sense s, selected from a predefined set of possibilities, usually known as *sense inventory*. JIGSAW takes as input a document encoded as a list of h words in order of their appearance, and returns a list of k MultiWordNet synsets ($k \leq h$), in which each synset s is obtained by disambiguating the target word w, by exploiting the *context* C in where w is found. The context C for w is defined as a set of words that precede and follow w. In the proposed algorithm, the sense inventory for w is obtained from MultiWordNet.

JIGSAW is based on the idea of combining three different strategies to disambiguate nouns, verbs, adjectives and adverbs. The main motivation behind our approach is that the effectiveness of a WSD algorithm is strongly influenced by the POS tag of the target word. An adaptation of Lesk dictionary-based WSD algorithm has been used to disambiguate adjectives and adverbs [11], an adaptation of the Resnik algorithm has been used to disambiguate nouns [12], while the algorithm we developed for disambiguating verbs exploits the nouns in the *context* of the verb as well as the nouns both in the glosses and in the phrases that MultiWordNet utilizes to describe the usage of a verb. The complete description of the adopted WSD strategy adopted is published in [10].

The WSD procedure implemented in the Content Analyzer allows to obtain a synset-based vector space representation, called bag-of-synsets (BOS), that is an extension of the classical bag-of-words (BOW) model. In the BOS model, a synset vector, rather than a word vector, corresponds to a document. The text in each slot is represented by the BOS model by counting separately the occurrences of a synset in the slots in which it occurs.

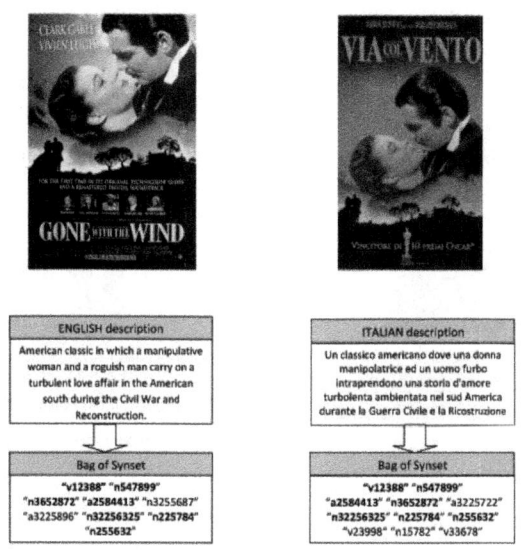

Fig. 2. Example of synset-based document representation

Figure 2 provides an example of representation for the movie by Victor Fleming *Gone with the wind*, corresponding to the Italian translation *Via col vento*. The textual description of the plot is provided both for the Italian and English version. The *Content Analyzer* produces the BOS containing the concepts extracted from the plot. The MultiWordNet-based document representation creates a bridge between the two languages: in a classical keyword-based approach the two plots would share none terms, while the adoption of the synset-based approach would allow a greater overlapping (seven shared synsets).

4   Profile Learner

The generation of the cross-language user profile is performed by the *Profile Learner*, which infers the profile as a binary text classifier. Therefore, the set of categories is restricted to c_+, the positive class (*user-likes*), and c_- the negative one (*user-dislikes*).

The induced probabilistic model is used to estimate the *a posteriori* probability, $P(c|d_j)$, of document d_j belonging to class c. The algorithm adopted for inferring user profiles is a Naïve Bayes text learning approach, widely used in content-based recommenders, which is not presented here because already described in [13]. What we would like to point out here is that the final outcome of the learning process is a probabilistic model used to classify a new document, written in any language, in the class c_+ or c_-.

Figure 3 provides an example of cross-language recommendations provided by MARS. The user likes a movie with an Italian plot and concepts extracted from the plot are stored in her synset-based profile. The recommendation step exploits this representation to suggest a movie whose plot is in English. The classical matching between keywords is replaced by a matching between synsets, allowing to suggest movies even if their descriptions do not contain shared terms.

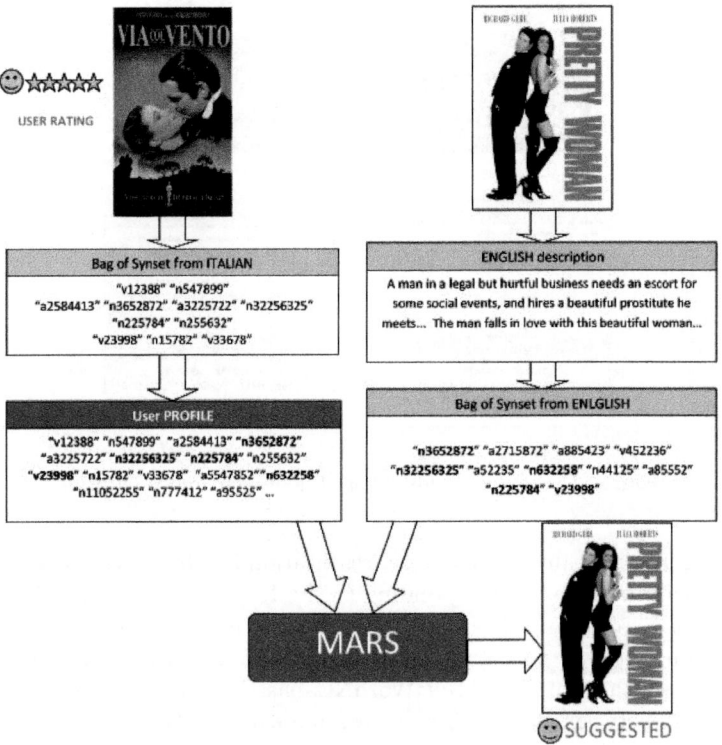

Fig. 3. Example of a cross-language recommendation

5   Experimental Evaluation

The goal of the experimental evaluation was to measure the predictive accuracy of language-independent (cross language) user profiles built using the BOS model. More specifically, we would like to test: 1) whether user profiles learned using examples in a specific language can be effectively exploited for recommending items in a different language; 2) whether the accuracy of the cross-language recommender system is comparable to that of the monolingual one. Experiments were carried out in a movie recommendation scenario, and the languages adopted in the evaluation phase are English and Italian.

5.1   Users and Dataset

The experimental work has been carried out on the MovieLens dataset[1], containing 100,000 ratings provided by 943 different users on 1,628 movies. The original dataset does not contain any information about the content of the movies. The content information for each movie was crawled from both the English and Italian version of Wikipedia. In particular the crawler gathers the *Title* of the movie, the name of the *Director*, the *Starring* and the *Plot*.

5.2   Design of the Experiment

User profiles are learned by analyzing the ratings stored in the MovieLens dataset. Each rate was expressed as a numerical vote on a 5-point Likert scale, ranging from 1=strongly dislike, to 5=strongly like. MARS is conceived as a text classifier, thus its effectiveness was evaluated by classification accuracy measures, namely *Precision* and *Recall*. F_β measure, a combination of Precision and Recall, is also used to have an overall measure of predictive accuracy (β sets the relative degree of importance attributed to Pr and Re. In this work we set β as 0.5) In the experiment, an item is considered *relevant* for a user if the rating is greater than or equal to 4, while MARS considers an item relevant for a user if the a-posteriori probability of the class *likes* is greater than 0.5.

We designed two different experiments, depending on 1) the language of items used for learning profiles, and 2) the language of items to be recommended:

- Exp#1 – *ENG-ITA*: profiles learned on movies with English description and recommendations provided on movies with Italian description;
- Exp#2 – *ITA-ENG*: profiles learned on movies with Italian description and recommendations produced on movies with English description.

We compared the results against the accuracy of classical monolanguage content-based recommender systems:

- Exp#3 – *ENG-ENG*: profiles learned on movies with English description and recommendations produced on movies with English description;
- Exp#4 – *ITA-ITA*: profiles learned on movies with Italian description and recommendations produced on movies with Italian description.

---

[1] http://www.grouplens.org

We executed one experiment for each user in the dataset. The ratings of each specific user and the content of the rated movies have been used for learning the user profile and measuring its predictive accuracy, using the aforementioned measures. Each experiment consisted of:

1. selecting ratings of the user and the description (English or Italian) of the movies rated by that user;
2. splitting the selected data into a training set Tr and a test set Ts;
3. using Tr for learning the corresponding user profile by exploiting the:
   - English movie descriptions (EXP#1);
   - Italian movie descriptions (EXP#2);
4. evaluating the predictive accuracy of the induced profile on Ts, using the aforementioned measures, by exploiting the:
   - Italian movie descriptions (EXP#1);
   - English movie descriptions (EXP#2);

In the same way, a single run for each user has been performed for computing the accuracy of monolingual recommender systems, but the process of learning user profiles from Tr and evaluating the predictive accuracy on Ts has been carried out using descriptions of movies in the same language, English or Italian. The methodology adopted for obtaining Tr and Ts was the 5-fold cross validation.

5.3   Discussion of Results

Results of the experiments are reported in Table 1, averaged over all the users.

Table 1. Experimental Results

Experiment	Pr	Re	F_β
EXP#1 – ENG-ITA	59.04	96.12	63.98
EXP#2 – ITA-ENG	58.77	95.87	63.70
EXP#3 – ENG-ENG	60.13	95.16	64.91
EXP#4 – ITA-ITA	58.73	96.42	63.71

The main outcome of the experimental session is that the strategy implemented for providing cross-language recommendations is quite effective. More specifically, user profiles learned using examples in a specific language, can be effectively exploited for recommending items in a different language, and the accuracy of the approach is comparable to those in which the learning and recommendation phase are performed on the same language. This means that the goal of the experiment has been reached. The best result was obtained by running Exp#3, that is a classical monolanguage recommender system using English content both for learning user profiles and providing recommendations. This result was expected, due to the highest accuracy of the JIGSAW WSD algorithm for English with respect to Italian. This means that the error introduced in the disambiguation step for representing documents as bag-of-synsets hurts the performance of the Profile Learner. It is worth to note that the result of Exp#4 related to movies whose description is in Italian is quite satisfactory. The result of the second experiment, in which Italian movie descriptions are used for learning profiles that are then exploited for recommending English movies is also satisfactory. This confirms the goodness of the approach designed for providing cross-language personalization.

6   Related Work

Up to our knowledge, the topic of Cross-Language and Multilanguage Information Filtering is still not properly investigated in literature.

Recently, the Multilingual Information Filtering task at CLEF 2009[2] has introduced the issues related to the cross-language representation in the area of Information Filtering. Damankesh et al. [3], propose the application of the theory of Human Plausible Reasoning (HPR) in the domain of filtering and cross language information retrieval. The system utilizes plausible inferences to infer new, unknown knowledge from existing knowledge to retrieve not only documents which are indexed by the query terms but also those which are plausibly relevant.

The state of the art in the area of cross-language Information Retrieval is undoubtedly richer, and can certainly help in designing effective cross-language Information Filtering systems. Oard [4] gives a good overview of the approaches for cross-language retrieval.

Ballesteros et al. [5] underlined the importance of phrasal translation in cross-language retrieval and explored the role of phrases in query expansion via local context analysis and local feedback.

The most recent approaches to Cross-Language Retrieval mainly rely on the use of large corpora like Wikipedia. Potthast et al. [6] introduce CL-ESA, a new multilingual retrieval model for the analysis of cross-language similarity. The approach is based on Explicit Semantic Analysis (ESA) [7], extending the original model to cross-lingual retrieval settings.

7   Conclusions and Future Work

This paper presented a semantic content-based recommender system for providing cross-language recommendations. The key idea is to provide a bridge among different languages by exploiting a language-independent representation of documents and user profiles based on word meanings, called bag-of-synsets. The assignment of the right meaning to words is based on a WSD algorithm that exploits MultiWordNet as sense repository. Experiments were carried out in a movie recommendation scenario, and the main outcome is that the accuracy of cross-language recommmmendations is comparable to that of classical

---

[2] http://www.clef-campaign.org/2009.html

(monolingual) content-based recommendations. In the future, we are planning to investigate the effectiveness of MARS on different domains and datasets. More specifically, we are working to extract cross-language profiles by gathering information from social networks, such as Facebook, LinkedIn, Twitter, etc., in which information are generally in different languages.

References

1. Pazzani, M.J., Billsus, D.: Content-Based Recommendation Systems. In: Brusilovsky, P., Kobsa, A., Nejdl, W. (eds.) Adaptive Web 2007. LNCS, vol. 4321, pp. 325–341. Springer, Heidelberg (2007)
2. Bentivogli, L., Pianta, E., Girardi, C.: Multiwordnet: developing an aligned multilingual database. In: First International Conference on Global WordNet, Mysore, India (2002)
3. Damankesh, A., Singh, J., Jahedpari, F., Shaalan, K., Oroumchian, F.: Using human plausible reasoning as a framework for multilingual information filtering. In: CLEF 2008: Proceedings of the 9th Workshop of the Cross-Language Evaluation Forum, Corfu, Greece (2008)
4. Oard, D.W.: Alternative approaches for cross-language text retrieval. In: AAAI Symposium on Cross-Language Text and Speech Retrieval. AAAI (1997)
5. Ballesteros, L., Croft, W.B.: Phrasal translation and query expansion techniques for cross-language information retrieval. In: SIGIR 1997: Proceedings of the 20th Annual International ACM SIGIR Conference on Research and Development in Information Retrieval, pp. 84–91. ACM, New York (1997)
6. Martin Potthast, B.S., Anderka, M.: A wikipedia-based multilingual retrieval model. In: Advances in Information Retrieval, pp. 522–530 (2008)
7. Gabrilovich, E., Markovitch, S.: Computing semantic relatedness using wikipedia-based explicit semantic analysis. In: Veloso, M.M. (ed.) IJCAI, pp. 1606–1611 (2007)
8. Manning, C., Schütze, H.: Foundations of Statistical Natural Language Processing. In: Text Categorization, ch. 16, pp. 575–608. MIT Press, Cambridge (1999)
9. Basile, P., de Gemmis, M., Gentile, A., Iaquinta, L., Lops, P., Semeraro, G.: META - MultilanguagE Text Analyzer. In: Proceedings of the Language and Speech Technnology Conference - LangTech 2008, Rome, Italy, February 28-29, 2008, pp. 137–140 (2008)
10. Basile, P., de Gemmis, M., Gentile, A., Lops, P., Semeraro, G.: UNIBA: JIGSAW algorithm for Word Sense Disambiguation. In: Proceedings of the 4th ACL 2007 International Workshop on Semantic Evaluations (SemEval 2007), Prague, Czech Republic, June 23-24, 2007. Association for Computational Linguistics, pp. 398–401 (2007)
11. Banerjee, S., Pedersen, T.: An adapted lesk algorithm for word sense disambiguation using wordnet. In: Gelbukh, A. (ed.) CICLing 2002. LNCS, vol. 2276, pp. 136–145. Springer, Heidelberg (2002)
12. Resnik, P.: Disambiguating noun groupings with respect to WordNet senses. In: Proceedings of the Third Workshop on Very Large Corpora. Association for Computational Linguistics, pp. 54–68 (1995)
13. de Gemmis, M., Lops, P., Semeraro, G., Basile, P.: Integrating Tags in a Semantic Content-based Recommender. In: Proceedings of the 2008 ACM Conference on Recommender Systems, RecSys 2008, Lausanne, Switzerland, October 23-25, pp. 163–170 (2008)

Towards Effective Recommendation of Social Data across Social Networking Sites

Yuan Wang[1], Jie Zhang[2], and Julita Vassileva[1]

[1] Department of Computer Science, University of Saskatchewan, Canada
{yuw193,jiv}@cs.usask.ca
[2] School of Computer Engineering, Nanyang Technological University, Singapore
zhangj@ntu.edu.sg

Abstract. Users of Social Networking Sites (SNSs) like Facebook, MySpace, LinkedIn, or Twitter, are often overwhelmed by the huge amount of social data (friends' updates and other activities). We propose using machine learning techniques to learn preferences of users and generate personalized recommendations. We apply four different machine learning techniques on previously rated activities and friends to generate personalized recommendations for activities that may be interesting to each user. We also use different non-textual and textual features to represent activities. The evaluation results show that good performance can be achieved when both non-textual and textual features are used, thus helping users deal with cognitive overload.

1 Introduction

Social Networking Sites (SNSs) have changed how people communicate: nowadays, people prefercommunication via SNSs over emails [1]. With the explosion of SNSs, it is also common that a user may engage with multiple SNSs. These users of multiple SNSs see a great number of status updates and other kinds of social data generated by their network friends everyday. This causes a significant information overload to users. One way to deal with information overload is by providing recommendations for interesting social activities, which allows the user to focus her attention more effectively.

In this paper, we present an approach for recommending social data in a dashboard application called "SocConnect", developed in our previous work and described in [2], that integrates social data from different SNSs (e.g. Facebook, Twitter), and also allows users to rate friends and/or their activities as favourite, neutral or disliked. We compare several machine learning techniques that can be used to learn users' preferences of activities to provide personalized recommendations of activities that are interesting to them. In the machine learning process, we use several different non-textual and textual features to represent social activities. The evaluation results show that some of the machine learning techniques achieve good performance (above 80% of correct recommendations on average). Both non-textual and textual features should be used for representing activities.

D. Dicheva and D. Dochev (Eds.): AIMSA 2010, LNAI 6304, pp. 61–70, 2010.

2   Related Work

2.1   Recommender Systems

There is a lot of research in the area of recommender systems dating back from the mid 1990ies. There are two main types of recommender systems: content-based (or feature-based) and collaborative (social). Content-based recommenders analyze features of the content in the set and match them to features of the user (e.g. preferences, interests), based on a user model developed by analyzing the previous actions of the user. Collaborative or social recommenders work by statistically correlating users based on their previous choices. Based on the assumption that people who have behaved similarly in the past will continue to do so, these recommenders suggest content, rated highly by a user, to similar users who have not seen the content yet. Collaborative (social) recommender systems are widely used to recommend movies, books, or other shopping items in e-commerce sites. More recently, recommender systems have been applied in SNSs, but there are still relatively few academic works in this area. SoNARS [3] recommends Facebook groups. It takes a hybrid approach, combining results from collaborative filtering and content-based algorithms. Dave Briccetti developed a Twitter desktop client application called TalkingPuffin (talkingpuffin.org). It allows users to remove "noise" (uninteresting updates) by manually muting users, retweets from specific users or certain applications. Many existing SNSs use social network analysis to recommend friends to users. This, however, does not help in dealing with information overload, on the contrary. So our research focuses on recommending status updates. Status update is different from items like movies, books, or shopping goods in two ways: first, the number of status updates arrive in large volumes, and are only relevant for very short time; second, a status update is more personal and aimed at a small audience. Due to these two features, a collaborative recommendation approach is not a good solution: collaborative filtering works well for a large group of similar users and requires previous ratings.

We focus on status updates recommendation that is content-based. It use machine learning techniques to make predictions based on the user's previous choices and generate personalized recommendations.

2.2   Text Recommendation

Our research shares similarity with text recommendation in the field of Information Retrieval and Personal Information Management, since each status update can be considered as one document. Text recommendation usually has four steps [4]: (1) recognizing user interest and document value; (2) representing user interest; (3) identifying other documents of potential interest; and (4) notifying the user - possibly through visualization. Our work follows these four steps.

Vector space is the most common method for modelling document value. A vector space represents a document or documents by the terms occurring in the document with a weight for each term. The weight represents the importance of the term in the given document. The most common two ways to calculate the weight are Term Frequency (TF) and Term Frequency - Inverse Document Frequency (TF-IDF).

TF is simply counting how many times each term occurs in the given document, defined as:

$$\text{TF}_i = \frac{N_i}{\sum_i N_i} \tag{1}$$

TF-IDF takes into account not only the importance of the term in the given document but also the general importance of the term across all documents, based on the number of documents containing this term. It can be defined as:

$$\text{TF-IDF}_i = \text{TF}_i \times \lg \frac{|A|}{|A_i|} \tag{2}$$

where $|A|$ is the total number of documents, and $|A_i|$ is the number of documents containing the term.

3   Personalized Recommendations in SocConnect

To relieve the information overload, SocConnect provides personalized recommendations of activities to individual users according to a prediction generated using their ratings on previous social data. Thus, our approach is content-based recommendation, rather than collaborative. In this section, we propose a list of potential non-textual and textual features for representing each activity and we present several machine learning techniques that were used to predict users' preferences on activities from the social networks Twitter and Facebook.

3.1   Learning User Preferences on Activities

Users directly express their preferences on activities and friends by using the function of rating activities as "favourite" or "disliked". The users' ratings of their friends are also used in predicting users' interests in activities posted by these friends. Based on the ratings, SocConnect can learn users' preferences and predict whether they will be interested in new similar activities from friends. Machine learning techniques are often used for learning and prediction. Soc-Connect applies the classic techniques of Decision Trees, Support Vector Machine [5], Bayesian Networks, and Radial Basis Functions [?]. In brief, Decision Tree learning is one of the most widely used techniques to produce discrete prediction about whether a user will find an activity interesting. It classifies an instance into multiple categories. Bayesian Belief Networks is a commonly used Bayesian learning technique. The method of Radial Basis Functions belongs to the category of instance-based learning to predict a real-valued function. Support Vector Machines have shown promising performance in binary classification problems. A performance analysis of these techniques (as implemented in Weka) on learning users' preferences on their social network activities will be presented in Section 4.

3.2   Features for Representing Activities

All machine learning techniques listed above require a set of features describing the data. We identify both non-textual and textual features that are potentially useful for learning.

Non-textual Features. Table 1 summarizes a list of relevant non-textual features and some of their possible values. Each activity has an actor (creator). SocConnect allows a user to rate friends as "favourite" or "disliked". Using these two features, we will be able to learn whether a user tends to be always interested in some particular friends' activities or activities from a particular type of friends (i.e. favourite or disliked friends). Each activity has a type. We also take into account the SNS sources of activity, such as Facebook and Twitter, since often users have a particular purpose for which they predominantly use a given SNS, e.g. Facebook for fun, Twitter for work-related updates. From this feature, we can find out whether a user is only interested in activities from particular SNS sources. Different applications used to generate those activities are also useful to consider. For example, if a user's friend plays "MafiaWars" on Facebook but the user does not, the status updates generated from the "MafiaWars" application may be annoying to the user.

Table 1. Non-Textual Features of Activities for Learning

Non-Textual Features	A Set of Possible Values
Actor	actor's SNS account ID
Actor Type	favourite; neutral; disliked
Activity Type	upload album; share link; upload a photo; status upload; use application; upload video; reply; twitter retweet; etc
Source	Facebook; Twitter; etc
Application	foursquare; FarmVille; etc

The above non-textual features of activities can be obtained through the APIs offered by SNSs. In our work, we also consider the textual content of activities, even though many activities, such as video uploads, do not have any textual content. The purpose of having these features is to investigate whether text analysis will contribute to the personalized recommendation of social activities.

Textual Features. In the text analysis part, we first remove the stop words and URL links in each activity. Two vector spaces are then calculated for each activity, one is using TF and another one is using TF-IDF. The reason of using both algorithms is to investigate whether the commonality (IDF value) of terms plays a role in the data mining process in the context of analysis social data.

Having the vector spaces for each activity and given training data containing a set of activities rated by a user as favourite, neutral or disliked, we sum up the weight values for each term in all the favourite, neutral and disliked activities, respectively. The results are three vectors over the training data, for the favourite, neutral and disliked activity sets respectively. Each vector consists of the total weight of each term in all activities of the corresponding set (either favourite, neutral or disliked activity set). We then calculate the cosine similarity between a vector representing each activity and the three vectors representing the favourite, neutral and disliked activity sets, denoted as S_F, S_N and S_D, respectively. Each of these similarity values can represent a textual feature for activities.

We can also use one combined textual feature C for an activity. Two ways can be used to generate a value for this feature. One way is to use the difference between the two similarity values, $C = S_F - S_D$. Another way is to map the difference into the three interest levels, favourite, neutral and disliked, as follows:

$$C = \begin{cases} \text{favourite} & \text{if } 0.33 < S_F - S_D \leq 1 \\ \text{neutral} & \text{if} -0.33 \leq S_F - S_D \leq 0.33 \\ \text{disliked} & \text{if } -1 \leq S_F - S_D < -0.33 \end{cases} \qquad (3)$$

In summary, we can have four potential textual features for representing activities, including S_F, S_N, S_D and the combined one C, as listed in Table 2. Note that the combined feature C can have a continuous value ($S_F - S_D$) or a discrete one (mapped interest levels). Also note that the values of each feature summarized in Table 2 can be calculated based on either TF or TF-IDF. The performance of the different features and the different ways of calculating feature values will be evaluated and compared in Section 4.

Table 2. Textual Features of Activities for Learning

Textual Features	Possible Values
S_F	$\in [0, 1]$
S_N	$\in [0, 1]$
S_D	$\in [0, 1]$
C	$S_F - S_D \in [-1, 1]$; or Mapped interest levels: $\in \{\text{favourite, neutral or disliked}\}$

After learning from a user-annotated list of activities from his or her friends, each of which is represented by a set of the feature values, a learning algorithm is able to predict whether a new activity from a friend will be considered as "favourite", "neutral" or "disliked" by the user.

4   Evaluation

We carried out experiments to evaluate 1) the performance of the four machine learning techniques for learning user preferences on social activities and 2) the performance of personalized recommendations when different features are used to represent social activities. Social data streams from ten subjects were used in the evaluation. Five of the subjects are from Saskatoon, Canada, and the other five are from New Jersey, USA. Half of them are students and the other half are workers. Six of the subjects are experienced users of Facebook and Twitter. For each of these subjects, we collected from Facebook and Twitter 200 recent activities of their friends. The other four subjects are relatively new users of Facebook and Twitter. For each of them, we collected around 100 recent activities of friends. Thus, in total, we collected around 1,600 user activities. We asked all subjects to rate their friends and activities. On average, they rated 38% of their friends as favourite or disliked friends and 45% of the activities as favourite or disliked. Thus, the data sample is quite diverse. A 10-fold cross validation was performed on the collected data from each subject, and the average performances of the machine learning techniques over the activities of all subjects are reported in the following sections.

4.1   Performance When Using only Non-textual Features

We first used only the set of non-textual features summarized in Table 1. Fig. 1 shows the performance of the four machine learning techniques. Although the performance difference among these techniques is not significant, support vector machine (SVM) provides the best performance, and it correctly classifies 69.9% of instances in the testing data. RBF performs the worst (68.4%). The performance of Decision Tree and that of Bayesian Belief Networks are about the same, which is around 69.5%. So, these machine learning techniques generally do not show good performance when only the non-textual features are used for representing activities.

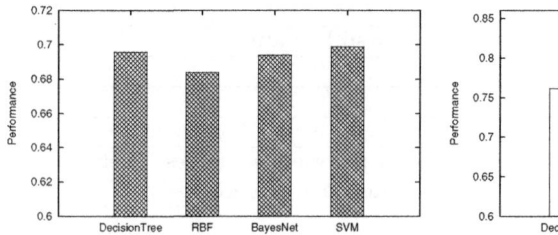
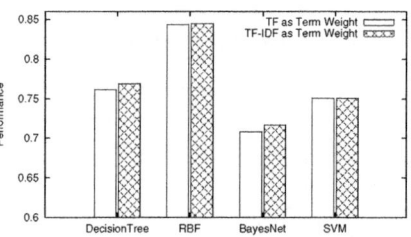

Fig. 1. Performance when only Non-Textual Features are Used

Fig. 2. Performance when Three Textual Features are Used

4.2   Performance When Using Only Textual Features

We then evaluated the performance of personalized recommendations on social activities when only the textual features summarized in Table 2 are used. In this set of experiments, we first tested the performance when the combined feature C is used. All the four machine learning techniques perform the same and achieve 64.9% of correct prediction. In addition, there is no difference when

TF or TF-IDF is used as term weight. Using this feature alone shows even worse performance than using the non-textual features.

We then tested the performance when the other three textual features (S_F, S_N and S_D) are used. The results are plotted in Fig. 2 when TF and TF-IDF are calculated for term weight respectively. We can see that now RBF performs the best (84.5% of correct prediction). RBF is known as generally showing good performance when the values of features are continuous, as it predicts a real-valued function. Decision Tree is the second best and has the performance of 76.9%. SVM is better than Bayesian Belief Network in this case. We can also see that there is still no much performance difference between TF and TF-IDF. From the evaluation results presented in this section, it is also clear that the performance when the three textual features are used is significantly better than that when the combined textual feature C is used and also better than the performance when non-textual features are used.

4.3   Using Both Non-textual and Textual Features

We further evaluated the performance of personalized recommendations on social activities when non-textual and textual features are both taken into account. We first use the combined feature C and the non-textual features. As described in Table 2, four different ways can be used to calculate the value for the feature C of an activity, listed as follows:

- TF+noMap: weight of term is calculated using TF and feature value is calculated by $S_F - S_D$;
- TF+Map: weight of term is calculated using TF and feature value is calculated by mapping $S_F - S_D$ to one of the three interest levels;
- TF-IDF+noMap: weight of term is calculated using TF-IDF and feature value is calculated by $S_F - S_D$;
- TF-IDF+Map: weight of term is calculated using TF-IDF and feature value is calculated by mapping $S_F - S_D$ to interest levels.

The performance of each method is summarized in Table 3. We can see that the methods without mapping to interest levels produce the better performance than those with mapping. There is no much difference between "TF-IDF+noMap" and "TF+noMap" or between "TF-IDF+Map" and "TF+Map". Thus, calculating term weight using TF-IDF does not provide much contribution to the personalized recommendation of social data. The performance when using both the combined feature C and the non-textual features (up to 79.4%) is much better

Table 3. Performance when Using C and Non-Textual Features

Methods	DecTree	RBF	BayesNet	SVM
TF+noMap	0.777	0.793	0.773	0.764
TF+Map	0.712	0.704	0.711	0.716
TF-IDF+noMap	0.780	0.794	0.761	0.749
TF-IDF+Map	0.718	0.698	0.713	0.718

than that using each alone (up to 69.9% with non-textual features and 64.9% with only the combined feature C).

We then use the combination of the three textual features (S_F, S_N and S_D) and the non-textual features. The results are plotted in Fig. 3 when TF and TF-IDF are calculated for term weight respectively. Again, there is no much performance difference between TF and TF-IDF. RBF performs the best (81.4%). Decision Tree and SVM perform similarly (around 80%). Bayesian Belief Network is the worst in this case (around 75.2%).

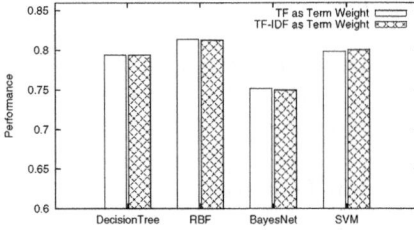
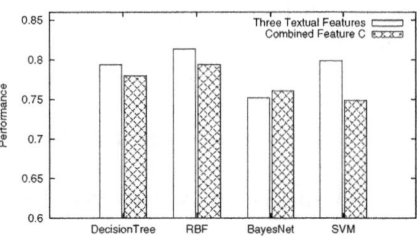

Fig. 3. Using S_F, S_N, S_D and Non-Textual Features

Fig. 4. Performance Comparison between Textual Features

We compare the performance between different textual features when the textual features are integrated with the non-textual features. In this comparison, we choose the best performance of the combined feature C. The result obtained is similar as that when only textual features are used, as shown in Fig. 4. In most of the cases, the three textual features provide better results than the combined feature. Bayesian Belief Network is the exception. The result concludes that it is generally better to use the three features separately instead of combining them.

4.4 More Analysis

To further analyze the obtained evaluation results, we also plot the performance of personalized recommendations when using only non-textual features, when using only textual features of S_F, S_N and S_D, and when using both, respectively in Fig. 5. We can see that in general, the best performance of the machine learning algorithms is produced when both non-textual and textual features are used. Thus, both non-textual and textual features contribute to the personalized recommendations of social activities. Note that RBF is exceptional. Its performance when using both non-textual and textual features is worse than that when using only textual features. Integrating discrete values of non-textual features degrades its performance. We analyzed the evaluation results using two factor ANOVA (analysis of variance) test with replication with 0.05 p-value, and the analysis shows that the difference between the performance of the combined approach and the other two approaches (textual and non-textual) is statistically significant. The ANOVA analysis did not show significant difference in the performance of the four tested machine learning algorithms. The combined text and non-text features approach yielded significantly better results with all four algorithms.

Using Weka's feature selection function, we can see which features are more important for individual users. We summarize in Fig. 6 the number of subjects for whom each feature was the most important one in the prediction. In this experiment, non-textual features and the three textual features (S_F, S_N and S_D) are used because they produce the best performance for most of the machine learning algorithms.

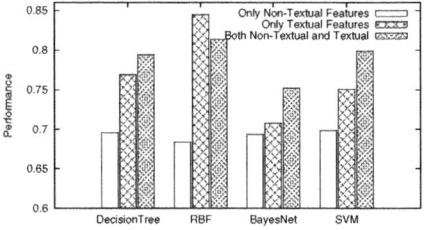
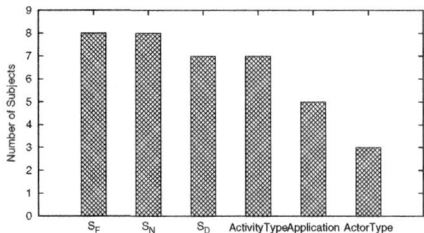

Fig. 5. Performance Comparison for Different Features

Fig. 6. The Most Important Features for Learning

For most of the users, the three textual features are important. This implies that most of the users are interested in the textual content of their friends' activities. "Activity Type" is also important for most of the users. For half of the users, "Application" is important. "Actor Type" is important for three users. The source of activities (i.e. whether they come from Twitter or Facebook) turns out to be not important. This interesting difference represents the diversity of social networking users' criteria in judging whether an activity is interesting to them, reflected in their ratings. Some users mainly care about the textual content of activities. Some users care about the type of their friends' activities. Some users care more about the applications that generate the activities, which are usually the games they are playing. And, some users care about their close friends' activities. The implication is that learning the user type would be useful in selecting the best suitable set of features for personalized recommendation of activities. We leave this for future work.

4.5   Conclusions from the Evaluation Results

Several important conclusions can be drawn from the evaluation results presented in the previous sections: a) both non-textual and textual features contribute to the personalized recommendation of social activities; the combination of textual and non-textual features performs significantly better than only textual or only non-textual features across all four algorithms; b) the best performance (84.5%) is produced by RBF using only the textual data, indicating that good performance can be achieved for the personalized recommendation of social activities; c) calculating term weight using TF-IDF does not show much advantage for textual features; and d) learning user types would be useful for further improving the performance of the personalized recommendations of activities.

5   Contribution and Future Work

Our work shows that it is possible to generate effective recommendations of social data using machine learning. Moreover, we found that both textual and non-textual features have to be taken into account and the results then the results would be comparably good for four machine learning algorithms. For future work, we are interested in exploring more deeply the relative importance of different features of SNS activities, to further improve the performance of the personalized recommendation. Other features that may be worth looking at include the targeted friends in activities (e.g in comments, responses, likes). Our immediate next step will be to conduct user studies to evaluate the quality of recommendations from the user point of view.

References

1. Chisari, M.: The future of social networking. In: Proceedings of the W3C Workshop on the Future of Social Networking (2009)
2. Wang, Y., Zhang, J., Vassileva, J.: SocConnect: A user-centric approach for social networking sites integration. In: Proceedings of the International Conference on Intelligent User Interface (IUI)Workshop on User Data Interoperability in the Social Web. (2010)
3. Carmagnola, F., Vernero, F., Grillo, P.: Sonars: A social networks-based algorithm for social recommender systems. In: Procecddings of the 17th International Conference on User Modeling, Adaptation, and Personalization (2009)
4. Claypool, M., Le, P., Waseda, M., Brown, D.: Implicit interest indicators. In: Intelligent User Interfaces, pp. 33–40. ACM Press, New York (2000)
5. Platt, J.C.: Fast training of support vector machines using sequential minimal optimization. In: Schoelkopf, B., Burges, C., Smola, A. (eds.) Advances in Kernel Methods: Support Vector Learning. MIT Press, Cambridge (1999)
6. Mitchell, T.M.: Machine Learning. McGraw-Hill, New York (1997)

Term Ranking and Categorization
for Ad-Hoc Navigation

Ondrej Ševce, Jozef Tvarožek, and Mária Bieliková

Slovak University of Technology, Faculty of Informatics and Information Technologies,
Ilkovičova 3, 842 16 Bratislava, Slovakia
ondrej.sevce@gmail.com, {jtvarozek,bielik}@fiit.stuba.sk

Abstract. Processing information in web pages and navigation on the web can take significant amount of time for users, requiring them to employ higher cognitive processes such as generalization and categorization. Providing users with annotated entities and terms contained in the text, and adaptive navigation based on these terms could help with the comprehension and better their orientation in the information space. In this paper, we present a method for ad-hoc navigation based on automatic terms retrieval, ranking and categorization. Recognized terms and categories are used as keywords for search in available content offering information spaces. Retrieved hyperlinks can be browsed by the user, while terms and categories gained from the last analyzed page are still available. Finally, the method includes user profiling, which enables grouping of the users based on their preferred terms and categories. Our results show that ad-hoc navigation can ease access to relevant related content on the web.

Keywords: term, category, navigation, conceptual user profile.

1 Introduction and Related Work

Comprehension and interpretation of the text in web pages and navigation in the web information space take significant amount of time for many users (lost in hyperspace problem [6]). In particular, news articles typically contain various entities (persons, places, events), each having its own context that is easily recalled by humans by re-collecting their previous personal experiences regarding these entities, posing a great challenge for machine processing. Systems for entity extraction from unstructured text are either domain specific, for example Essie which operates in medical domain [5], or domain independent, for example the user-friendly Wikify! System [11], which provides descriptions of entities gathered from Wikipedia, producing a "wikified" page to the user.

When extracting entities and terms, one of the issues to deal with is entity disambiguation. Entities and terms appear in the text in their "surface form", which may refer to various interpretations of the entity. This ambiguity can be eliminated by considering contextual evidence (words or other entities that describe or co-occur with the entity) and category tags (which describe topics to which the entity belongs to) [12], or by machine learning on large data sample [7]. Category tags can be operationally

D. Dicheva and D. Dochev (Eds.): AIMSA 2010, LNAI 6304, pp. 71–80, 2010.
© Springer-Verlag Berlin Heidelberg 2010

retrieved from available folksonomies using graph algorithms, also providing the corresponding tag hierarchies [9]. The number of available entity extraction tools is increasing, and latest approaches tend to employ more than one extraction system, thereby increasing both entity recall (more systems recognize more entities) and precision [10].

When providing entity or term extraction results to a user, it may be valuable to assign relevance rating to entity, or to sort entities in order of relevance. Term rating and term ranking are tight together, as the higher rating of the term leads to its position closer to the top of the list. There are approaches that rank terms based on semantic techniques, like for example term expansion used along with terms and documents mapping into L_2 space, and computing the inner product of this space to express similarity [4]. To adjust terms similarity, the sets of terms senses are compared. In [8] the term relevance scoring computation is based on considering term to document relations and also term to term relations. The method involves creation of indexed ontology, which provides valuable metadata for search refinement.

Search behavior of users shows that when navigating to the target, instead of using keywords, users navigate with small, local steps using their contextual knowledge as a guide [13]. Adaptive navigation that is based on retrieved entities and terms can thus have positive impact on user's sense of orientation in the web environment. There are several techniques which support adaptive navigation, such as annotation, sorting, hiding or generating of hyperlinks. Our approach is based on generating hyperlinks; in particular, it provides dynamic recommendation of relevant links [2].

In this paper, we propose an ad-hoc navigation method which relies on short-time user preferences. Terms and categories recognized in the text selected by the user are used as keywords for search in available content offering information spaces. Terms and categories are retrieved using shallow linguistic processing, which proved sufficient results for the purpose of keywords extraction. Based on keywords identified, the method provides links extracted from tweets and bookmarks retrieved from popular online systems Twitter and Delicious. The user can browse the web information space, while the context of the last analyzed page (represented by extracted concepts and recommended links) is still available. The ad-hoc navigation is engaged by explicit user's action, behaving as on-demand service.

Our approach frees users from devising relevant keywords, and gives them a stable context, which can be used as a basis in the web navigation. The difference between our proposed method and other existing methods for term extraction is in the ranking of retrieved terms, which in our case focuses the user's attention to the most important terms available in processed text. Another aspect of our method is that it is tightly integrated with the navigation. The emphasis was put on minimal user's effort simultaneously with providing wide range of navigation possibilities. We proceeded towards this goal also by integrating user interface to the web browser, what enables easy access to the methods results. Further, our method includes user profiling which enables grouping of users based on their preferred terms and categories.

Following our method, we developed a system called *Marquess*, which includes web service capable of processing texts and returning machine-ranked terms and categories. It also supports user profiling based on principles of [1]. For the client side, a Mozilla Firefox add-on was developed, which enables communication with *Marquess* and other web services directly from browser window.

2   Method of Ad-Hoc Navigation Using Term Ranking

We propose a method for automatic term retrieval, ranking, categorization, and ad-hoc navigation, which employs also user profiling. In the following text, we use words *term*, *category* and *concept*, as follows: *term* – the surface form [12] of abstract or concrete entity, as it occurs in the text (for example Barrack Obama); *category* – the surface form of some common characteristic or some generalization of related entities (for example Presidents of the United States); *concept* – term or category.

The typical use case of our method consists of the following steps (see Figure 1):

1. Select a web page or part of its text and send it for processing.
2. Browse retrieved terms and categories.
3. Add / remove terms and categories to / from the user profile.
4. Navigate the user to other pages upon selected concept(s).

The user looks up interesting content on the current web page. She may choose to send the whole page or a part of it for term extraction. The set of terms and categories is then ordered by ranking, and the user is enabled to pick preferred concepts into her profile, or select concept(s) and make it the basis for requesting navigational paths from other information spaces. Concepts are selected explicitly by the user, who is motivated by the need of additional information about the concept. Retrieved navigational links are provided to the user, and their target pages can become sources for subsequent term extraction. The process may be repeated multiple times (see Figure 2) until the user is satisfied with results.

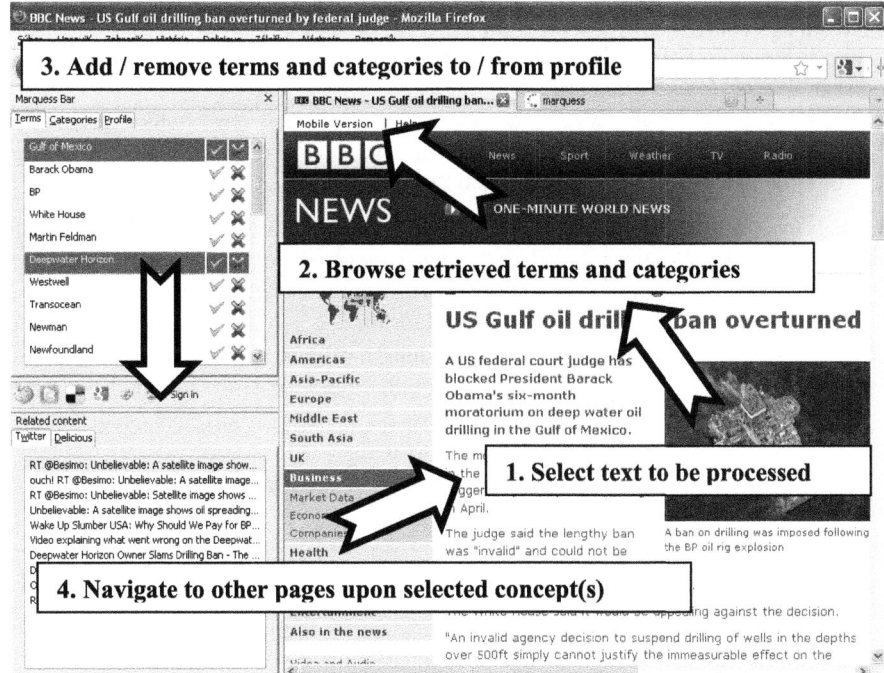

Fig. 1. Method use case – four steps, which enable concept based ad-hoc navigation

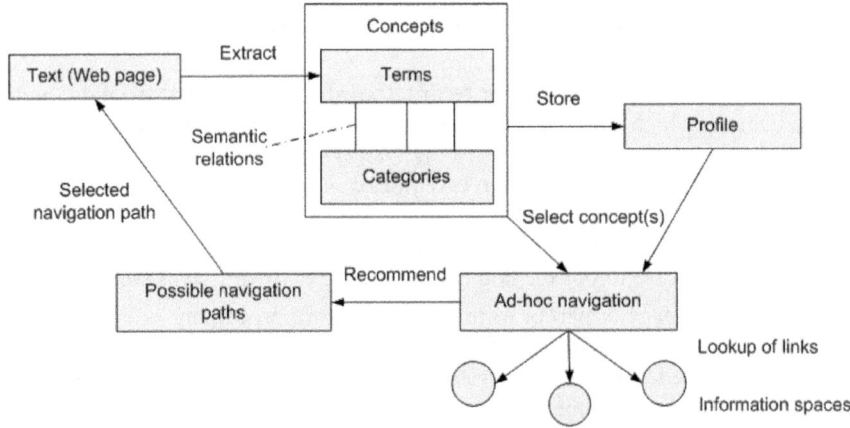

Fig. 2. Navigation recommendation cycle

2.1 Retrieving Terms

To retrieve terms, we employ shallow linguistic analysis in which we distinguish between original and nested occurrence of a term. For example, the text „*The White House Office of Health Reform said the process was going really well*" contains terms "*White House Office of Health Reform*", "*White House*", and "*Health Reform*". We distinguish two types of term occurrences:

- *Original occurrence of a term* – the term in the text is not part of any other term and it logically fits into the context of text. In the stated example, it would be the term "*White House Office of Health Reform*".
- *Nested occurrence of a term* – the term is part of other term, which gives more detailed information and fits better into the context of text. In the stated example, it would be the terms "*White House*" and "*Health Reform*".

Similar approach was used in [14] where author considers occurrence of nested "candidate" terms, which are included in "longer candidate" terms. In our approach, the machine term ranking requires counting of original and nested occurrences of terms in the text, for which we propose following algorithm: t – Vector of tokens (analyzed text), T – Text words count, N – Maximal length of term (words count), o – Occurrence vector of length T; each position contains index of term, in which the token occurred at last, c – Vector of retrieved terms, oo – Vector of original occurrences of terms, no – Vector of nested occurrences of terms.

The algorithm for term retrieving and occurrence ranking is as follows:

```
t = input_text.tokenize(), T = t.lenght
o = vector(T), oo = vector(T), no = vector(T)
c = vector(0)

for (i = 1 to T) {
  potential_term = t[i] + t[i+1] ... t[i+N]
  for (j = N to 0) {
    if (j != N) removeLastToken(potential_term)
```

```
if (isTerm(potential_term)) {
  if (not c.contains(potential_term))
    c.addNewTerm(potential_term)
  C = c.indexOf(potential_term)
  if (o[i] == -1 and o[i+j] == -1)  oo[C] += 1
  else if (o[i] == o[i+j])
    { no[C] += 1; c[C].setNestedIn(c[o[i]]) }
  for (k = i to i+j) o[k] = C
} } }
```

The searching for terms starts at the first token. The `potential_term` is initialized to N subsequent tokens. If `potential_term` is recognized as a new term it is added to the vector of retrieved terms. Next, the occurrence vector is checked (on positions of marginal tokens of current term), whether the actual term was already included in some other term. If not, the original occurrence of the actual term is increased, otherwise (if the actual term is enclosed in other term) the nested occurrence is increased and a relation between the terms is recorded. Next, the occurrence vector of every token included in the term is set to the index of the current term. In subsequent iterations, the last token of `potential_term` is removed until only the first token is the `potential_term`. Recognizing of terms continues in this way beginning with each token of the analyzed text.

We use the DBPedia dataset consisting of Wikipedia articles labels as the primary source of terms [3], i.e. as each term is directly related with article about itself. The retrieving of terms is based on string matching. The DBPedia dataset consists of more than three million articles labels (in English version).

To enable real-time dataset search, we index the dataset's content using a hash map where key is the first word of article label, and value is the position of the first occurrence of this word in the dataset. Then, during the search for `potential_term` in the dataset, the position of first word of `potential_term` can be easily looked up in the hash map, and subsequently, the articles labels beginning with this word are compared with `potential_term`. If a match is found the `potential_term` is added to the list of retrieved terms.

2.2 Ranking Terms and Categories

The rating of a term is estimated by presented formula 1 (considering term occurrences and word count). Devising the weight coefficients presented in the formula is explained in the Section 3.

$$T_i = (W_{wc} \cdot wc_i) + (W_{oo} \cdot oo_i) + (W_{no} \cdot no_i) \tag{1}$$

where T_i – relevance rating of term, wc_i – word count of term, oo_i – original occurrences of term, no_i – nested occurrences of term, W_{wc} – weight of the word count of term, W_{oo} – weight of original occurrences of term, W_{no} – weight of nested occurrences of term.

Articles in Wikipedia are grouped in more than 400,000 categories. DBPedia offers the dataset of categories and relations between articles and categories currently containing about two million records. We take advantage of the human-made relations when looking up categories of a particular term. Each category gains relevance rating based on the ratings of its related terms (the rating of category is the sum of ratings of related terms, see Formula 2). Categories are presented to the user in a separate list, ranked by machine-computed relevance, and simultaneously providing the user with more general information related to the analyzed text.

$$C = \sum\nolimits_{j=1}^{n} T_j \qquad (2)$$

where C – relevance rating of category, n – number of terms which occur in the text, and are related to the category, T_j – rating of term.

2.3 Ad-Hoc Navigation

Based on client's interactions with *Marquess*, information spaces are searched for additional content by using their online interfaces. The ad-hoc navigation is affected by user's choice of a page to be analyzed, and subsequently, by picking the concept(s) to be looked up in other information spaces. User is enabled to pick one or more concepts at the time. When multiple concepts are chosen the search string is built using these concepts in order of their ranking.

Client side of the implemented system is interacting with popular micro-blogging system Twitter, bookmarking system Delicious, provides simple Google search feature, and direct links to Wikipedia articles (for terms and categories). Twitter and Delicious services return tweets and bookmarks ordered by creation time (recent results appear first); we present them to users in this same order. The system extracts links from tweets and enables user to read the tweet, view target page of the link or home page of person who published the tweet. Opening the links included in tweets can be a direct way to obtain recent news and information. Delicious bookmarks usually provide more time insensitive content, guiding users to general information about the selected concept(s).

2.4 Profiling

In the proposed method, profiling partly depends on user's interactions with the system. After collecting and ranking of terms and categories, these are presented to the user in temporary lists. The user may interact with these lists by adding or removing terms and categories to or from her profile. When accumulated in profile, concepts create a base of user's preferences, and have universal usage, such as keywords for search engine queries, keywords for searching in folksonomies, matching RSS feeds, and so on. The profile contains also data about user's interactions with the system (as proposed in [1]) and relations between terms and analyzed documents, what makes it the base for further content recommendation.

3   Evaluation

To evaluate the proposed method, we performed three experiments. The goal of the first experiment was to set the weights coefficients of the formula 1 so that we would gain satisfactory term ranking. Second experiment evaluated weights coefficients gained in the first experiment by comparing human and machine rating of categories relevance. In the third experiment, we evaluated the relevance of links provided by the ad-hoc navigation.

In the first experiment, we evaluated different combinations of weights presented in formula 1. The machine rating of terms is based on original and nested occurrences of the term and on word count of the terms (*oo*, *no*, *wc*). These parameters are multiplied by weights coefficients and their sum is the machine rating of the term (which implies the ranking of the term). To set the values of weights coefficients, we involved human experts in the experiment, and used their relevance rating to optimize weights values. Figure 3 shows the dependence of MSE on weights combinations (explained further).

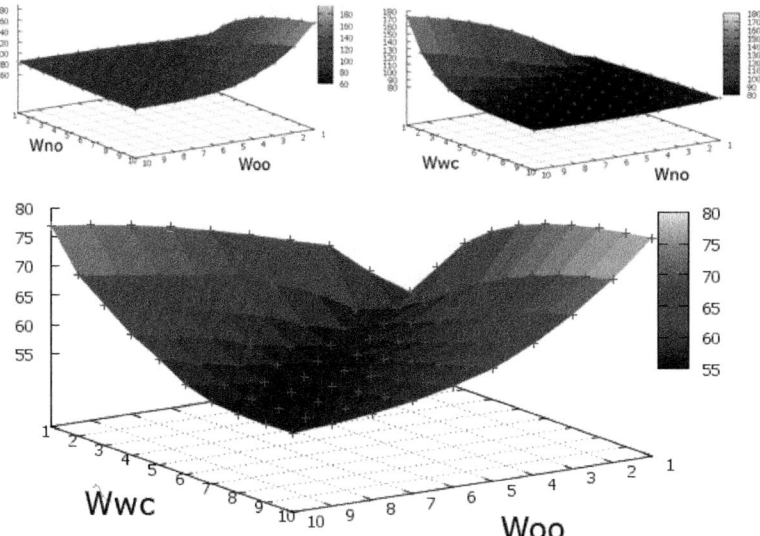

Fig. 3. Values of MSE computed from combinations of W_{wc}, W_{no} and W_{oo}. Lower MSE means smaller difference between human and machine terms rating. The graphs show that certain combinations of weight coefficients W_{wc} and W_{oo} gives results closest to the human rating.

Seven human experts rated relevance of terms found in six various articles from BBC news on the scale from zero to ten. Average human rating of each term was considered as the real relevance of the term. The system was optimized by testing combinations of weights (1,331 combinations, each weight W_{oo}, W_{no}, W_{wc} was represented 11 times, on the scale from 0 to 10) to gain ratings closest to the average human ratings. The weight scale measure was set to one, as more fine scale measurement didn't significantly affect the results.

The quality of machine rating was evaluated by MSE (MSE was calculated for each article as the sum of the square roots of differences between average human rating of term and the machine rating of term, for each term retrieved from the article). The z-dimension of graphs in Figure 3 shows computed MSE for given weights coefficients combination. The coefficients combination that gave closest rating to the average human rating was $W_{oo} = 3$, $W_{no} = 4$, $W_{wc} = 2$, MSE = 45.894.

In the second experiment, we evaluated optimized weights coefficients. We let the same group of human experts rate the relevance of categories related to retrieved terms. The relevance of category was marked by the expert by saying "yes, it is relevant to the text", or "no, it is not relevant to the text". Each category was rated by seven human experts, and the relevance of category was proportional to the number of "yes, it is relevant to the text" choices.

Table 1. Difference between normalized human and machine rating of categories related to the terms extracted from analyzed documents

Article	Average rating MSE	Categories	Standard deviation of MSE
1	7.610	15.0	7.963
2	4.391	26.0	4.173
3	5.955	22.0	11.061
4	2.563	5.0	1.674
5	14.936	10.0	11.979
6	9.732	15.0	12.089
Average	**7.531**	**15.5**	**8.156**

Totally, average MSE for one category rating was 7.18, which means that the average difference between human and machine rating was 2.68 on the scale from 0 to 10.

In the third experiment, we evaluated the relevance of links provided by the ad-hoc navigation. The navigation links were looked up by selecting one, two or three terms with highest ranking. We evaluated total number of navigation links retrieved, number of relevant navigation links and finally number of off topic navigation links.

Table 2. Evaluation of ad-hoc navigation

	Twitter			Delicious		
	1 term	2 terms	3 terms	1 term	2 terms	3 terms
Avg. keyw. count	1.71	2.57	3.85	1.71	2.57	3.85
Total tw./bkm.	8.86	11.43	8.29	12.86	7.53	2.64
Relevant to article	4.71	5.67	3.52	5.71	3.25	2.14
Relevant to terms	4.71	5.86	3.95	10.14	5.48	2.32
Off topic	1.14	0.29	0.00	0.68	0.30	0.11

Obtained results suggest the following findings about ad-hoc navigation: (i) Twitter reacts very fast on news articles, as the relevance of discovered links is nearly identical for terms and for the article (terms "gain" the relevance from being stated in related articles); (ii) Number of off topic navigation links is decreasing while the number of used terms increases (both Twitter and Delicious); (iii) Searching two highest-ranked terms brought the highest number of navigation links (total, relevant to article, relevant to terms) from Twitter; in the case of Delicious, using only the single highest-ranked term brought the most navigation links.

4 Summary and Discussion

In this paper, we have presented and evaluated a method for automatic terms retrieval, ranking, categorization and ad-hoc navigation. One of the key aspects of our method – term ranking is based on shallow linguistic analysis which appears to be sufficient for the purpose of ad-hoc navigation. When rating terms, three weight coefficients are used (for each measured parameter). These weight coefficients were optimized by adapting machine rating of terms to average human relevance rating of terms.

Selecting and evaluating various combinations of weights (where one of weights was always set to zero) demonstrated that the importance of original occurrences and word count of term is higher than the importance of nested occurrences, although the best result was achieved while nested occurrences weight was higher than the other two weights. When rating categories with optimized weight coefficients, we gained quite unbalanced variation between human and machine ratings, although categories were related to the text via semantic relations with retrieved terms. By using human-made semantic relations, every category offered to a human expert for rating was relevant to the text. The unresolved question is, if the experts knew about the relation of this category to the article; the relevance rating of categories may be influenced by this "nescience" of human experts. However, user's interest in categories and their related content is partially based on personal preferences, so the precise relevance order of categories may not be the crucial issue of the proposed method.

The ad-hoc navigation showed both strengths and weaknesses. The positive aspects of this kind of navigation are its context dependency and adaptability. These emerge from its integration with dynamic folksonomies providing data processed by collective intelligence. These data are not perfect, for example Twitter contains high amount of noise and redundancy. The noise is represented by off topic tweets reflecting current events or terms related to these events. Noise can be partly eliminated by filtering tweets without links.

The redundancy is difficult to discover, and thus not easy to eliminate. By frequent use of link proxies in tweets, many links in fact point to the same article. On the other hand, links retrieved from Delicious were affected by less noise and contained less redundancy. Delicious links also proved higher relevance when requesting bookmarks for articles regarding more significant or long-term topics, while some topics were ignored by Delicious users. These results confirmed that Twitter is a good resource of links for current topics, while Delicious provides links with more long-term usability. Therefore our ad-hoc navigation method could be useful for users demanding various information about actual events, or users with deeper interests in particular topics.

In the future, we plan to improve the navigation method and users profiling by discovering of similarities between users' profiles. Information stored in profiles should be used for relating users via compliant terms, categories and documents. These relations may allow for a more sophisticated content recommendation. There are also unresolved issues about terms ambiguity, for example names of persons often refer to different persons. These issues should be eliminated by integrating additional services and gathering meta-data about terms retrieved from the texts.

Acknowledgments. This work was supported by the Scientific Grant Agency of SR, grant No. VG1/0508/09, the Cultural and Educational Grant Agency of SR, grant No.

028-025STU-4/2010, and it is a partial result of the Research & Development Operational Program for the project Support of Center of Excellence for Smart Technologies, Systems and Services II, ITMS 25240120029, co-funded by ERDF.

References

1. Pazzani, M., Billsus, D.: Content-based Recommendation Systems. In: Brusilovsky, P., Kobsa, A., Nejdl, W. (eds.) Adaptive Web 2007. LNCS, vol. 4321, pp. 325–341. Springer, Heidelberg (2007)
2. Brusilovsky, P., Maybury, M.T.: From adaptive hypermedia to the adaptive web. Communications of the ACM 45(5) (2002)
3. Auer, S., Bizer, C., Lehmann, J., Kobilarov, G., Cyganiak, R., Ives, Z.: Dbpedia: A nucleus for a web of open data. In: Aberer, K., Choi, K.-S., Noy, N., Allemang, D., Lee, K.-I., Nixon, L.J.B., Golbeck, J., Mika, P., Maynard, D., Mizoguchi, R., Schreiber, G., Cudré-Mauroux, P. (eds.) ASWC 2007 and ISWC 2007. LNCS, vol. 4825, pp. 722–735. Springer, Heidelberg (2007)
4. Wittek, P., Darányi, S., Tan, C.L.: Improving text classification by a sense spectrum approach to term expansion. In: Proc. of the 13th Conf. on Computational Natural Language Learning, pp. 183–191 (2009)
5. Ide, N.C., Loane, R.F., Demner-Fushman, D.: Essie: A Concept-based Search Engine for Structured Biomedical Text. American Medical Informatics Assoc. 14(3), 253–263 (2007)
6. Otter, M., Johnson, H.: Lost in hyperspace: metrics and mental models. Interacting with Computers 13, 1–40 (2000)
7. Milne, D., Witten, H.I.: Learning to link with Wikipedia. In: Proc. of the 17th ACM Conf. on Information and Knowledge Management, pp. 509–518 (2008)
8. Šimko, M., Bieliková, M.: Improving Search Results with Lightweight Semantic Search. In: Proc. of the Workshop on Semantic Search, SemSearch 2009 at the 18th Int. World Wide Web Conf., WWW 2009. CEUR, vol. 491, pp. 53–54 (2009)
9. Barla, M., Bieliková, M.: On Deriving Tagsonomies: Keyword Relations coming from the Crowd. In: Nguyen, N.T., Kowalczyk, R., Chen, S.-M. (eds.) ICCCI 2009. LNCS (LNAI), vol. 5796, pp. 309–320. Springer, Heidelberg (2009)
10. Iacobelli, F., Birnbaum, L., Hammond, J.K.: Tell me more, not just "more of the same". In: Proc. of the 14th Int. Conf. on Intelligent User Interfaces, pp. 81–90 (2010)
11. Mihalcea, R., Csomai, A.: Wikify! linking documents to encyclopedic knowledge. In: CIKM 2007: Proc. of the 16th ACM Conf. on Information and Knowledge Management, pp. 233–242. ACM, New York (2007)
12. Cucerzan, S.: Large-Scale Named Entity Disambiguation Based on Wikipedia Data. In: Proc. of Empirical Methods in Natural Language Processing, pp. 708–716 (2007)
13. Teevan, J., Alvarado, C., Ackerman, S.M., Karger, D.R.: The perfect search engine is not enough: a study of orienteering behavior in directed search. In: Proc. of the SIGCHI Conf. on Human Factors in Computing Systems, pp. 415–422 (2004)
14. Fahmi, I.: C-value method for multi-word term extraction. In: Lecture for Seminar in Statistics and Methodology, Alfa-informatica, RuG (2005)

Time Optimized Algorithm for Web Document Presentation Adaptation

Rong Pan and Peter Dolog

IWIS – Intelligent Web and Information Systems,
Department of Computer Science, Aalborg University,
Selma Lagerlöfs Vej 300, DK-9220 Aalborg, Denmark
{rpan,dolog}@cs.aau.dk

Abstract. Currently information on the web is accessed through different devices. Each device has its own properties such as resolution, size, and capabilities to display information in different format and so on. This calls for adaptation of information presentation for such platforms. This paper proposes content-optimized and time-optimized algorithms for information presentation adaptation for different devices based on its hierarchical model. The model is formalized in order to experiment with different algorithms.

Keywords: adaptation, hierarchical model, web document.

1 Introduction

Currently the mobile multimedia network is developing fast, and many more choices are being provided for people to access Internet resources. A large number of resources are available on the web. However, one fixed page size fits all approach does not fit mobile web very well mainly because of their different capabilities to exhibit information comparing with personal computer (PC) monitors, such as the size, resolution and the color quality of a device. For example, displaying a web page on a PC might turn out a mess on the screen of a PDA or a mobile phone: texts are jam-packed due to the limited screen size. Pictures are fragmented and presented with inappropriate resolution, requiring scrolling both horizontally and vertically necessary to read the whole page. With the development of various interconnected devices, the situation is worsening.

An adaptive presentation approach of a web document might relieve such a problem. That is, the mobile device should be able to choose the best way to display the content automatically. Our former work [11] proposed a method of retaining some redundancy on the web documents by repeating the same information in different scales of detail like size and color, to allow the terminal (which means the PDA, mobile phone's screen) to choose the most suitable scale for it. In the previous example, a traditional web page can be repeated in different grains of detail, together with some meta-tags for the terminal to identify. When it is browsed on a PDA, the quality of pictures will be reduced to fit the screen. If it is shown on an ordinary mobile phone,

D. Dicheva and D. Dochev (Eds.): AIMSA 2010, LNAI 6304, pp. 81–90, 2010.
© Springer-Verlag Berlin Heidelberg 2010

only some text-based descriptions will be provided to facilitate the even smaller display. To sum up, the web document will permit the devices to display auto-adaptive according to their abilities, by choosing the content's appropriate form, and hence alleviate the misrepresentation.

This work is a progress of a series of former works [7, 9, 10, 11]. Previously a simpler hierarchical model was introduced as a special case of the one defined in Pan et al [11], in which the model is generalized from binary trees to more practical un-ranked ones. Building on this ongoing work, this paper's main contributions are as follows:

- For the hierarchical model, it provides an algorithm for time-optimized solutions which is more feasible with its detailed analyses.
- The comparison among other two algorithms for content-optimized solutions and time-optimized solutions; discussion about the advantages and disadvantages of each solution.

The remainder of this paper is organized as follows. In Section 2 we discuss related work. Section 3 will recall the hierarchical model and the formal problem related to it, then Section 4 will propose the algorithms for such problem, analyze it and compare the algorithms. Section 5 introduces the evaluation for three algorithms, and Section 6 presents the conclusion and future works.

2   Related Works

The ontology or knowledge based method [1, 2, 6] is a main approach to creating an adaptive or pervasive computing environment. The knowledge to be presented is structured upon its formal semantics and reasoned about using some reasoning tools, based upon certain axioms and rules. This is suitable to derive deeper facts beyond known knowledge and their exhibition. Compared with their way, the work presented in this paper is more pages implementation-related and fit for simple format-based hypertext to be displayed. In [13], traditional bipartite model of ontology is extended with the social dimension, leading to a tripartite model of actors, concepts and instances, and illustrates ontology emergence by two case studies, an analysis of a large scale folksonomy system and a novel method for the extraction of community-based ontology from web pages.

[3] presents a statemachine based approach to design and implement adaptive presentation and navigation. The work in this paper can be used as a target model for a generator from a state machine.

Phanouriou [14] and Ku et al [5] also proposed markup languages to interconnect different devices. The former one proposed a comprehensive solution to the problem of building device-independent (or multi-channel) user interfaces promoting the separation of the interface from the application logic. It introduced an interface model to separate the user interface from the application logic and the presentation device. It also presented User Interface Markup Language 2 to realize the model. The latter proposed the device-independent markup language that generated automatically and thus unified the interfaces of home appliances, while interfaces generated by the proposed transformation engine would also take into account interfaces previously generated for the user and create single combined interfaces for multiple connected appliances. These two mainly focus on both user and application interfaces.

Er-Jongmanee [4] took XML and XSLT to create user interfaces which is also a possible way of implementation, but her work concentrates on design time adaptation while the work presented in this paper focus runtime adaptation.

Wang and Sajeev [15] provided a good conclusion for the state-of-the-art in such field. They studied abstract interface specification languages as the approach for de-vice-independent application interface design. Then they classified their design into three groups, discussed their features and analyzed their similarities and differences.

3 Hierarchical Model

This section aims at a brief recollection of the hierarchical model defined in Pan et al [11] to make this paper more self-contained. Detailed descriptions of this model can be found in [11].

Informally, our approach of web document presentation adaptation is to provide dif-ferent abstraction levels of the content, by the following cutting-and-condensing method. First, divide a web document into N_0 segments. Then condense neighboring segments into a briefer one by abstracting words and/or reducing picture quality, viz. losing some type of content details. This is performed recursively to form different levels of more general segments. Each segment corresponds to several adjacent ones in a lower level, forming a complete tree structure with all leaves on the same level. In the achieved tree structure, each segment serves as a node, and the relationship of abstrac-tion forms parents and children. At last, the whole document is condensed as one segment, that is, the root of the tree structure (the title of the web page, for example). Suppose abstracted segments never need more capability of the terminal (in the light that they never take more area on the terminal) than their children, then our problem can be formalized as finding proper nodes on the tree to display, whose capability consump-tion suits the terminal's limit. Let us now describe formally the model.

First is a mapping to assign a serial number for each segment

$$f : Infoparts \rightarrow \mathbb{N}$$

in which we make a sequence of all the segments of all levels, from the n-th level (most condensed; root level) to the 0-th level (most detailed; leaf level). We assign 0 as the root's mapping and denote the number of nodes on the i-th level as N_i. Hence each segment's mapping can be deducted according to its position in the tree.

Definition 1. For $m, n \in \mathbb{N}$, denote m as n's *parent* if $f^{-1}(m)$ is a directly compressed part of $f^{-1}(n)$ and some of its neighbors; if m is n's parent, then n is m's *child*.

From Definition 1 we obtain two mappings ch and pa:

$ch : \mathbb{N} \rightarrow 2^{\mathbb{N}}$, mapping a number to all its children;

$pa : \mathbb{N} \rightarrow \mathbb{N}$, mapping a number to its parent.

And for the convenience of statement, we indicate a lifting of ch as

$$CH : 2^{\mathbb{N}} \rightarrow 2^{\mathbb{N}}, S \mapsto \{ch(n) \mid n \in S\}.$$

Definition 2. For $m, n \in \mathbb{N}$, denote m as n's *ancestor* if a sequence $m, m_1, m_2, \ldots, m_k, n$ exists, such that in each adjacent pair, the left is the parent of the right; if m is n's ancestor, then n is m's *descendant*.

Based on Definition 2, another two mappings *de* and *an* are defined:

$de : \mathbb{N} \rightarrow 2^{\mathbb{N}}$, mapping a number to all its descendants;

$an : \mathbb{N} \rightarrow 2^{\mathbb{N}}$, mapping a number to all its ancestors.

We can define the *weight* of each node as its exhibited size decided by the layout manager of the browser. When the layout manager generates the nodes, each node has its own size, which can be calculated as the *weight* of the consumed resources.

Definition 3. Denote w_i as i's *weight*, where w_i is defined by an existing mapping $we : \mathbb{N} \rightarrow \mathbb{N}$, and $w_i = we(i)$.

With these definitions, the problem can be changed to a new form: with a given sequence of weights $w_0, w_1, \ldots w_N$, where $w_i \leq \sum_{j \in ch(i)} w_j$, construct a set $S \subseteq \{0, 1, \ldots, N\}$, satisfying that for any a, $N - N_0 \leq a \leq N$, $\exists b \in S$, $b = a$ or $b \in an(a)$, and $\sum_{i \in S} w_i \lessdot w$, where w is given.

Definition 4. Denote a tree T as a *hierarchical model*, if each node of T has been assigned a natural number as an identifier (the assignment is in breadth-first order and begins at zero), and each node i has unique weight w_i, where any parent node i and its children set $ch(i)$ have relationship $w_i \leq \sum_{j \in ch(i)} w_j$ for their weights.

Definition 5. Set S is defined to be a *cover set* of a hierarchical model T, if $\forall i \in S, i \in T$ holds for any node i, and for any leaf node $i (N - N_0 \leq i \leq N)$ in T, on the path from it to the root node 0, there exists one and only one node s, such that $s \in S$. (For any empty hierarchical model, its cover set is defined as empty set.)

The problem of auto-adaptation of web document can be formalized as:

Definition 6. For certain hierarchical model T and constant w, their *hierarchical problem* is to find a cover set S of T, such that provided the sum of weights of the nodes in S does not exceed w, $|S|$ should be maximized.

For any given hierarchical model T and constant w, a natural means to find a best (or largest) cover set S is to try every cover set of T, and choose the one with the most nodes and not exceeding the weight limit w.

The cover set containing only the root, namely, $\{0\}$, is a cover set. Besides this one, there are still many others, which all follow the rule that each of them is built up by some smaller cover sets of the root's children trees. It is not difficult to prove that only these two types of cover sets exist. So an equivalent definition may be stated as follows.

Definition 7. Set S is defined to be a *cover set* of a hierarchical model T, if $S = \{0\}$, or $S = \bigcup_{i=1}^{N_{n-1}} S_i$, where $S_i \cdots S_{N_{n-1}}$ are cover sets of 0's child trees, respectively. (For any empty hierarchical model, its cover set is defined as empty set.)

This recursive definition reveals the essence of cover sets. It also shows a direct way to find all the cover sets of a hierarchical model: first find all the cover sets of all root's children trees, then make a combination with each cover set of each child tree, plus the one containing only the root.

4 Algorithms for the Hierarchical Model

4.1 Two Related Algorithms for Content-Optimized Solutions

In Pan et al [11] we discussed two algorithms for the hierarchical model; and the content-optimized solutions are to reveal as much content as possible (which might be infeasible in computation) while time-optimized plays the other role. One of the algorithms is based on direct search which has both a local set to catch the current cover set and a global repository to store all of them. It first tries to add each whole level of nodes into the current cover set, then saves the cover set into the repository, and attempts to replace some of the nodes with the children next.

Another algorithm is based on dynamic plan. The goal is to find and store the number of nodes in the best cover set of each node as the root of a small hierarchical model, which is part of the original one, and combine these cover sets with the numbers stored to form the solution to the original problem.

The first algorithm's time complexity is $O(m^N)$. The second one's time complexity is $O(\binom{w-1}{m-1}wN)$, with N as the number of nodes in the hierarchical model, for each node has exactly m children, each of which has a weight at most w.

4.2 An Algorithm for Time-Optimized Solutions

For the hierarchical problem of a model, there is no formal definition for what a time-optimized solution is; hence it can be achieved in many different ways. Here the greedy principle is adopted to conduct a local search for a possible solution to the problem. It restricts the search within two adjacent levels of the model, namely, for given weight limit w and hierarchical model T, denoting CH^i as the i-th composition of CH, if

$$\sum_{i \in CH^{n-k-1}(\{0\})} w_i \leq w \leq \sum_{i \in CH^{n-k}(\{0\})} w_i$$

holds, say, the k-th level of T has a sum of weight no less than w and the $(k+1)$-th level has a weight sum no more than w, then the search will base on all the nodes on the k-th level, whose total weight just exceeds the limit, and some of these nodes will be replaced with their parents on the $(k+1)$-th level to reduce the total weight to a value below the limit.

The procedure of our algorithm can be divided into two steps. The first is to find a proper level as the search base, and the second is to replace some nodes on the base level with their parents. In the second step another greedy strategy is used, that on the search base level, the weight which is the sum of children's weight subtracting their parents', and divided by the number of the children minus one, is sorted descendently. The children with larger quotients above are replaced to enforce a faster decrease of total weight. The greedy strategy used here is to prevent the selection of replacing order from drowning in a KNAPSACK-like problem. At last, also it can be seen that the solutions are restricted in the two adjacent levels.

procedure TIME-OPTIMIZED (*h_model*, *w*) **returns** the time-optimized cover set of *h_model*

    inputs: *h_model*, the hierarchical model

          w, the weight limit

  $i \leftarrow n$; *result* $\leftarrow \{n_0\}$

  while $0 \leq i$

    weight $\leftarrow$ the total weight of the nodes on *level_i*

    if *weight* $\leq w$ **then**

      result $\leftarrow$ {all nodes on *level_i*}

    else if $i=n$ **then**

      result $\leftarrow$ **empty**; **exit while**

    else

      DESC_SORT($SA = \{(\sum_{k \in ch(j)} w_k - w_j) / (|ch(k)| - 1) \mid j \in CH^{n-i}(\{0\})\}$)

      $j \leftarrow 0$

      for $j \leftarrow 0$ **to** $N_{i-1} - 1$ **step** 1

        REPLACE(*result*, *ch* (*j*-th node of *SA*), *j*-th node of *SA*)

        RENEW(*weight*)

        if *weight* $\leq w$ **then**

          exit for and **while**

        end if

      end for

    end if

    $i \leftarrow i - 1$

  end while

  return *result*

end TIME-OPTIMIZED

The greedy idea restricts the search within the nodes: each level is just scanned once at most, and even each node is also considered no more than once. So the complexity of the algorithm is highly improved, that is, $O(N \log N)$ (where N is the number of nodes). This is feasible enough for most applications.

However, the solution gained by this means is generally a possible one. If we take the number of nodes in the cover set as the only standard to judge the quality of a solution, then sometimes this solution can have a considerable gap from the best one, due to the imbalance of the weights on different branches of trees. Below is a some-what extreme example to illustrate this.

Example 1. Suppose the nodes of a hierarchical model have their weights listed as below, and the weight limit is 7. The time-optimized solution algorithm for this will find t result $S' = \{n_1, n_2, n_8, n_9\}$, shown in Figure 1 (with the numbers beside nodes to represent their weights).

This algorithm's time complexity perfectly fits the needs of web document presentation in runtime, especially in the mobile multimedia network environment. Mean-while, the nodes that this algorithm returns are placed on the two adjacent levels of

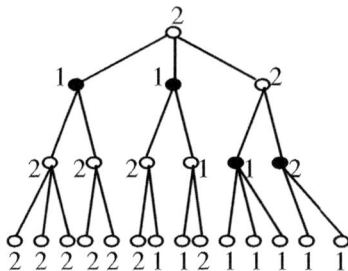

Fig. 1. An Example of the Time-optimized Solutions

the model, so they are similar in abstraction scale of the web document, and are thus more convenient for display, in support of this practical algorithm.

5   Evaluation

For this time-optimized algorithm and the two for content-optimized solutions in Pan et al [9], we have performed experiments over many different sorts of web documents and done statistics with their models.

5.1   Motivation

As far as evaluation is concerned, we focus on two aspects, viz. time consumption of the algorithms and quality of nodes in the cover sets they provide, respectively. Here "quality of nodes" consists of two criteria, i.e. the sum of nodes' weight and the similarity of levels the nodes are in. Hence we have three perspectives of observation for our experiments: time consumption, sum of nodes' weight, and level variance. We will compare and explain these three aspects in the following sections.

5.2   Method

The idea of the experiment is as follows. 1. Select a group of HTML pages as a basic test page. 2. Process the various elements in the page as the nodes in the hierarchical model. 3. Assign the weight of each node based on the consumed resources, where the terminal maximum weight limit is based on the issuance capacity. 4. Use the above algorithm for the cover sets. 5. Finally, the nodes in the model obtained are organized as a new re-mix, and make comparison among the three algorithms according to the returned nodes set and the time consumption. Thus the original web page is completed for adaptive information issuance.

5.3   Dataset

As for experiment dataset, we have built respective 1,000 and 10,000 hierarchical models from two sets of web documents. When we build those models, the web documents' layout is removed and we only consider the number of characters to calculate the weight of each node. We want to show the differences of results when choosing the various numbers of hierarchical models. For both sets of 1,000 and 10,000 test cases we still have two dimensions as scales of the data: level numbers of models and weight limits of each node. The level numbers are 3, 4 and 5 and the weight limits are 3, 4 and 6, dividing each dataset as 9 groups. We also want to show the stability of the algorithm in time consuming and the quality of cover sets by choosing the different number of levels and weight limits. The observed time consumption is based on the average time of executing the same experiment ten times over the same dataset (to obtain more accurate result).

5.4   Experimental Result

Of the entire 10,000 hierarchical models constructed, in the condition of the maximal weight limit is 6 and for the 5-levels, the two algorithms for content-optimized solution both return 57,981 nodes, while the one for time-optimized gave 53,277; the ratio of the two types of solutions is about 1.09:1. It is found that, on a very large hierarchical model (which is the only one), the content-optimized solution has 10 more nodes than the time-optimized one; in other situations the difference in nodes of two types of solutions has not exceeded 5, and mostly there are only one in difference.

With a little gap in nodes quantity, these three algorithms have distinct time performance. For all the 10,000 cases, the direct search algorithm took 399.487 seconds, the dynamic plan version is done in 0.257 seconds, and it is just 0.018 for the time-optimized one. The first is over 22,194 times more than the last one. The comparison on time consumption is shown in Table 1.

Another observation on our experiments is the other evaluation of cover set quality: the level variance of nodes, which is the sum of all cases' square roots of variance concerning the level difference of nodes in each cover set. The smaller the value is, the nearer levels are the nodes in the cover sets on. Since the time-optimized solution algorithm always finds solutions in adjacent levels, its level variance is just about half of the former two's. On this aspect the time-optimized algorithm performs better than the other two due to its greedy essence. On average, the time-optimized algorithm finds cover sets with slightly more than half level variance of that of the results from the other two methods, say 2043.37 compared with 3571.19 (57.22% in proportion). This trend is more evident for hierarchical models with more levels, as this data is 626.01 vs. 988.74 (63.31%) for 3-level hierarchical models whereas 731.13 vs. 1428.43 (51.18%) for 5-level ones. This might be attributed to the fact that the nodes' levels tend to vary more in models with more levels (and hence the other two algorithms tend to find worse results albeit they have more total nodes). We conjecture this trend would be more obvious for more complicated hierarchical models where time-optimized algorithm would yield more practical results for issuance.

Through the comparison, the direct search algorithm should be eliminated from practical use. The dynamic plan can find better cover sets within not long time, but the abstraction degree of nodes is not as good as the one for time-optimized solutions, whose execution time is also the least. To conclude, in a practical environment, these two algorithms should be evaluated more precisely according to the application's real needs.

Table 1. Comparison of the three algorithms On Time Consumption

Algorithm	1000 Cases				10000 Cases			
	Weight Limit	3- Levels	4- Levels	5- Levels	Weight Limit	3- Levels	4- Levels	5- Levels
Direct Search	3	0.637	3.473	64.893	3	2.339	9.690	381.250
	4	0.781	3.561	67.815	4	2.351	10.310	390.120
	6	0.820	3.667	69.375	6	2.397	11.375	399.487
Dynamic Plan	3	0.000	0.001	0.016	3	0.031	0.072	0.156
	4	0.001	0.002	0.032	4	0.046	0.087	0.188
	6	0.016	0.002	0.038	6	0.049	0.110	0.257
Time- Optimized	3	0.001	0.001	0.001	3	0.006	0.007	0.009
	4	0.001	0.001	0.001	4	0.006	0.008	0.010
	6	0.001	0.001	0.002	6	0.006	0.016	0.018

6 Conclusion and Future Works

In this paper we proposed a time-optimized algorithm based on the HTML hierarchical model for adaptive web document presentation, and compared it with our former works. The comparison indicates that our previous content-optimized solutions can find more comprehensive cover sets; however, the time-optimized solution algorithm finds solutions in more adjacent levels and has great advantage in time complexity, which is therefore regarded better.

Our future works include designing an adaptive extension of XHTML recommendation finding both other strategies for solutions to the problem, and more potential facets of the device's capabilities to help to extend the XHTML. Meanwhile, as the algorithms rely on a weighting of the nodes in the hierarchical model, we will study different weighting algorithms for document modeling in the future work.

References

1. Cui, G., Sun, D., et al.: WebUnify: An Ontology-based Web Site Organization and Publication Platform for Device Adaptation. In: Proceedings of SNPD 2004 International Conference (July 2004)
2. Dolog, P., Nejdl, W.: Semantic Web Technologies for the Adaptive Web. In: Brusilovsky, P., Kobsa, A., Nejdl, W. (eds.) The Adaptive Web: Methods and Strategies for Web Personalization, pp. 697–719. Springer, Heidelberg (2007)

3. Dolog, P., Nejdl, W.: Using UML and XMI for Generating Adaptive Navigation Sequences in Web Based Systems. In: Stevens, P., Whittle, J., Booch, G. (eds.) UML 2003. LNCS, vol. 2863, pp. 205–219. Springer, Heidelberg (2003)
4. Er-Jongmanee, T.: XML-Driven Device Independent User Interface–build Rich Client Applications Using XML. Master thesis, Lehrstuhl für Informatik V (2005)
5. Ku, T., Park, D., Moon, K.: Device Independent Authoring Language. In: Fourth Annual ACIS International Conference on Computer and Information Science, pp. 508–512 (2005)
6. Lewis, D., Conlan, O., et al.: Managing Adaptive Pervasive Computing using Knowledge-based Service Integration and Rule-based Behavior. In: Proceedings of IFIP/IEEE Network Operations and Management Systems, Seoul, Korea (April 2004)
7. Luo, C., Pan, R., Wang, S.: A Hierarchical Model for Auto-adjusting of Information Issuance. International Journal of Computer Science and Network Security, IJCSNS (December 2006)
8. Ma, W., Bedner, I., Chang, G., et al.: Framework for adaptive content delivery in heterogeneous network environments. In: Proceedings of Multimedia Computing and Networking, San Jose (January 2000)
9. Pan, R.: The PLCH Binary Tree Model of the Auto-adaptation of Web Information Issuance. Journal of Computer Applications (May 2005)
10. Pan, R.: A Mathematical Model for the Hierarchization of Web Information and Its Second-Best Algorithm. Computer Engineering and Science (October 2005)
11. Pan, R., Wei, H., Wang, S., Luo, C.: Auto-adaptation of Web Content: Model and Algorithm. In: ICWMMN 2008, Beijing, China, October 12-15, 2008, pp. 507–511 (2008)
12. Pemberton, S., et al. (eds.): XHTML 1.0 Specification, W3C Recommendation (August 2002), http://www.w3.org/TR/xhtml1/
13. Mika, P.: Ontologies are us: A unified model of social networks and semantics. Web Semantics: Science, Services and Agents on the World Wide Web 5(1), 5–15 (2007)
14. Phanouriou, C.: UIML: A Device-Independent User Interface Markup Language, PhD thesis, Virginia Polytechnic Institute and State University (2000)
15. Wang, L., Sajeev, A.S.M.: Abstract Interface Specification Languages for Device Independent Interface Design: Classification, Analysis and Challenges. In: Proceedings of First International Symposium on Pervasive Computing and Applications, Xinjiang, China, August, pp. 241–246. IEEE Press, Los Alamitos

Discrepancy-Based Sliced Neighborhood Search

Fabio Parisini, Michele Lombardi, and Michela Milano

D.E.I.S., University of Bologna, Italy
{fabio.parisini,michele.lombardi2,michela.milano}@unibo.it

Abstract. Limited Discrepancy Search (LDS) is one of the most widely used search strategies in Constraint Programming; given a solution suggested by a search heuristics, LDS explores the search space at increasing discrepancy with respect to such solution. In optimization problems, LDS is used as a way to explore the k-distance neighborhood of an incumbent solution using constraint propagation and tree search. However, for large problems, the size of the resulting neighborhoods limits the k-distance (i.e. the number of discrepancies) that can be efficiently explored. If the first solution is far from the optimal one, exploring limited neighborhood leads to small improvements.

Therefore, we propose a variant of LDS that samples the LDS space by exploring slices of (possibly very large) discrepancy based neighborhoods. Instead of deciding only the number of variables that can change (the k-distance) we decide which $n-k$ variables should be fixed to the value they have in the incumbent solution. We present results on hard Asymmetric Traveling Salesman Problem with Time Windows (ATSPTW) instances to show the effectiveness of the approach.

Keywords: Constraint Programming, Local Search, LDS, LNS.

1   Introduction

It is a common practice to use tree search methods to solve Combinatorial Optimization Problems, exploiting carefully tuned successor ordering heuristics to guide the search towards promising regions. When the heuristics guides the search to a failure state, backtracking is performed up to an open decision point, where the choice performed by the heuristics is reverted. We call this alternate choice a "wrong turn".

Limited Discrepancy Search (LDS) [5] addresses the problem of limiting the number of "wrong turns" (i.e. discrepancies) along the way; it explores portions of the search space at increasing discrepancy values w.r.t. a given successor ordering heuristics. We will focus on a variation of LDS, Improved Limited Discrepancy Search (ILDS) [7], which first explores the portions of the search tree which can be reached by making exactly one choice different from the heuristics, then allows exactly two differences (discrepancies), and so on, till a termination condition is reached or the desired solution has been found. These regions at increasing discrepancy k are the so-called k-distance neighborhoods of the solution proposed by the heuristics. In the following we will use LDS to refer to ILDS.

D. Dicheva and D. Dochev (Eds.): AIMSA 2010, LNAI 6304, pp. 91–100, 2010.

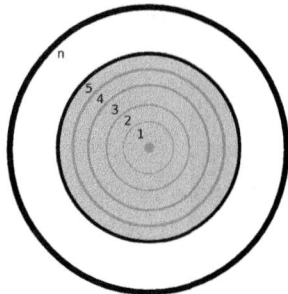

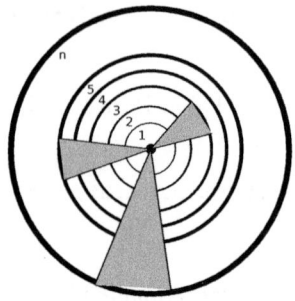

Fig. 1. Basic LDS space exploration scheme

Fig. 2. SNS space exploration scheme

We deal with optimization problems for which we have an incumbent solution $\bar{x}$ having cost $Cost_{\bar{x}}$; LDS tries to reduce the cost of the incumbent solution, by exploring the k-distance neighborhood of $\bar{x}$ via tree search. The drawback of LDS is that for large problems, exploring neighborhood at high discrepancy (high values of k) is very time consuming. As a consequence, if the incumbent solution is far from the optimal one, small neighborhoods are likely to contain low quality solution as well.

For this reason, we propose a very simple yet effective variant of LDS that is able to explore slices of large discrepancy neighborhoods efficiently; we call it Sliced Neighborhood Search, SNS. It is heavily based on restarts each fixing a (possibly large) discrepancy k and randomly selecting $n - k$ variables that should be maintained fixed w.r.t. the incumbent solution. In this way we do not explore exhaustively the k-distance neighborhood of the incumbent solutions but only slices of it.

Pictorially, we can imagine our incumbent solution $\bar{x}$ as a point in a two-dimensional space. The Euclidean distance between two points on the two-dimensional space corresponds to the discrepancy between two solutions.

Figure 1 shows the behavior of basic LDS. The central point is the incumbent solution $\bar{x}$, while the concentric circumferences enclose areas of the search space at a discrepancy from $\bar{x}$ smaller than the corresponding number. For large problems, LDS may efficiently explore only low discrepancy areas.

Figure 2, instead, shows the main idea underlying SNS; the central point and the circumferences have the same meaning as in figure 1. The grayed out slices in the circles represent the search areas explored at each iteration of SNS; we expect to be able to explore portions of the search space which are at a higher discrepancy value than those explored via LDS within the same time limit.

We provide experimental results on large ATSPTW instances showing the effectiveness of our solution w.r.t. standard LDS.

2   Extending LDS

In the following we introduce basic concepts on LDS for the solution of Combinatorial Optimization Problems; let us start with some notations:

- the problem variables are identified as $x = [x_1, x_2 \ldots x_n]$.
- an incumbent solution $\bar{x} = [\bar{x}_1, \bar{x}_2 \ldots \bar{x}_n]$ is a feasible assignment of values to the problem variables.

In optimization problems we are looking for feasible solutions which are better in cost than an incumbent solution $\bar{x}$. LDS can be used to limit the search space to the k-distance neighborhood of $\bar{x}$; in this setting we count a discrepancy each time a variable is assigned to a value which is different from the value the same variable takes in the incumbent.

Definition 1 (Discrepancy). *The discrepancy between $\bar{x}$ and x can be computed as*

$$k = \sum_{i=1}^{n} d_i \qquad where \qquad d_i = \begin{cases} 1 \ if \ x_i \neq \bar{x}_i; \\ 0 \ otherwise. \end{cases} \qquad (1)$$

When exploring the neighborhood of an incumbent solution $\bar{x}$ looking for a solution x, we can set an "at most-k" discrepancy bound by posting $n - k$ equality constraints of the kind $x_i = \bar{x}_i$. Similarly, we can set an "at least-k" discrepancy bound by posting k difference constraints of the kind $x_i \neq \bar{x}_i$. Both bounds can be set at the same time and different values of k can be used for each bound.

We introduce two predicates to offer a compact graphical representation of the search space explored by LDS.

Definition 2. *Predicate $fix(L, \bar{x})$ sets a subset L of the problem variables to the value they had in the incumbent solution:*

$$fix(L, \bar{x}) \iff \forall x_i \in L, \quad x_i = \bar{x}_i$$

Definition 3. *Predicate $change(L, \bar{x})$ removes the incumbent solution value from the domain of a subset L of the problem variables:*

$$change(L, \bar{x}) \iff \forall x_i \in L, \quad x_i \neq \bar{x}_i$$

One can see how to set at-most and at-least discrepancies by simply setting equality or difference constraints on x; the search space explored by LDS at each discrepancy value can be written as follows.

Definition 4. *The search space explored by LDS at a given discrepancy k from the incumbent solution $\bar{x}$ is obtained by setting at-least and at-most discrepancy bounds equal to k.*

Figure 3 shows the standard LDS search tree using the incumbent solution $\bar{x}$ as heuristics. It can be viewed as a three level search strategy.

1. First level: the search space is split by increasing discrepancy values k, in figure 3 from $k = 1$ up to a given value i.

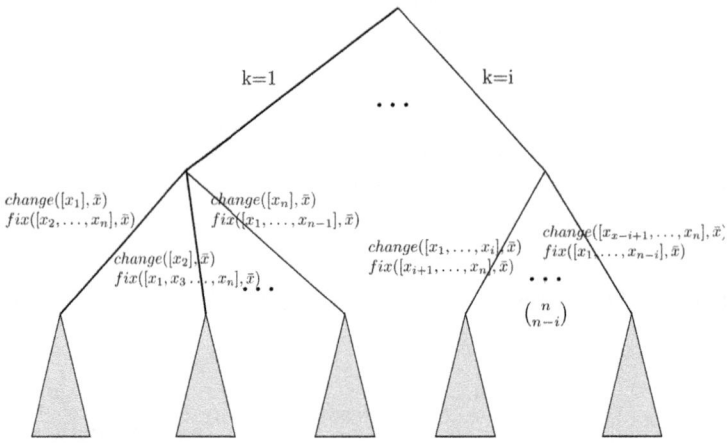

Fig. 3. LDS as neighborhood exploration of an incumbent solution $\bar{x}$

2. Second level: for each of the given discrepancy values k we enumerate all the possible combinations of subsets of k variables over the total n variables. Then we force k those variables to take a value different from the one they had in the incumbent solution; the remaining $n - k$ variables are set to the value they had in the incumbent solution.
3. Third level: for each constraint set at the second level, a search is performed on the still unbound k variables; for a fixed value of discrepancy k there are $\binom{n}{n-k}$ level 3 sub-trees. This level is represented by the bottom sub-trees in figure 3.

LDS explores the search tree depicted in figure 3 in a Depth First Search (DFS) fashion.

2.1   Limitations of LDS

When we use LDS to explore a neighborhood around an incumbent solution we first explore portions of the search space which are close to the solution, i.e. at low discrepancy values, increasing the discrepancy value during the search, until we reach a maximum discrepancy value or some other termination condition is encountered.

This kind of approach is troublesome when we deal with problem instances with a high number of variables. In those settings LDS does not manage to explore high discrepancy values in a reasonable computation time, limiting the search space to be very close to the incumbent solution. If the incumbent solution has a large optimality gap, it is very likely that small neighborhoods contain low quality solutions.

3   Sliced Neighborhood Search

To overcome LDS limitations we propose a simple, yet effective variant of the standard search strategy called Sliced Neighborhood Search, SNS.

We take inspiration from Large Neighborhood Search LNS, that uses the incumbent solution $\bar{x}$ and iteratively relaxes a fragment of $\bar{x}$ and then re-optimizes it using tree search. LNS has been proved successful on a number of applications [8] by relying on carefully tuned fragment selection and optimization. LNS applications so far are highly problem dependent. We propose instead a problem-independent search strategy whose aim is to explore larger neighborhoods than those explored by LDS, but maintaining the problem independence.

SNS iteratively explores neighborhoods of an incumbent solution $\bar{x}$ by triggering a sequence of restarts; at each restart a randomly chosen transversal slice of a neighborhood is chosen and explored. SNS works by setting extra constraints on a subset of k variables of the n problem variables, namely setting at-least and at-most discrepancy bounds. The constraints are used to ensure that the search space we explore is a neighborhood of the incumbent solution $\bar{x}$; in the remainder of this section we assume to just use equality constraints, as we will justify in Section 4.2.

For example, let's take an incumbent solution $\bar{x}$ of a problem instance having 6 variables $x_1 \ldots x_6$ taking values $[2, 4, 9, 5, 2, 8]$ in the incumbent solution. With LDS we would first explore exhaustively the search space at discrepancy value 1, where just 1 of the 6 variables can change at a time, then at discrepancy 2 and so on. SNS, instead, would fix a certain number of variables to the incumbent solution value, then the real search would start; for example, we could set $x_2 = 4$, $x_3 = 9$ and $x_6 = 8$ and perform a standard tree search just on x_1, x_4 and x_5.

Our approach relies on restarts and randomization to ensure the considered neighborhood is sufficiently well explored. SNS performs many randomized iterations, each iteration exploring just a "slice" of the limited discrepancy space w.r.t. $\bar{x}$, choosing each time a different set of variables and using small time limits for each iteration.

To have a clearer idea of what kind of space the SNS is exploring, we can examine figure 3 once again. LDS is exploring all the search tree depicted in the figure in a DFS manner; one can see that our strategy is doing something similar to sampling this search tree, i.e. randomly choosing some of the third level search sub-trees and exploring them. For the sake of precision, sampling that tree would mean searching at discrepancy *exactly* k. Conversely SNS explores a search space having at-most discrepancy k from $\bar{x}$; the search tree which SNS is sampling is thus the one depicted in figure 4. In such a configuration the third level sub-trees at discrepancy i are subsumed by the ones at discrepancy $i + 1$; that is not a problem for SNS, as the number of iterations is much smaller than the number of third level sub-trees, so the probability to explore overlapping search spaces is negligible. A formal definition of the approach follows.

Definition 5. *Sliced Neighborhood Search is a search approach designed to explore a limited discrepancy neighborhood of an incumbent solution $\bar{x}$ looking for*

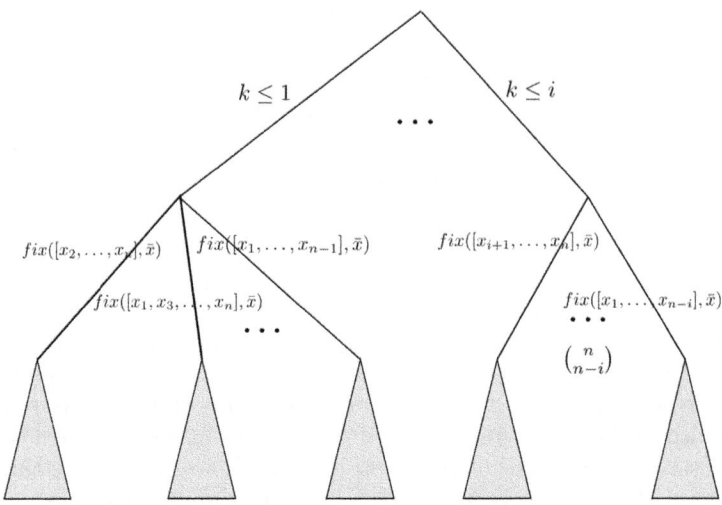

Fig. 4. Sliced Neighborhood Search sampling search tree

improving solutions. The approach has an iterative structure; it explores a randomly chosen search space characterized by a given at-least and at-most discrepancy from $\bar{x}$ at each iteration.

SNS has two main parameters which need to be carefully tuned:

- At-least and at-most discrepancy bounds to be used in each iteration; in particular, these values could change at each iteration and they could be related to the problem size n.
- time allocated to each iteration: it could be computationally expensive to perform a complete exploration of the sampled neighborhood. In our configuration a time limit is set on the iterations.

4   Experimental Results

In this section, we provide some experimental results to showcase the benefits of the SNS approach against the basic LDS. Experiments are conducted using ILOG Solver 6.3 and ILOG Scheduler 6.3 on a 2Ghz Pentium IV running Linux with 1 GB RAM.

4.1   Traveling Salesman Problem with Time Windows

The *Travelling Salesman Problem with Time Windows* (TSPTW) is the problem of finding a minimum cost path visiting a set of cities exactly once, where each city must be visited within a specific time window. TSPTW has important applications in routing and scheduling. It is therefore extensively studied in Operations Research and CP (see, e.g., [2,3]). In the following, we focus on the asymmetric TSPTW (ATSPTW) where the cost from a city i to another city j may not be the same as the cost from j to i.

In the TSPTW two main components coexist, namely a *Travelling Salesman Problem* (TSP) and a scheduling problem.

In TSPs, *optimization* usually results to be the most difficult issue: although the feasibility problem of finding a Hamiltonian tour in a graph is NP-hard, in the specific applications it is usually "easy" to find feasible solutions, while the optimal one is very "hard" to determine. On the other hand, scheduling problems with release dates and due dates usually set serious *feasibility* issues since they may involve disjunctive, precedence and capacity constraints at the same time. The solution of scheduling problems is probably one of the most promising Constraint Programming (CP) area of application to date.

We have adopted the model of [3] in which each $Next_i$ variable gives the city to be visited after city i. Each city i is associated with an activity and the variable $Start_i$ indicates the time at which the service of i begins. The model embeds an *alldifferent(X)* constraint (posted on the $Next$ variables) and a cost function $C(Next)$. We applied the cost-based filtering described in [4]; additionally, to improve the quality of the computed bounds on discrepancy-limited neighborhoods, a discrepancy constraint on the $Next$ variables is inserted in the additive bounding chain, together with the Assignment Problem relaxation (see [6] for details).

The adopted tree search schema builds a path starting from the initial city by assigning the $Next$ variables; at each step the search heuristics we adopted selects an activity j in the domain of $Next_i$; in detail, the activity having the smallest reduced-cost value $\bar{c}_{ij}$, (obtained from the Assignment Problem relaxation) is chosen and assigned. To allow a fair comparison, the same heuristics is used both in basic LDS and in each of the iterations of our approach.

4.2   Experiments

The experiments were conducted on the set of ATSPTW instances proposed by Ascheuer [1]; in particular we used the largest instances (with more than 50 cities). The incumbent solutions were chosen from a pool of candidates, found via CP-Based Local Branching [6].

We set a time limit of 300 seconds on each test run (referred to as global time limit). Additionally, a smaller time limit (namely, 10% of the global one) was set on each SNS iteration; this allows the (possibly partial) exploration of several SNS samples and is basically required for the method to be effective.

The SNS performance is mainly influenced by 3 elements, namely A) the effectiveness of the search within the third level sub-trees; then B) the actual presence of improving solutions in the discrepancy range identified by at-least and at-most bounds. Finally, C) the quality of the sampling process performed by SNS by randomly choosing portions of search space. In deeper detail:

Search effectiveness in the third level sub-trees. We refer as third level search sub-tree to the search space explored in Figure 4 after discrepancy bounds are enforced by posting equality and difference constraints. Search effectiveness in the sub-trees strongly depends on the propagation performed by such constraints, and clearly has a big part in SNS performance. While equality constraints enable strong propagation, we expect difference constraints to be less effective.

Solution Density in the selected discrepancy range. Regardless of how efficiently the selected discrepancy range is explored, the success of SNS depends on the actual presence of improving solutions in such range and their number. This in turn depends on the problem structure, and on the selected at-least and at-most discrepancy bounds.

Effectiveness of the Sample Space exploration. As SNS is basically sampling the the search space depicted in figure 4, the effectiveness of such sapling process has a strong impact on the method performance. Such effectiveness is measured by the number of collected samples (i.e. SNS iterations) over the size of the Sample Space.

Let Δ_{least} be the at-least discrepancy bound, Δ_{most} the at-most discrepancy bound and let Δ_{least} and Δ_{most} be fixed throughout the whole SNS process; then the size of the Sample Space (i.e. the overall number of third level sub-trees) is given by:

$$\binom{n}{n - \Delta_{most}}\binom{\Delta_{most}}{\Delta_{least}}$$

For a fixed number of SNS iterations, the larger the Sample Space, the least effective we expect the method to be. Note that such size is minimum for $\Delta_{most} = 0$ or n and $\Delta_{least} = 0$ or n, and quickly grow as Δ_{most} approaches $n/2$ and Δ_{least} approaches $\Delta_{most}/2$.

The best results obtained for SNS are given in table 1; here we compare LDS with a configuration of our approach having at-most discrepancy set to 40% of the problem size n. We also report results from what we call the *Random Restart* approach, in which we randomize the search heuristics and perform a sequence of fail limited restarts. We did not take into account LDS variations working on the sorting of branch exploration at a given discrepancy value, such as Depth-bounded Discrepancy Search [9], as such sorting in our implementation is handled by the search heuristics once the discrepancy constraints have been set. The instances are ordered by increasing number of cities and the instance name contains such number. The column labeled "Best Cost" reports the cost of the best known solution from the literature. The column "Ref Cost" reports the cost of an incumbent solution.

For all the search strategies the cost of the best solution found within the global time limit is reported; the discrepancy at which that solution is found follows, together with the percentage of the cost improvement. Since the incumbent solutions offer an initial cost value which is quite near to the best one, the percentage improvement provides a better evaluation of the search strategy ability to cover a large part of the gap. This percentage is computed as

$$\%Impr = 100 * (Cost - Ref\ Cost)/(Ref\ Cost - Best\ Cost)$$

Where *Cost* is the cost value found by the considered configuration, while *RefCost* and *BestCost* have the meaning explained above.

The table shows that SNS is able to get results consistently better than basic LDS; our approach is able to find improving solutions at discrepancy values which are much higher than the ones which can be explored by LDS. The chosen at-most discrepancy value is small enough to allow an effective exploration of the search sub-trees and big enough to keep a sufficiently high solution density within the sub-trees. Random Restart performs better than SNS only on the biggest instances; this is due to the fact that SNS is constrained to explore the neighborhood of initial solutions whose quality worsens while the instance size grows.

We have also run experiments on different Minimum and Maximum Discrepancy values; we report some results in table 2.

In this table we use the notation $\Delta(\Delta_{least}, \Delta_{most})$ to identify a SNS configuration having at-least discrepancy Δ_{least} and at-most discrepancy Δ_{most} from $\bar{x}$. We compare the best SNS configuration, $\Delta(0\%, 40\%)$, with configurations having

Table 1. Big instances, more than 50 cities, 300 CPU seconds time limit

Instance	Best Cost	Ref Cost	basic LDS				$\Delta_{max}=40\%$				Random Restart	
			Cost	Sol Disc	% Impr		Cost	Sol Disc	% Impr		Cost	% Impr
rbg050a	414	430	430	-	0.00		424	5	37.50		430	0.00
rbg050b	527	570	570	-	0.00		570	-	0.00		570	0.00
rbg050c	536	563	563	-	0.00		545	7	66.67		563	0.00
rbg055a	814	814	814	-	0.00		814	-	0.00		814	0.00
rbg067a	1048	1056	1051	4	62.50		1051	7	62.50		1056	0.00
rbg092a	1093	1245	1208	3	24.34		1150	8	62.50		1245	0.00
rbg125a	1409	1762	1706	3	15.86		1632	20	36.83		1762	0.00
rbg132	1360	1934	1882	3	9.06		1815	13	20.73		1934	0.00
rbg132.2	1083	2164	2152	3	1.11		2040	31	11.47		1969	18.04
rbg152	1783	2415	2371	3	6.96		2336	23	12.5		2414	0.16
rbg152.3	1539	2570	2570	-	0.00		2548	42	2.13		2539	3.01
rbg172a	1799	2991	2942	3	4.11		2873	31	9.9		2904	7.30
rbg193	2414	3401	3360	3	4.15		3266	31	13.68		3389	1.21
rbg193.2	2093	3330	3290	3	3.23		3233	45	7.84		3131	16.09
rbg201a	2189	3748	3694	3	3.46		3650	24	6.29		3748	0.00
rbg233.2	2304	4142	4103	3	2.12		4059	49	4.52		3834	16.76

Table 2. Big instances, more than 50 cities, 300 CPU seconds time limit

Instance	Best Cost	Ref Cost	$\Delta(0\%, 40\%)$		$\Delta(10\%, 50\%)$		$\Delta(10\%, 40\%)$		$\Delta(25\%, 55\%)$	
			Cost	% Impr	Cost	% Impr	Cost	% Impr	Cost	% Impr
rbg050a	414	430	424	37.50	429	6.25	429	6.25	430	0.00
rbg050b	527	570	570	0.00	570	0.00	570	0.00	570	0.00
rbg050c	536	563	545	66.67	563	0.00	563	0.00	563	0.00
rbg055a	814	814	814	-	814	-	814	-	814	-
rbg067a	1048	1056	1051	62.50	1048	100	1051	62.50	1051	62.50
rbg092a	1093	1245	1150	62.50	1203	27.63	1203	27.63	1217	18.42
rbg125a	1409	1762	1632	36.83	1682	22.66	1721	11.61	1762	0.00
rbg132	1360	1934	1815	20.73	1925	1.57	1916	3.14	1934	0.00
rbg132.2	1083	2164	2040	11.47	2063	9.34	2035	11.93	2108	5.18
rbg152	1783	2415	2336	12.5	2377	6.01	2365	7.91	2410	0.79
rbg152.3	1539	2570	2548	2.13	2555	1.45	2499	6.89	2528	4.07
rbg172a	1799	2991	2873	9.90	2918	6.12	2932	4.95	2991	0.00
rbg193	2414	3401	3266	13.68	3278	12.46	3401	0.00	3401	0.00
rbg193.2	2093	3330	3233	7.84	3330	0.00	3330	0.00	3330	0.00
rbg201a	2189	3748	3650	6.29	3748	0.00	3748	0.00	3748	0.00
rbg233.2	2304	4142	4059	4.52	4142	0.00	4142	0.00	4142	0.00

$\Delta_{least} \neq 0$, i.e. $\Delta(10\%, 50\%)$, $\Delta(10\%, 40\%)$ and $\Delta(25\%, 55\%)$. We observe that, even if using a non null at-least discrepancy bound further reduces the size of the sub-trees to be explored, the overall performances of the SNS configurations worsen. This is due to the fact that, when $\Delta_{least} \neq 0$, the size of the sampling space grows quickly, lowering the quality of the sampling process performed by SNS.

5   Conclusions and Future Work

We have developed a general and effective search technique to heuristically explore the neighborhood of an incumbent solution $\bar{x}$ up to high discrepancy values, hybridizing elements coming from LDS and LNS. The tests run can be effectively read in terms of size of sampling space, search sub-tree size and solution density, supporting with experimental results the idea that the best configurations, obtaining consistently better results than LDS, have a at-most discrepancy bound around 40% of the problem size. Following these guidelines, an interesting direction of further development is the introduction of learning processes during the SNS iterations, to automatically tune the SNS parameters during the search process.

References

1. Ascheuer, N.: Hamiltonian path problems in the on-line optimization of flexible manufacturing systems. PhD thesis, Technische Universität Berlin (1995)
2. Desrosiers, J., Dumas, Y., Solomon, M.M., Soumis, F.: Time constrained routing and scheduling. In: Network Routing, pp. 35–139 (1995)
3. Focacci, F., Lodi, A., Milano, M.: A hybrid exact algorithm for the tsptw. Informs Journal on Computing 14(4), 403–417 (2002)
4. Focacci, F., Lodi, A., Milano, M.: Optimization-oriented global constraints. Constraints 7, 351–365 (2002)
5. Harvey, W., Ginsberg, M.L.: Limited discrepancy search. In: Proc. of IJCAI 1995, pp. 607–615. Morgan Kaufmann, San Francisco (1995)
6. Kiziltan, Z., Lodi, A., Milano, M., Parisini, F.: Cp-based local branching. In: Bessière, C. (ed.) CP 2007. LNCS, vol. 4741, pp. 847–855. Springer, Heidelberg (2007)
7. Korf, R.E.: Improved limited discrepancy search. In: Proceedings of AAAI 1996, pp. 286–291. MIT Press, Cambridge (1996)
8. Shaw, P.: Using constraint programming and local search methods to solve vehicle routing problems. In: Maher, M.J., Puget, J.-F. (eds.) CP 1998. LNCS, vol. 1520, pp. 417–431. Springer, Heidelberg (1998)
9. Walsh, T.: Depth-bounded discrepancy search. In: Proceedings of IJCAI 1997, pp. 1388–1393 (1997)

Constraint Models for Reasoning on Unification in Inductive Logic Programming

Roman Barták

Charles University in Prague, Faculty of Mathematics and Physics
Malostranské nám. 2/25, 118 00 Praha 1, Czech Republic
bartak@ktiml.mff.cuni.cz

Abstract. Inductive Logic Programming (ILP) deals with the problem of find-ing a hypothesis covering all positive examples and excluding negative exam-ples. One of the sub-problems is specifying the structure of the hypothesis, that is, the choice of atoms and position of variables in the atoms. In this paper we suggest using constraint satisfaction to describe which variables are unified in the hypotheses. This corresponds to finding the position of variables in atoms. In particular, we present a constraint model with index variables accompanied by a Boolean model to strengthen inference and hence improve efficiency. The efficiency of models is demonstrated experimentally.

Keywords: constraint modelling, inductive logic programming, meta-reasoning.

1 Introduction

Inductive logic programming (ILP) is a subfield of machine learning which uses first-order logic as a uniform representation for examples, background knowledge and hypotheses [10]. In the core form, ILP deals with the problem of finding a hypothesis that covers all positive examples and excludes all negative examples. More formally, hypothesis and examples are supposed to be clauses and the hypothesis is required to logically entail all *positive examples* and no *negative example*. Entailment is checked using θ-*subsumption* [11] which is a decidable restriction of logical entailment. For simplicity of notation, we will assume clauses to be expressed as sets of literals, and, without loss of generality, we will only work with positive literals, that is, non-negated atoms. All terms in the learning examples (hypotheses, respectively) are constants (variables) written in the lower (upper) cases. For instance, E = {arc(a,b), arc(b,c), arc(c,a)} is an example and H = {arc(X,Y), arc(Y,Z)} is a hypothesis. Hypothesis H *subsumes* example E, if there exists a substitution θ of variables such that Hθ ⊆ E. In the above example, substitution θ = {X/a, Y/b, Z/c} implies that H subsumes E. Maloberti and Sebag [9] suggested an efficient algorithm *Django* for checking θ-subsumption based on formulating the problem as a binary constraint satisfaction problem. Django brought dramatic speed-up for θ-subsumption and consequently for the entire ILP system. This motivates us to exploit constraint satisfaction techniques in other parts of the ILP system.

D. Dicheva and D. Dochev (Eds.): AIMSA 2010, LNAI 6304, pp. 101–110, 2010.
© Springer-Verlag Berlin Heidelberg 2010

In this paper we focus on a *template consistency problem* that belongs among two computationally equivalent basic ILP problems (the other problem is the bounded consistency problem) [7]. In this problem, a template T is given for the hypothesis and hypothesis H is obtained by applying certain substitution σ to template T, that is, H = Tσ. In our case, the template is a set of positive literals where all terms are different variables, for example T = {arc(A,B), arc(C,D)}. Substitution σ = {C/B} defines the hypothesis H mentioned above (after renaming the variables). As the terms in both the template and the hypothesis are always variables, we can see the substitution σ as a description of unification of variables in T, in our example B = C. In other words, we are solving the problem of finding a unification of variables in the template that gives us a *consistent hypothesis* subsuming all positive examples and no negative example. In this paper we suggest a constraint model and a solving approach for the above problem. This model is naturally integrated with the constraint model for θ-subsumption which is based on the ideas of Django (we use the standard constraint satisfaction techniques, namely maintain arc consistency, and we do not restrict to binary constraints).

The paper is organized as follows. We will first abstract from the ILP template consistency problem and formally define the problem purely in terms of constraint satisfaction. Then we will describe the base solving approach for finding the unification of variables. The main part of the paper will be about various constraint models supporting the solving approach. These are novel models that exploit some already used principles such as symmetry breaking, channeling, and global reasoning. Finally, we will present the experimental comparison of these models using random structured problems.

2 Problem Formalization

Recall that a *constraint satisfaction problem* (CSP) is a triple (X, D, C), where X is a finite set of decision variables, for each $x_i \in$ X, $D_i \in$ D is a finite set of possible values for the variable x_i (the domain), and C is a finite set of constraints [4]. A constraint is a relation over a subset of variables (its *scope*) that restricts the possible combinations of values to be assigned to the variables in its scope. The constraints can be expressed in extension using a set of compatible value tuples. A *solution to a CSP* is a complete instantiation of variables such that the values are taken from respective domains and all constraints are satisfied. Constraint satisfaction techniques are frequently based on the combination of inference techniques (called consistency techniques) and search.

We can reformulate the ILP template consistency problem [7] as follows. Assume that we have a finite set of constraint satisfaction problems, where each problem C_i is defined over *n* variables $\{X_{i,1}, X_{i,2}, ..., X_{i,n}\}$. In terms of ILP, each problem C_i defines the problem whether the hypothesis with variables $\{X_{i,1}, X_{i,2}, ..., X_{i,n}\}$ subsumes the example E_i. The constraint satisfaction problem C_i can be derived from the example E_i as described in [9]. In general, each predicate symbol *p* with some arity *k* appearing in the example is represented as *k*-ary constraint c_p where the set of compatible *k*-tuples defining the constraint is specified by arguments of atoms of predicate *p* in the example. Hence from the example $E = \{\text{arc}(a,b), \text{arc}(b,c), \text{arc}(c,a)\}$ we define the binary constraint c_{arc} as a set of pairs of compatible values $\{(a,b), (b,c), (c,a)\}$. Now, for the hypothesis H we can formulate the θ-subsumption problem as a constraint satisfaction problem in such a way that we take the variables from the hypothesis, the domains of variables consist of all constants in the example, and for each atom $p(Y_1, Y_2,\ldots, Y_k)$ in the hypothesis we use constraint c_p over the variables in the atom. For example, for hypothesis $H = \{\text{arc}(A,B), \text{arc}(C,D)\}$ and for the above example E we formulate the θ-subsumption problem as a constraint satisfaction problem with variables $\{A,B,C,D\}$, all with domain $\{a,b,c\}$, and constraints $\{c_{arc}(A,B), c_{arc}(C,D)\}$. Clearly, a solution to such a CSP, that is, the instantiation of the variables satisfying the constraints, defines the substitution θ of variables such that $H\theta \subseteq E$. Hence, the hypothesis subsumes the example if and only if the corresponding CSP has a solution. Note that each example requires a separate set of variables because the subsuming substitutions (instantiations of the variables) may be different between the examples.

Now, let us assume that the set of problems C_i is divided into two disjoint sets *Pos* and *Neg*. This clearly corresponds to the positive and the negative examples in ILP. The task is to find a set *Uni* of pairs (k,l), where $k,l \in \{1,2,\ldots,n\}$, such that the following two conditions hold:

$$\forall\ C_i \in\ Pos \quad C_i \cup \{\ X_{i,k} = X_{i,l} \mid (k,l) \in\ Uni\} \text{ has a solution,}$$
$$\forall\ C_i \in\ Neg \quad C_i \cup \{\ X_{i,k} = X_{i,l} \mid (k,l) \in\ Uni\} \text{ has no solution.}$$

By the solution we mean the instantiation of variables $\{X_{i,1}, X_{i,2},\ldots, X_{i,n}\}$ satisfying all the specified constraints. In other words, we have a set of constraint satisfaction problems that "share" some equality constraints defined by set *Uni*.

The above task corresponds exactly to the template consistency problem. We have a template T with n different variables and the set *Uni* describes which variables in the template should be unified to obtain a consistent hypothesis. This is because the constraint satisfaction problem $C_i \cup \{X_{i,k} = X_{i,l} \mid (k,l) \in\ Uni\}$ defines the subsumption problem for example E_i, where the added equality constraints $X_{i,k} = X_{i,l}$ describe how the hypothesis is obtained from the template.

3 Solving Approach

In the previous section we fully abstracted from the template consistency problem so we can focus on the formal task specified there, which is finding the set *Uni* satisfying the specified properties. Notice that the set *Uni* adds constraints to problems C_i so the main reason for including a pair (k,l) in *Uni* is to break satisfaction of some "negative problem" $C_i \in\ Neg$ (we do not need these equality constraints to satisfy the "positive problems"). It would be possible to model the problem fully as a Quantified CSP (QCSP) [2] supporting the description of a non-existence of a solution for the negative problems. We rather decided to mimic the solving ideas behind a QCSP in an ad-hoc search procedure implemented on top of existing constraint solvers which gives us the power of inference of the constraint solvers. In particular, we explore the solutions of the negative problems and break these solutions by adding equality constraints for the variables assigned to different values in the solution as the following pseudo-code shows.

```
Uni <- {}
for each Cᵢ∈Pos do post_constraint(Cᵢ)
for each Cⱼ∈Neg do
  for each solution Sol of Cⱼ ∪ { Xⱼ,ₖ = Xⱼ,₁ | (k,l)∈Uni} do
    choice point {
      select (k,l) s.t. Xⱼ,ₖ ≠ Xⱼ,₁ in Sol
      Uni <- Uni ∪ {(k,l)}
      for each Cᵢ∈Pos do post_constraint(Xᵢ,ₖ = Xᵢ,₁)
    }
for each Cᵢ∈Pos do instantiate(Xᵢ,₁, Xᵢ,₂,…, Xᵢ,ₙ)
```

The solver works as follows. First, we post the constraints for all positive problems. This is because these constraints must be satisfied independently on the set *Uni*. Posting these constraints allows us to exploit the inference techniques behind these constraints (maintaining arc consistency) and hence to detect earlier if any such constraint is violated and hence any problem from *Pos* cannot be satisfied after adding some equality constraints. Then we incrementally build the set *Uni* by taking the negative problems one by one and trying to find the solution for them assuming the equality constraints from the current set *Uni* (initially empty). If the negative problem has no solution then everything is fine as this is what we require. Otherwise, we take the found solution *Sol* of the negative problem $C_j \in Neg$, that is, the instantiation of variables $\{X_{j,1}, X_{j,2}, \ldots, X_{j,n}\}$, and we "break" this solution by adding an equality constraint between a pair of variables, say $X_{j,k}$, $X_{j,l}$, that are instantiated to different values in *Sol*. By adding the pair (k,l) to *Uni* we ensure that *Sol* is no more a solution to the problem $C_j \cup \{ X_{i,p} = X_{i,q} \mid (p,q) \in Uni\}$. Note that there might be more such pairs (k,l) for a given solution *Sol* so we introduce a choice point here and when a failure is detected later we backtrack to the closest choice point and select another pair for *Uni*. Assuming the example from the previous section E = {arc(a,b), arc(b,c), arc(c,a)} with hypothesis H = {arc(A,B), arc(C,D)}, the solution could be {A/a, B/b, C/a, D/b)} and we can break this solution by adding one of the constraints A=B, A=D, B=C, or C=D. Notice that as soon as we add the pair (k,l) to *Uni* we post the corresponding equality constraints to all positive problems so the underlying inference techniques can find out if there is any inconsistency. These constraints are removed, if the pair (k,l) is removed from *Uni* upon backtracking. If we process all negative examples, we ensure that no negative example has a solution. However, as the inference techniques are usually incomplete, we need to validate that all positive examples have a solution by instantiating their variables. If this is not the case, we backtrack to the closest choice point and select some alternative unification there. For solving both positive and negative examples, we use a standard constraint satisfaction approach that instantiates the variables using min-dom heuristic (the variables with the smallest domains are instantiated first) and maintains arc consistency during search. The values for variables are tried in the order of appearance in the problem.

4 Constraint Models for Unification

The above-described solving approach uses a base constraint model (the constraints from the "positive" problems) that is extended by the equality (unification) constraints derived from the solutions of "negative" problems. In each decision (choice) point we solve another CSP (a negative problem) whose solution defines the possible options for the search algorithm. Notice that if we refuse the unification of some variables X_k and X_l at some step then it may still happen that the solver tries this unification again and again to break other solutions of the same problem or of other negative problems. This is clearly undesirable behaviour as it decreases the time efficiency of search by exploring branches that are known to fail.

Note that the idea of semantic branching [6] cannot be directly applied to prevent the above inefficiency – if X_k and X_l cannot be unified then posting the negated constraint in the form $X_{i,k} \neq X_{i,l}$ among the corresponding variables in all positive problems cannot be done because the variables $X_{i,k}$ and $X_{i,l}$ can be instantiated to identical value in some positive problems C_i but not in all. The negation of the conjunction of equality constraints is a disjunction of inequality constraints. General disjunctive constraints do not propagate well unless a dedicated filtering algorithm is proposed for a particular type of disjunction (see for example the filtering algorithms for the constraints modelling unary resources [8]). We suggest using meta-reasoning on unifications based on keeping the information about which variables were unified and which variables were decided not to be unified in the hypothesis.

4.1 Index Model

Our first model is based on the observation that if a set of variables is unified then we can take the variable with the smallest index to represent this set and all other variables in the set are mapped to this variable. For example, unification $X_2 = X_3$ can be represented by mapping X_3 to X_2. The proposed constraint model uses index variable I_i for the i-th variable in the template to describe the mapping. The domain of I_i is $\{1,\ldots,i\}$. Each time, a couple (k,l) is added to *Uni*, the constraint $I_k = I_l$ is posted to describe that these two variables are unified with (mapped to) the same variable. If it is decided that this couple is not unified (the alternative branch in the choice point), constraint $I_k \neq I_l$ is posted. Hence, the semantic branching can be applied to the index variables. Note that this is correct as constraint $I_k \neq I_l$ only says that these variables should not be unified in the future but it does not force the corresponding variables in the positive examples to be different.

To ensure that each variable is mapped to the first variable in the set of unified variables we use a constraint $\forall i=1,\ldots,n$ $element(I_i, [I_1,\ldots, I_n], I_i)$, where n is the total count of variables. The semantics of $element(X,List,Y)$ is as follows: Y unifies/equals to the X-th element of List. For example, $[1,1,2]$ is not a valid list of indexes (it represents $X_1 = X_2$ and $X_2 = X_3$) as it violates the constraint $element(2,[1,1,2],2)$. The correct representation of this unification should be [1,1,1]. The *element* constraints thus ensure that each set of unifications is represented by exactly one list of indexes. Other *element* constraints are also used as a channel connecting the index model with each positive problem C_i as follows: $\forall j=1,...,n$ *element*$(I_j, [X_{i,1},.., X_{i,n}], X_{i,j})$. These *channeling constraints* allow propagation of information from the positive problems to the index model and vice versa.

The index model can also exploit the structure of the template. Typically, the template represents a set of atoms, for example $\{arc(X_1,X_2), arc(X_3,X_4), arc(X_5,X_6)\}$. However, the template itself is modeled as a list (variables are ordered) so swapping atoms generates a different unification set. Moreover, some unification may collapse the template to a smaller hypothesis which is not desirable (when looking for a consistent hypothesis without knowing apriori its structure, the templates can be generated incrementally in the increasing size so a smaller template was proved not to provide a consistent hypothesis before). In our example, $X_1 = X_3$ and $X_2 = X_4$ collapses the first two atoms to one. To remove this deficiency we suggest the following constraint. We collect the tuples of index variables belonging to variables in atoms with the same name, in our example, we obtain pairs (I_1,I_2), (I_3,I_4), (I_5,I_6). For each atom and a corresponding list of variable tuples L, we post a constraint *lex*(L) forcing the variable tuples in list L to be lexicographically ordered [3]. In our example, it means $(I_1,I_2) < (I_3,I_4) < (I_5,I_6)$ which ensures that no atom *arc* is unified with another atom *arc* in the template (and hence no atom will disappear from T). This constraint can be seen as a form of *symmetry breaking*. The following example demonstrates the complete index model (without the channeling constraints):

template:
   $arc(X_1,X_2)$, $arc(X_3,X_4)$, $arc(X_5,X_6)$, $red(X_7)$, $red(X_8)$, $red(X_9)$, $green(X_{10})$

index model:
   variables
      $I_1, ..., I_{10}$
   domains
      $\forall i=1,...,10 \; D_i = \{1,...,i\}$
   constraints
      $\forall i=1,...,10 \; element(I_i, [I_1,.., I_{10}], I_i)$
      $lex([(I_1,I_2), (I_3,I_4), (I_5,I_6)])$
      $lex([(I_7), (I_8), (I_9)])$

4.2 Boolean Model

One of the problems of the index model is that it cannot easily detect conflicting constraints $I_i \neq I_j$ and $I_i = I_j$ unless the domains are singleton. This deficiency of local inference, namely arc consistency, can be resolved by applying global reasoning. Assume that we have n variables in the template. We propose using a matrix U of size $n \times n$ to describe which variables are unified (1) and where the unification is forbidden (0). Initially, the matrix consists of unbounded variables $U_{i,j}$ such that $U_{i,j}$ is identical to (unified with) $U_{j,i}$ and $U_{i,i} = 1$ (bounded). When a couple (i,j) is added to *Uni*, we simply unify rows i and j of the matrix U, that is, $\forall k\ U_{i,k} = U_{j,k}$ (and also $U_{k,i} = U_{k,j}$). In particular we obtain $U_{i,j} = U_{j,i} = 1$. This unification operation implicitly keeps the transitive closure of the equality constraints between the variables (if $U_{i,l} = 1$ for some l then after the unification $U_{j,l} = 1$ also holds and vice versa). When a couple (i,j) is decided not to be unified then we set $U_{i,j} = 0$. If this is not possible because of $U_{i,j} = 1$ then we can immediately deduce a failure (we have a conflict between requiring X_i and X_j to be unified and not to be unified at the same time). Similarly, if sometime later we deduce that $X_i = X_j$ either directly via posting this constraint or indirectly via the transitive closure of equality (unification) constraints then we can also deduce a failure. The Boolean matrix for template $\{arc(X_1,X_2), arc(X_3,X_4), arc(X_5,X_6)\}$, where we already decided that $X_1 \neq X_2$, $X_3 \neq X_4$, $X_5 \neq X_6$ (see the next section on hints) will look like this:

1	0	A	B	C	D
0	1	E	F	G	H
A	E	1	0	I	J
B	F	0	1	K	L
C	G	I	K	1	0
D	H	J	L	0	1

Now, if we decide during search that $X_2 = X_3$ then we obtain the following table by unifying the second and third rows (columns):

1	0	0	B	C	D
0	1	1	0	G	H
0	1	1	0	G	H
B	0	0	1	K	L
C	G	G	K	1	0
D	H	H	L	0	1

Notice that we can immediately deduce that for example $X_2 \neq X_4$ so if we ever try unification $X_2 = X_4$ it will immediately fail. Or if we decide that $X_2 \neq X_5$ ($G = 0$) then we immediately get also $X_3 \neq X_5$.

The global reasoning behind the Boolean model can be implemented as a special global constraint but the above operations over the Boolean matrix do the same job. Note also that the Boolean model can be used independently of the index model or both models can be combined to exploit their complementary strengths (each model represents some form of local inference). It may seem that the Boolean model infers more information than the index model, but it is not clear how to implement the symmetry breaking from the index model there. Hence both models have their value (see the section on experiments).

4.2.1 Hints
We already explored some structural information in the symmetry breaking constraints of the index model, but we can do more. Assume that certain constraint in some positive problem C_i forbids a pair of variables $X_{i,k}$ and $X_{i,l}$ to be instantiated

to the same value, for example the constraint allows pairs $\{(a,b), (b,c), (c,a)\}$. If we post constraint $X_{i,k} = X_{i,l}$ then arc consistency does not deduce a failure immediately (domains of both variables are $\{a,b,c\}$) and the failure is only detected when we instantiate the variables or prune their domains. Nevertheless, it is possible to identify such pairs $(X_{i,k}, X_{i,l})$ of non-unifiable variables in positive problems in advance and express this information in the Boolean model by setting $U_{k,l} = 0$. We call it a *hint* and in our experiment we deduce hints only from individual constraints. More precisely, $U_{k,l} = 0$ if variables $X_{i,k}$ and $X_{i,l}$ appear in the same constraint in some positive problem C_i and there is no tuple satisfying that constraint such that $X_{i,k} = X_{i,l}$. In terms of ILP, from the positive example $E = \{arc(a,b), arc(b,c), arc(c,a)\}$ we can deduce that $X \neq Y$ for any pair of variables appearing in any atom $arc(X,Y)$ of the template. Note finally that such hints can be found fully automatically just by exploring the value tuples in each constraint.

5 Experimental Results

Searching for unifications in the template is a novel approach to specify hypothesis in ILP. As we are not aware about another published approach for solving the template consistency problems, we provide experimental results comparing the efficiency of the proposed models only. In particular, the experiments demonstrate the contribution of individual models/constraints to overall efficiency. The constraint models were implemented in SICStus Prolog 4.0.8 and the experiments run on 2.53 GHz Core 2 Duo processor with 4 GB RAM under Windows 7 Professional system.

We used the problems of identifying common structures in randomly generated structured graphs. In particular, we generated random structured graphs with 20 nodes where the new nodes are attached to the graph with 3 arcs according to the Barabási-Réka model [1]. Each dataset consisted of ten positive and ten negative examples representing the graphs and the template consisted of 5 nodes (10 atoms and 15 variables for each dataset). Table 1 shows the results, in particular, the runtimes in milliseconds. The experiments clearly show that the Boolean model is not satisfactory and that symmetry breaking constraints of the index model contribute significantly to the efficiency. Moreover, when both models are used together (combined model), channeling constraints add non-necessary overhead and they can be removed (decoupled model). To show the influence of symmetry breaking *lex* constraints, we also removed them from the decoupled model (no SB) which decreased efficiency significantly (this is basically the Boolean model with 20-30% overhead from the index model without symmetry breaking that does not prune more the search space). There are two problems where the Boolean model beats all other models. The reason is that the random graphs may contain a common structure that is a subset of the template. Recall that the *lex* constraints forbid collapsing atoms in the template and hence in this case, the Boolean model allows a smaller hypothesis (easier to find) than the other models ("no SB" column confirms it). The last column in Table 1 shows that hints are really useful for solving the complicated problems and if they are removed then efficiency degrades.

Table 1. Comparison of runtimes (milliseconds) for identifying the common structures in graphs (Barabási-Réka). The bold numbers indicate the best runtime.

Index	Boolean	Combined	Decoupled		
			full	no SB	no hints
32	202	47	**31**	250	32
250	7956	343	250	10312	**249**
54351	85192	12495	**6365**	108780	54163
49421	14679	52541	**6552**	18362	49858
>600000	531670	132288	**38579**	>600000	>600000
>600000	**38735**	347804	108717	50171	>600000
>600000	>600000	>600000	**158308**	>600000	>600000
>600000	**65302**	>600000	378005	84817	>600000

We also did experiments with other data sets namely we generated random structured graphs using the Erdős-Rényi model [5]. Each dataset again consisted of ten positive and ten negative examples (graphs). The random graphs contain 20 nodes with the density of arcs 0.2 and the template contains 5 nodes. Table 2 shows the results, in particular, it describes the size of the template (the number of atoms and variables) and the runtimes in milliseconds. The results confirm that the conclusions from previous paragraph about the contribution of symmetry breaking and hints to overall efficiency and about the importance of decoupling the models are valid. Again, the Boolean model is better for one problem due to the same reason as before.

Table 2. Comparison of runtimes (milliseconds) for identifying the common structures in graphs (Erdős-Rényi). The bold numbers indicate the best runtime.

#atoms	#vars	Index	Boolean	Combined	Decoupled		
					full	no SB	no hints
7	9	**0**	15	16	**0**	16	**0**
8	11	**0**	16	15	**0**	15	**0**
8	11	31	281	31	**21**	343	32
8	11	94	343	109	**78**	421	**78**
9	13	94	1046	125	**93**	1373	94
9	13	328	1810	436	**312**	2309	**312**
9	13	1232	9626	1606	**1170**	12324	1185
10	15	>600000	**86425**	>600000	236514	110981	>600000

6 Conclusions

The paper suggests and compares several constraint models for describing unification of variables in ILP templates. In particular, it shows that the model based on indexes and the Boolean model can be successfully combined and accompanied by hints derived from data to overcome some deficiencies of local consistency techniques. In particular, the index model is useful to express some symmetry breaking constraints while the Boolean model does global reasoning over equality (unification) and inequality constraints and allows natural expression of hints about existence of such constraints. The experiments also showed that it might be useful to use the models independently without channelling information between them. This shows that the models do not exploit information from each other and the channelling constraints only add overhead.

As far as we know this is a completely novel approach to the ILP template consistency problem. Actually, we are not aware about any other approach for solving this problem that we can compare with. We built the solving approach on top of the constraint satisfaction technology which proved to be very efficient in solving the θ-subsumption problem. Though the motivation is going from ILP, the presented ideas, especially the Boolean model accompanied by hints, are generally applicable to problems where equality and inequality constraints need to be combined.

Acknowledgements. The research is supported by the Czech Science Foundation under the project 201/08/0509. I would like to thank Filip Železný and Ondřej Kuželka for introducing me to the area of ILP and providing a generator of random problems for experiments.

References

1. Barabási, A.-L., Réka, A.: Emergence of scaling in random networks. Science 286, 509–512 (1999)
2. Bordeaux, L., Monfroy, E.: Beyond NP: Arc-consistency for quantified constraints. In: Van Hentenryck, P. (ed.) CP 2002. LNCS, vol. 2470, pp. 371–386. Springer, Heidelberg (2002)
3. Carlsson, M., Beldiceanu, N.: Arc-Consistency for a Chain of Lexicographic Ordering Constraints. SICS Technical Report T2002-18 (2002)
4. Dechter, R.: Constraint Processing. Morgan Kaufmann, San Francisco (2003)
5. Erdős, P., Rényi, A.: On Random Graphs I. Publicationes Mathematicae 6, 290–297 (1959)
6. Giunchiglia, F., Sebastiani, R.: Building decision procedures for modal logics from propositional decision procedures - the case study of modal K. In: McRobbie, M.A., Slaney, J.K. (eds.) CADE 1996. LNCS (LNAI), vol. 1104, pp. 583–597. Springer, Heidelberg (1996)
7. Gottlob, G., Leone, N., Scarcello, F.: On the complexity of some inductive logic programming problems. New Generation Computing 17, 53–75 (1999)
8. Le Pape, C., Baptiste, P., Nuijten, W.: Constraint-Based Scheduling: Applying Constraint Programming to Scheduling Problems. Kluwer Academic Publishers, Dordrecht (2001)
9. Maloberti, J., Sebag, M.: Fast Theta-Subsumption with Constraint Satisfaction Algorithms. Machine Learning 55, 137–174 (2004)
10. Muggleton, S., De Raedt, L.: Inductive logic programming: Theory and methods. Journal of Logic Programming 19, 629–679 (1994)
11. Plotkin, G.: A note on inductive generalization. In: Meltzer, B., Michie, D. (eds.) Machine Intelligence, vol. 5, pp. 153–163. Edinburgh University Press, Edinburgh (1970)

Coalition Structure Generation with GRASP

Nicola Di Mauro, Teresa M.A. Basile, Stefano Ferilli, and Floriana Esposito

University of Bari "Aldo Moro", Italy
{ndm,basile,ferilli,esposito}@di.uniba.it

Abstract. The coalition structure generation problem represents an active research area in multi-agent systems. A coalition structure is defined as a partition of the agents involved in a system into disjoint coalitions. The problem of finding the optimal coalition structure is NP-complete. In order to find the optimal solution in a combinatorial optimization problem it is theoretically possible to enumerate the solutions and evaluate each. But this approach is infeasible since the number of solutions often grows exponentially with the size of the problem. In this paper we present a greedy adaptive search procedure (GRASP) to efficiently search the space of coalition structures in order to find an optimal one.

1 Introduction

An active area of research in multi-agent systems (MASs) is the coalition structure generation (CSG) of agents (equivalent to the complete set partitioning problem). In particular it is interesting to find coalition structures maximizing the sum of the values of the coalitions, that represent the maximum payoff the agents belonging to the coalition can jointly receive by cooperating. A coalition structure is defined as a partition of the agents involved in a system into disjoint coalitions. The problem of finding the optimal coalition structure is $\mathcal{NP}$-complete [1,2]. Coalition generation shares a similar structure with a number of common problems in theoretical computer science, such as in combinatorial auctions [3], in which bidders can place bids on combinations of items; in job shop scheduling; and as in set partitioning and set covering problems.

Sometimes in MASs there is a time limit for finding a solution, the agents must be reactive and they should act as fast as possible. Hence for the specific task of CSG it is necessary to have approximation algorithms able to quickly find solutions that are within a specific factor of an optimal solution. Hence, the goal of this paper is to propose a new algorithm for the CSG problem able to quickly find a near optimal solution.

The problem of CSG has been studied in the context of characteristic function games (CFGs) in which the value of each coalition is given by a characteristic function, and the values of a coalition structure are obtained by summing the value of the contained coalitions. The problem of coalition structure generation is $\mathcal{NP}$-hard, indeed as proved in [2], given n the number of agents, the number of possible coalition structures than can be generated is $O(n^n)$ and $\omega(n^{n/2})$. Moreover, in order to establish any bound from the optimal, any algorithm must search at least $2^n - 1$ coalition structures. The CSG process can be viewed as being composed of three activities [2]: a) *coalition structure generation*, corresponding to the process of generating coalitions such that agents within each coalition coordinate their activities, but agents do not coordinate between

D. Dicheva and D. Dochev (Eds.): AIMSA 2010, LNAI 6304, pp. 111–120, 2010.

coalitions. This means partitioning the set of agents into exhaustive and disjoint coalitions. This partition is called a coalition structure (CS); b) *optimization*: solving the optimization problem of each coalition. This means pooling the tasks and resources of the agents in the coalition, and solving this joint problem; and c) *payoff distribution*: dividing the value of the generated solution among agents.

Even if these activities are independent of each other, they have some interactions. For example, the coalition that an agent wants to join depends on the portion of the value that the agent would be allocated in each potential coalition. This paper focuses on the coalition structure generation in settings where there are too many coalition structures to enumerate and evaluate due to costly or bounded computation and limited time. Instead, agents have to select a subset of coalition structures on which to focus their search.

Specifically, in this work we adopted a stochastic local search procedure [4], named GRASP [5], to solve the problem of coalition structure generation in CFGs. The main advantage of using a stochastic local search is to avoid exploring an exponential number of coalition structures providing a near optimal solution. Indeed, our algorithm does not provide guarantees about finding the global optimal solution. In particular the questions we would like to pose are:

- **Q1**) can the metaheuristic GRASP be used as a valuable anytime solution for the CSG problem? In many cases, as in CSG, it is necessary to terminate the algorithm prior to completion due to time limits and to reactivity requirements. In this situation, it is possible to adopt anytime algorithms (i.e. algorithms that may be terminated prior to completion and returning an approximation of the correct answer) whose quality depends on the amount of computation.
- **Q2**) can the metaheuristic GRASP be adopted for the CSG problem to find optimal solution faster than the state of the art exact algorithms for CSG problem? In case of optimization combinatorial problems, stochastic local search algorithms have been proved to be very efficient in finding near optimal solution [4]. In many cases, they outperformed the deterministic algorithms in computing the optimal solution.

2   Definitions

Given a set $A = \{a_1, a_2, \ldots, a_n\}$ of n agents, $|A| = n$, called the *grand coalition*, a *coalition* S is a non-empty subset of the set A, $\emptyset \neq S \subseteq A$. A *coalition structure* (CS) $C = \{C_1, C_2, \ldots, C_k\} \subseteq 2^A$ is a partition of the set A, and k is its size, i.e. $\forall i, j$: $C_i \cap C_j = \emptyset$ and $\cup_{i=1}^{k} C_i = A$. For $C = \{C_1, C_2, \ldots, C_k\}$, we define $\cup C \triangleq \cup_{i=1}^{k} C_i$. We will denote the set of all coalition structures of A as $\mathcal{M}(A)$.

As in common practice [2,6], we consider coalition structure generation in *characteristic function games* (CFGs). In CFGs the value of a coalition structure C is given by $V(C) = \sum_{S \in C} v(S)$, where $v : 2^A \to \mathbb{R}$ and $v(S)$ is the value of the coalition S. Intuitively, $v(S)$ represents the maximum payoff the members of S can jointly receive by cooperating. As in [2], we assume that $v(S) \geq 0$. In case of negative values, it is possible to normalize the coalition values, obtaining a game strategically equivalent to the original game [7], by subtracting a lower bound value from all coalition values. Given a set of agents A, the goal is to maximize the social welfare of the agents by finding a coalition structure $C^* = \arg\max_{C \in \mathcal{M}(A)} V(C)$.

Given n agents, the size of the input to a CSG algorithm is exponential, since it contains the values $v(\cdot)$ associated to each of the $(2^n - 1)$ possible coalitions. Furthermore, the number of coalition structures grows as the number of agents increases and corresponds to $\sum_{i=1}^{n} Z(n, i)$, where $Z(n, i)$, also known as the Stirling number of the second kind, is the number of coalition structures with i coalitions, and may be computed using the following recurrence: $Z(n, i) = iZ(n - 1, i) + Z(n - 1, i - 1)$, where $Z(n, n) = Z(n, 1) = 1$. As proved in [2], the number of coalition structures is $O(n^n)$ and $\omega(n^{n/2})$, and hence an exhaustive enumeration becomes prohibitive.

In this paper we focus on games that are neither *superadditive* nor *subadditive* for which the problem of coalition structure generation is computationally complex. Indeed, for superadditive games where $v(S \cup T) \geq v(S) + v(T)$ (meaning any two disjoint coalitions are better off by merging together), and for subadditive games where $v(S \cup T) < v(S) + v(T)$ for all disjoint coalitions $S, T \subseteq A$, the problem of coalition structure generation is trivial. In particular, in superadditive games, the agents are better off forming the grand coalition where all agents operate together ($C^* = \{A\}$), while in subadditive games, the agents are better off by operating alone ($C^* = \{\{a_1\}, \{a_2\}, \ldots, \{a_n\}\}$).

3   Related Work

Previous works on CSG can be broadly divided into two main categories: exact algorithms that return an optimal solution, and approximate algorithms that find an approximate solution with limited resources.

A deterministic algorithm must systematically explore the search space of candidate solutions. One of the first algorithms returning an optimal solution is the dynamic programming algorithm (DP) proposed in [8] for the set partitioning problem. This algorithm is polynomial in the size of the input ($2^n - 1$) and it runs in $O(3^n)$ time, which is significantly less than an exhaustive enumeration ($O(n^n)$). However, DP is not an anytime algorithm, and has a large memory requirement. Indeed, for each coalition C it computes the tables $t_1(C)$ and $t_2(C)$. It computes all the possible splits of the coalition C and assigns to $t_1(C)$ the best split and to $t_2(C)$ its value. In [6] the authors proposed an improved version of the DP algorithm (IDP) performing fewer operations and requiring less memory than DP. IDP, as shown by the authors, is considered one of the fastest available exact algorithm in the literature computing an optimal solution.

Both DP and IDP are not anytime algorithms, they cannot be interrupted before their normal termination. In [2], Sandholm et al. have presented the first anytime algorithm, that can be interrupted to obtain a solution within a time limit but not guaranteed to be optimal. When not interrupted it returns the optimal solution. The CSG process can be viewed as a search in a coalition structure graph as reported in Figure 1. One desideratum is to be able to guarantee that the CS is within a worst case bound from optimal, i.e. that searching through a subset N of coalition structures, $k = \min\{k'\}$ where $k' \geq \frac{V(S^*)}{V(S_N^*)}$ is finite, and as small as possible, where S^* is the best CS and S_N^* is the best CS that has been seen in the subset N. In [2] has been proved that: a) to bound k, it suffices to search the lowest two levels of the coalition structure graph (with this

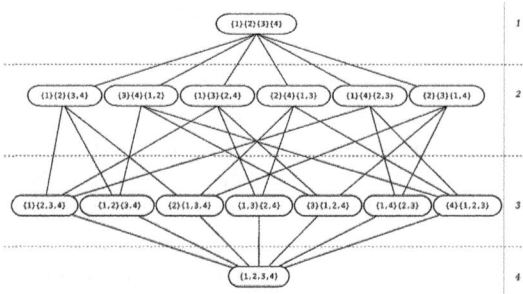

Fig. 1. Coalition structure graph for a 4-agent game

search, the bound $k = n$, and the number of nodes searched is 2^{n-1}); b) this bound is tight; and, c) no other search algorithm can establish any bound k while searching only 2^{n-1} nodes or fewer.

A new anytime algorithm has been proposed in [9], named IP, whose idea is to partition the space of the possible solutions into sub-spaces such that it is possible to compute upper and lower bounds on the values of the best CSs they contain. Then, these bounds are used to prune all the sub-spaces that cannot contain the optimal solution. Finally, the algorithm searches through the remaining sub-spaces adopting a branch-and-bound technique avoiding to examine all the solutions within the searched sub-spaces.

As regards the approximate algorithms, in [10] it has been proposed a solution based on a genetic algorithm, which performs well when there is some regularity in the search space. Indeed, the authors assume, in order to apply their algorithm, that the value of a coalition is dependent of other coalitions in the CS, making the algorithm not well suited for the general case. A new solution [11] is based on a Simulated Annealing algorithm [12], a widely used stochastic local search method. At each iteration the algorithm selects a random neighbour solution s' of a CS s. The search proceeds with an adjacent CS s' of the original CS s if s' yields a better social welfare than s. Otherwise, the search is continued with s' with probability $e^{(V(s')-V(s))/t}$, where t is the temperature parameter that decreases according to the annealing schedule $t = \alpha t$.

4   GRASP for CSG

The resource limits posed by MASs, such as the time for finding a solution, require to have approximation algorithms able to quickly find solutions that are within a specific factor of an optimal solution. In this section we present a new anytime algorithm for CSG based on a stochastic local search procedure, named GRASP-CSG.

A method to find high-quality solutions for a combinatorial problem is a two steps approach consisting of a greedy construction phase followed by a perturbative[1] local search [4]. Namely, the greedy construction method starts the process from an empty candidate solution and at each construction step adds the best ranked component according to a heuristic selection function. Successively, a perturbative local search

---

[1] A perturbative local search changes candidate solutions by modifying one or more of the corresponding solution components.

Algorithm 1. GRASP-CSG

Require: V: the characteristic function; A: the set of n agents; *maxiter*: maximum number of iterations

Ensure: solution $\widehat{C} \in \mathcal{M}(A)$

$\widehat{C} = \emptyset, V(\widehat{C}) = -\infty$

iter $= 0$

while iter $<$ maxiter **do**

    $\alpha = \text{rand}(0,1)$;

    $C = \emptyset; i = 0$

    /* construction */

    while $i < n$ **do**

        $\mathcal{S} = \{C'|C' = add(C, A)\}$

        $\overline{s} = \max\{V(T)|T \in \mathcal{C}\}$

        $\underline{s} = \min\{V(T)|T \in \mathcal{C}\}$

        RCL $= \{C' \in \mathcal{S}|V(C') \geq \underline{s} + \alpha(\overline{s} - \underline{s})\}$

        randomly select an element C from RCL

        $i \leftarrow i + 1$

    /* local search */

    $\mathcal{N} = \{C' \in neigh(C)|V(C') > V(C)\}$

    while $\mathcal{N} \neq \emptyset$ **do**

        select $C \in \mathcal{N}$

        $\mathcal{N} \leftarrow \{C' \in neigh(C)|V(C') > V(C)\}$

    if $V(C) > V(\widehat{C})$ **then**

        $\widehat{C} = C$

    iter $=$ iter $+ 1$

  return $\widehat{C}$

---

algorithm is used to improve the candidate solution thus obtained. Advantages of this search method, over other stochastic local search algorithms, are the much better solution quality and fewer perturbative improvement steps to reach the local optimum. Greedy Randomized Adaptive Search Procedures (GRASP) [5] solve the problem of the limited number of different candidate solutions generated by a greedy construction search methods by randomising the construction method. GRASP is an iterative process, in which each iteration consists of a construction phase, producing a feasible solution, and a local search phase, finding a local optimum in the neighborhood of the constructed solution. The best overall solution is returned.

Algorithm 1 reports the outline for GRASP-CSG included in the ELK system. In each iteration, it computes a solution C by using a randomised constructive search procedure and then applies a local search procedure to C yielding an improved solution. The main procedure is made up of two components: a constructive phase and a local search phase. The constructive search algorithm used in GRASP-CSG iteratively adds a solution component by randomly selecting it, according to a uniform distribution, from a set, named *restricted candidate list* (RCL), of highly ranked solution components with respect to a greedy function $g : C \to \mathbb{R}$. The probabilistic component of GRASP-CSG is characterized by randomly choosing one of the best candidates in the RCL. In our case the greedy function g corresponds to the characteristic function V presented in Section 2. In particular, given V, the heuristic function, and $\mathcal{C}$, the set of feasible solution components, $\underline{s} = \min\{V(C)|C \in \mathcal{C}\}$ and $\overline{s} = \max\{V(C)|C \in \mathcal{C}\}$ are computed. Then the RCL is defined by including in it all the components C such that $V(C) \geq \underline{s} + \alpha(\overline{s} - \underline{s})$. The parameter α controls the amounts of greediness and randomness. A value $\alpha = 1$ corresponds to a greedy construction procedure, while $\alpha = 0$ produces a random construction. As reported in [13], GRASP with a fixed nonzero RCL parameter α is not asymptotically convergent to a global optimum. The solution to make the algorithm asymptotically globally convergent, could be to randomly select the parameter value from the continuous interval $[0, 1]$ at the beginning of each iteration and using this value during the entire iteration, as we implemented in GRASP-CSG.

Given a set of nonempty subsets of n agents A, $C = \{C_1, C_2, \ldots, C_t\}$, such that $C_i \cap C_j \neq \emptyset$ and $\cup C \subset A$, the function $add(C, A)$ used in the construction phase returns a refinement C' obtained from C using one of the following operators:

1. $C' \rightarrow C \setminus \{C_i\} \cup \{C_i'\}$ where $C_i' = C_i \cup \{a_i\}$ and $a_i \notin \cup C$, or
2. $C' \rightarrow C \cup \{C_i\}$ where $C_i = \{a_i\}$ and $a_i \notin \cup C$.

Starting from the empty set, in the first iteration all the coalitions containing exactly one agent are considered and the best is selected for further specialization. At the iteration i, the working set of coalition C is refined by trying to add an agent to one of the coalitions in C or a new coalition containing the new agent is added to C.

To improve the solution generated by the construction phase, a local search is used. It works by iteratively replacing the current solution with a better solution taken from the neighborhood of the current solution while there is a better solution in the neighborhood. In order to build the neighborhood of a coalition structure C, $neigh(C)$, the following operators, useful to transform partitions of the grand coalition, have been used. Given a CS $C = \{C_1, C_2, \ldots, C_t\}$:

split: $C \rightarrow C \setminus \{C_i\} \cup \{C_k, C_h\}$, where $C_k \cup C_h = C_i$, with $C_k, C_h \neq \emptyset$;
merge: $C \rightarrow C \setminus \{C_i, C_j\}_{i \neq j} \cup \{C_k\}$, where $C_k = C_i \cup C_j$;
shift: $C \rightarrow C \setminus \{C_i, C_j\}_{i \neq j} \cup \{C_i', C_j'\}$, where $C_i' = C_i \setminus \{a_i\}$ and $C_j' = C_j \cup \{a_i\}$, with $a_i \in C_i$.
exchange: $C \rightarrow C \setminus \{C_i, C_j\}_{i \neq j} \cup \{C_i', C_j'\}$, where $C_i' = C_i \setminus \{a_i\} \cup \{a_j\}$ and $C_j' = C_j \setminus \{a_j\} \cup \{a_i\}$, with $a_i \in C_i$ and $a_j \in C_j$;
extract: $C \rightarrow C \setminus \{C_i\}_{i \neq j} \cup \{C_i', C_j\}$, where $C_i' = C_i \setminus \{a_i\}$ and $C_j = \{a_i\}$, with $a_i \in C_i$.

In the local search phase, the neighborhood of a coalition structure C is built by applying all the previous operators (split, merge, shift, exchange and extract) to C. As an example, in Table 1 is reported the application of the operators to the CS $\{\{12\}\{3\}\{4\}\}$. The problem in using more than the two classical merge and split operators corresponds to the fact of obtaining repetitions of the same CS. This problem deserves further attention, each operator must take into account how other operators works. Concerning the representation of the characteristic function and the search space, given n agents $A = \{a_1, a_2, \ldots, a_n\}$, we recall that the number of possible coalitions is $2^n - 1$. Hence, the characteristic function $v : 2^n \rightarrow \mathbb{R}$ is represented as a vector CF in the following way. Each subset $S \subseteq A$ (coalition) is described as a binary number $c_B = b_1 b_2 \cdots b_n$ where each $b_i = 1$ if $a_i \in S$, $b_i = 0$ otherwise. For instance, given $n = 4$, the coalition

Table 1. Operators applied to the CS $\{12\}\{3\}\{4\}$.

split	merge	shift	exchange	extract
$\{1\}\{2\}\{3\}\{4\}$	$\{123\}\{4\}$	$\{2\}\{13\}\{4\}$	$\{23\}\{1\}\{4\}$	$\{1\}\{2\}\{3\}\{4\}$
	$\{124\}\{3\}$	$\{2\}\{3\}\{14\}$	$\{24\}\{3\}\{1\}$	
	$\{12\}\{34\}$	$\{1\}\{23\}\{4\}$	$\{13\}\{2\}\{4\}$	
		$\{1\}\{3\}\{24\}$	$\{14\}\{3\}\{2\}$	

$\{a_2, a_3\}$ corresponds to the binary number 0110. Now, given the binary representation of a coalition S, its decimal value corresponds to the index in the vector CF where its corresponding value $v(S)$ is memorised. This gives us the possibility to have a random access to the values of the characteristic functions in order to efficiently compute the value V of a coalition structure.

Given a coalition structure $C = \{C_1, C_2, \ldots, C_k\}$, assuming that the C_i are ordered by their smallest elements, a convenient representation of the CS is a sequence $d_1 d_2 \cdots d_n$ where $d_i = j$, if the agent a_i belongs to the coalition C_j. Such sequences are known as *restricted growth sequences* [14] in the combinatorial literature. For instance, the sequence corresponding to the coalition structure $C = \{C_1, C_2, C_3\} = \{\{1, 2\}, \{3\}, \{4\}\}$ is 1123. Now in order to compute $V(C)$, we have to solve the sum $v(C_1) + v(C_2) + v(C_3)$, where C_1 corresponds to the binary number 1100, C_2 corresponds to the binary number 0010, and C_3 corresponds to the binary number 0001. Hence, $V(C) = v(C_1) + v(C_2) + v(C_3) = CF[12] + CF[2] + CF[1]$, where CF is the vector containing the values of the characteristic function.

5   Experimental Results

In order to evaluate our GRASP-CSG, we implemented it in the C language and the corresponding source code has been included in the ELK system[2]. We also implemented the algorithm proposed by Sandholm et al. in [2], DP [8], and IDP [6], whose source code has been included in the ELK system. GRASP-CSG has been compared to those algorithms and to the Simulated Annealing algorithm (SA) proposed in [11], kindly provided by the authors.

In the following we present experimental results on the behaviour of these algorithms for some benchmarks considering solution qualities and the runtime performances. We firstly compared GRASP-CSG to DP and IDP. Then we evaluated its ability in generating solutions anytime when compared to the SA and to the Sandholm et al. algorithms.

Instances of the CSG problem have been defined using the following distributions for the values of the characteristic function:

1. Normal if $v(C) \sim N(\mu, \sigma^2)$ where $\mu = 1$ and $\sigma = 0.1$;
2. Uniform if $v(C) \sim U(a, b)$ where $a = 0$ and $b = 1$.

The algorithms are executed on PC with an Intel Core2 Duo CPU T7250 @ 2.00GHz and 2GB of RAM, running Linux kernel 2.6.31.

---

[2] ELK is a system including many algorithms for the CSG problem whose source code is publicly available at http://www.di.uniba.it/$\sim$ndm/elk/

5.1   Optimal Solution

Given different numbers of agents, ranging from 14 to 27, we compared GRASP-CSG to DP and IDP reporting the time required to find the optimal coalition structure. As reported in Figure 2, where the time in seconds is plotted in a log scale, GRASP-CSG outperforms both DP and IDP for all the distributions. Note that there is one line for DP and IDP since they do not depend on the input distribution but only on the input dimension. Over the problems, the execution time for DP (resp. IDP) ranges from 0.124 (resp. 0.06) seconds to approximately 366586 (resp. 207433) seconds, while for GRASP-CSG it ranges from 0.011 seconds to 0.12 seconds on average. In particular, for 27 agents GRASP-CSG is 1728608 times faster than IDP. As regards DP and IDP, due to high time consuming, values for problems with a number of agents ranging from 23 to 27 have been analytically calculated since their complexity linearly depends on the problem size. As we can see from Figure 2 this improvement grows as the dimension of the problem grows. Even if we cannot make a direct comparison to IP, as the authors reported in [9], IP is 570 times better than IDP in the case of uniform distribution for 27 agents. Since GRASP-CSG is 1728608 times faster than IDP for the same problem, we can argue that GRASP-CSG is faster than IP.

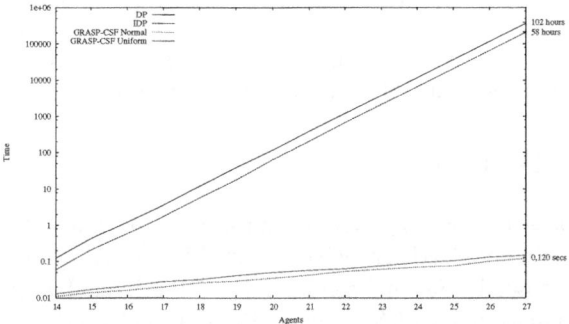

Fig. 2. Comparison between DP, IDP and GRASP-CSG

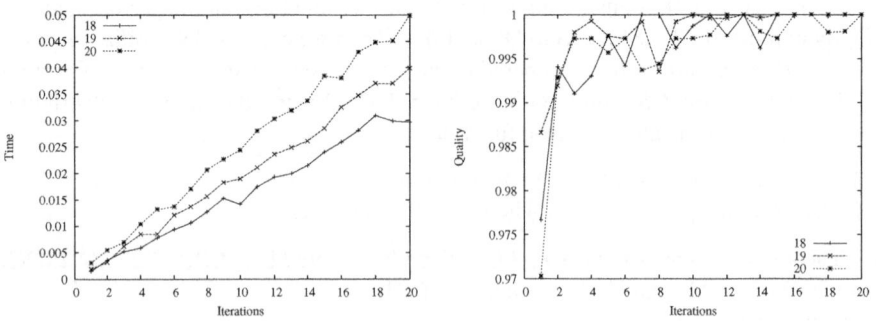

Fig. 3. Time (in seconds) and relative quality of GRASP-CSG obtained on the first 20 iterations

As regards GRASP-CSG we set the maxiter parameter to 20 even if in many cases the optimal CS has been found with fewer iterations, see the details in Figure 3. However, in this experiment this limit guarantees to find always the optimal coalition. Given the number of agents, 10 different instances of the problem for each distribution have been generated and time is averaged. Figure 3 reports an insight of the GRASP-CSG execution for three problems (18, 19 and 20 agents). For each iteration the graphs report the time and the relative quality of the solution averaged over 10 instances. The relative quality of a coalition structure C is obtained as $V(C)/V(C^*)$ where C^* is the optimal CS. As we can see, the quality of the obtained CSs is very high just in the first iterations computed with few milliseconds.

5.2 Approximate Solution

In this second experiment we compared the anytime characteristic of GRASP-CSG to the Sandholm et al. algorithm [2] and the Simulated Annealing algorithm [11]. We generated 10 instances for each problem with agents ranging from 14 to 20 and uniform distribution $U(0,1)$. For each problem we set a time limit to return a good solution and we recorded the relative error of the obtained solution S by each of the three algorithms computed as $e = 1 - V(S)/V(S^*)$, where S^* is the best CS. Table 2 reports the error averaged over the 10 instances for each problem. As we can see GRASP-CSG is always able to find a better CS than those obtained by Sandholm et al. and SA. With this second experiment we can conclude that GRASP-CSG quickly finds very good solutions.

Table 2. Comparison of Sandholm et al., Simulated Annealing (SA) and GRASP-CSG

Agents	Sandholm	SA	GRASP-CSG
14	0.1340	0.0320	0.0046
15	0.1862	0.0331	0.0000
16	0.1814	0.0205	0.0033
17	0.1514	0.0104	0.0000
18	0.1057	0.0001	0.0000
19	0.1393	0.0052	0.0005
20	0.1399	0.0021	0.0000

6   Conclusions

The paper presented the application of the stochastic local search GRASP to the problem of coalition structure generation. As reported in the experimental section the proposed algorithm outperforms some of the state of the art algorithms in computing optimal coalition structures.

As a future work it should be interesting to investigate the behaviour of the operators used to create the neighborhood of a coalition structure. In particular, an in deep study may be conducted in learning to choose the correct operators with respect to the distribution of the coalition values. Furthermore the application of shift, exchange and extract operators should generate repetitions of the same coalition structure obtained with the split and merge operators. Hence, an analysis on how to overcome this problem, avoiding to spend time and space resources, deserves more attention. Furthermore, we are planning to apply the proposed method to the more general games such as Partition Function Games [15], where the coalition's value may depend on the formation of another coalition, and to the task-based colaition formation problem [16].

References

1. Garey, M.R., Johnson, D.S.: Computers and Intractability: A Guide to the Theory of NP-Completeness. W. H. Freeman & Co., New York (1990)
2. Sandholm, T., Larson, K., Andersson, M., Shehory, O., Tohmé, F.: Coalition structure generation with worst case guarantees. Artificial Intelligence 111(1-2), 209–238 (1999)
3. Cramton, P., Shoham, Y., Steinberg, R. (eds.): Combinatorial Auctions. MIT Press, Cambridge (2006)
4. Hoos, H., Stützle, T.: Stochastic Local Search: Foundations & Applications. Morgan Kaufmann Publishers Inc., San Francisco (2004)
5. Feo, T.A., Resende, M.G.C.: Greedy randomized adaptive search procedures. Journal of Global Optimization 6, 109–133 (1995)
6. Rahwan, T., Jennings, N.R.: An improved dynamic programming algorithm for coalition structure generation. In: Prooceedings of AAMAS 2008, pp. 1417–1420 (2008)
7. Kahan, J.P., Rapoport, A.: Theories of Coalition Formation. Lawrence Erlbaum Associates Publisher, Mahwah (1984)
8. Yeh, D.Y.: A dynamic programming approach to the complete set partitioning problem. BIT 26(4), 467–474 (1986)
9. Rahwan, T., Ramchurn, S.D., Jennings, N.R., Giovannucci, A.: An anytime algorithm for optimal coalition structure generation. Journal of Artificial Intelligence Research 34(1), 521–567 (2009)
10. Sen, S., Dutta, P.S.: Searching for optimal coalition structures. In: Prooceedings of ICMAS 2000, pp. 287–292. IEEE Computer Society, Los Alamitos (2000)
11. Keinänen, H.: Simulated annealing for multi-agent coalition formation. In: Håkansson, A., Nguyen, N.T., Hartung, R.L., Howlett, R.J., Jain, L.C. (eds.) KES-AMSTA 2009. LNCS, vol. 5559, pp. 30–39. Springer, Heidelberg (2009)
12. Kirkpatrick, S., Gelatt Jr., C.D., Vecchi, M.P.: Optimization by simulated annealing. Science 220(4598), 671–680 (1983)
13. Mockus, J., Eddy, E., Mockus, A., Mockus, L., Reklaitis, G.V.: Bayesian Heuristic Approach to Discrete and Global Optimization. Kluwer Academic Publishers, Dordrecht (1997)
14. Milne, S.C.: Restricted growth functions, rank row matchings of partitions lattices, and q-stirling numbers. Advances in Mathemathics 43, 173–196 (1982)
15. Rahwan, T., Michalak, T., Jennings, N., Wooldridge, M., McBurney, P.: Coalition structure generation in multi-agent systems with positive and negative externalities. In: Proceedings of IJCAI 2009, pp. 257–263 (2009)
16. Dang, V.D., Jennings, N.R.: Coalition structure generation in task-based settings. In: Proceeding of ECAI 2006, pp. 210–214. IOS Press, Amsterdam (2006)

A Bayesian Model for Entity Type Disambiguation

Barbara Bazzanella[1], Heiko Stoermer[2], and Paolo Bouquet[3]

[1] University of Trento,
Dept. of Cognitive Science,
Trento, Italy
barbara.bazzanella@unitn.it
[2] Fondazione Bruno Kessler
DKM unit
Trento, Italy
hstoermer@gmail.com
[3] University of Trento
Dept. of Information Science and Engineering
Trento, Italy
bouquet@disi.unitn.it

Abstract. Searching for information about individual entities such as persons, locations, events, is a major activity in Internet search. General purpose search engines are the most commonly used access point to entity-centric information. In this context, keyword queries are the primary means of retrieving information about specific entities. We argue that an important first step to understand the information need that underly an entity-centric query is to understand what type of entity the user is looking for. We call this process Entity Type Disambiguation. In this paper we present a Naive Bayesian Model for entity type disambiguation that explores our assumption that an entity type can be inferred from the attributes a user specifies in a search query. The model has been applied to queries provided by a large sample of participants in an experiment performing an entity search task. The results of the model performance evaluation are presented and discussed. Finally, an extension of the model that includes the distribution of attributes inside the query is presented and its implications are investigated.

1 Introduction

In the transition from a "document web" to a "semantic web", one of the most significant changes in paradigm is the shift away from documents as the central element of information retrieval, towards something closer to the actual information need of the user. Neglecting navigational and transactional queries [4] in the context of our work, we follow the assumption that informational queries can be satisfied by identifying which individual entity a user is looking for information *about*. Studies have shown that user behaviour is often characterized by defining

D. Dicheva and D. Dochev (Eds.): AIMSA 2010, LNAI 6304, pp. 121–130, 2010.
© Springer-Verlag Berlin Heidelberg 2010

a certain context in which the desired information is most likely to be found [14], and from our perspective, an individual entity can be such a context.

Our work is thus concentrating on the question how to understand such a context, i.e. how to determine from a set of keywords whether there is a part that describes an individual entity, and which kind of entity it describes, in order to limit the search to information about this precise entity. This approach is particularly meaningful for searching in Semantic Web content, where "aboutness" is a central aspect of information modelling. Knowing about what we want to know something can help us limit the search space significantly and improve the quality of search results.

A first step in this direction was a study that we have performed in 2008, with the aim of asking people how they actually *describe* entities. Results of this study were published in [2], and provided us with a first hypothesis on the most important set of features commonly employed by users. This study has lead to the implementation of a novel algorithm for entity linkage [12], specially tailored for use-cases of the Semantic Web, as well as provided the background for the core data model in the Entity Name System [1]. The topic of this paper is a new study that has been conducted to gain insights into the same questions from a different perspective (user queries instead of descriptions), to see whether our initial findings can be confirmed, and to explore whether it is possible to identify an "entity part" in a keyword query.

The analysis of the outcomes of this study are significant for the Semantic Web community for several reasons. First, we confirm many of the findings of [2], which represents a useful contribution to the ontological modelling of entity types, because we provide an extensive list of the most common features used to describe entities; these can directly be re-used to create or even to evaluate an ontological model. Secondly, as mentioned, search on the Semantic Web is more and more going in the direction of question answering, and understanding which (type of) entity we are talking about can be important in this process. Finally, our findings can help us disambiguate an entity-related query; to a human, the term "Washington" in the two queries "George Washington" and "Washington USA" is clearly referring to different objects. We are hoping to give a contribution to the construction of new algorithms that also make this possible in a machine.

2   An Entity Search Experiment

Queries for specific entities represent a variation of the expressed information need that has been studied in many IR contexts [11, 13]. A query for a specific entity can be considered like a way to translate a human information need into a small number of attributes that the user considers relevant to identify the entity. Therefore, the analysis of real user queries should provide valuable insights into which kinds of attributes humans actually consider relevant to identify different types of entities during the search process.

As a first step towards a better understanding of this aspect of the query formulation process, we performed an experimental study. This study is part of an ongoing research to better understand how people represent and identify entities. In the previous experiment mentioned above, we adopted a bottom-up approach to investigate how people extensively describe individual entities belonging to a small set of entity types. The selection of these entity types was driven by a set of ontological requirements. In this current study we focus on the same collection of entity types, partly also to confirm our initial findings (see the technical report [2] accompanying the first study for more theoretical details).

The goal of this study was to investigate the process that leads users to organize and represent their information needs using simple queries, limiting the analysis to queries that look for specific type of entities (person, organization, event, artifact and location). This study explores two main issues: 1) to investigate which attributes are considered more relevant by people to identify specific types of entities in a query formulation task; 2) to identify significant patterns of attributes that reproduce recurrent strategies in entity searching.

2.1 Methodology

To answer our research questions we conducted an online experiment with a significant amount of users (301 participants with average age of 31.40 years). The experiment consists of ten query formulation tasks. Participants are presented with an entity type (e.g., person) and they are asked to imagine any individual entity of their choosing belonging to this type (e.g., Barack Obama). Once the individual entity is chosen, participants are asked to formulate a query with the intent to find the homepage of the entity considered. In our example a plausible query may be <Barack Obama president USA>. Every participant is asked to perform ten such tasks, submitting their queries through a mimicked search engine interface. Five tasks present entity types at a very high level of abstraction (*high-level entity types*: person, organization, event, artifact and location). All the participants were tested on all the high-level tasks. The other five tasks correspond to more specific entity types (*low-level entity types*), selected from a predefined set of possible subtypes for each high-level type (see table 1 for the complete list of types). Every participant performed only one low-level task for each high-level entity type. The task order was randomized between subjects.

Table 1. Entity types and subtypes

Person	Organization	Event	Artifact	Location
politician	company	conference	product	tourist location
manager	association	meeting	artwork	city
professor	university	exhibition	building	shop
sports person	government	show	book	hotel
actor	agency	accident	article of clothing	restaurant
		sports event		

2.2 A Naive Bayes Model of Attribute Relevance

The first goal of our research is to identify which kinds of attributes humans consider relevant to identify different types of entities during the search process. To answer this question we suggest to adopt a Naive Bayes Model of attribute relevance. This choice is motivated by two main reasons. The first is that quantifying the level of relevance of a feature for a category is a well-known approach in cognitive studies on human categorization [9]. Moreover, the Bayesian model of attribute relevance corresponds to one of the measures proposed in cognitive psychology [8] (cue validity) to quantify the relevance of a feature for general categories. A second reason is that the formalization of Bayesian statistics provides a middle ground where cognitive models and probabilistic models developed in other research fields (statistics, machine learning, and artificial intelligence) can find the opportunity for communication and integration.

In order to clarify the terms of our approach, we first introduce the basic tokens of the Naive Bayesian Model (NBM). We can represent a query Q as a set of unknown terms $T = (t_1, t_2, ..., t_n)$, each of which can be a single word or a combination of words. We assume that each term t specifies the value of an attribute a. Assume that $A = (a_1, a_2, ..., a_n)$ is a set of attribute types. We map every term t into one appropriate type in A. Finally, suppose that $E = (e_1, e_2, ..., e_m)$ is a small number of entity types. Our goal is to quantify the relevance of an attribute a for a given entity type e_i. In the NBM framework this corresponds to compute the posterior probability $p(e_i|a)$:

$$p(e_i|a) = \frac{p(e_i)*p(a|e_i)}{\sum_{i=1}^{m} p(e_i)*p(a|e_i)} \tag{1}$$

The NBM is a probabilistic model based on the assumption of strong independence between attributes that means that the presence (or absence) of a particular attribute is unrelated to the presence (or absence) of any other attribute. Under this assumption, we can extend the model to the case of multiple attributes $\mathbf{a} = (a_1, a_2, ..., a_s)$ as defined in Eq. 2.

$$p(e_i|\mathbf{a}) = \frac{p(e_i)*p(\mathbf{a}|e_i)}{\sum_{i=1}^{m} p(e_i)*p(\mathbf{a}|e_i)} \qquad \text{where} \qquad p(\mathbf{a}|e_i) = \prod_{j=1}^{s} p(a_j|e_i) \tag{2}$$

In this way, we can express the combined relevance of two or more types of attributes for a given entity type (e.g., the combined relevance of "name" and "surname" for "person") and detect the most likely type of the target entity. This corresponds to a function, $f : \mathbf{a} \rightarrow (E)$, that takes as argument a vector $\mathbf{a}$ of attributes and returns the most likely entity type for that combination of attributes. f is known as *Naive Bayes Classifier*.

Preprocessing. Before applying the Bayesian Model to our data we performed two steps of preprocessing (see table 2 for examples). The first step, (*syntactic preprocessing*), involved extracting the terms from the queries. In this phase we also cleaned the dataset from unusual queries such as blank queries (empty), strings with only punctuation marks or senseless queries. Once the terms have been extracted from the queries, they were mapped into the attribute type set A. This mapping corresponded to the second step of preprocessing (*semantic preprocessing*). The first step was conducted in a semiautomatic way (i.e., the deletion of empty queries and a rough tokenization by segmenting the text at each space were performed automatically but the assignment of words to terms

Table 2. Two-step Preprocessing

Query	Syntactic Preproc.	Semantic Preproc.
Q_1 =ISCW 2010 Shangai	t_1=ISWC t_2=2010 t_3=Shangai	$t_1 \Rightarrow$ event name $t_2 \Rightarrow$ date:year $t_3 \Rightarrow$ location (city)
Q_2= McCain Republican	t_1=McCain t_2=Republican	$t_1 \Rightarrow$ surname $t_2 \Rightarrow$ political party

was performed manually), whereas the semantic preprocessing was performed entirely manually.

3 Results

In our experiment we collected an amount of 4017 queries. The average query length was 2.04 terms (mode=2 and median=2), which is in line with the results reported in literature (see for example [6]). Over 35% contained only one term and less than 3% of the queries contained five or more terms. Almost none of the queries utilized Boolean operators (over 99%). In only ten queries the operator AND was used, whereas the use of other operators was inexistent. The analysis of the word frequency distribution showed a very limited usage of articles, prepositions, and conjunctions. The only word without content that appeared in the first 30 most frequently used words was the preposition "of".

3.1 Bayesian Relevance of Attribute Types

In Table 3(a) we report the results of applying the Bayesian Model described in Eq. 1 for the five high-level entity types addressed in our experiments. For each entity type we list the attributes with the highest relevance. These attributes satisfy two requirements: they are frequently used by subjects to formulate their queries about a specific entity type and at the same time they are rarely used to search for other entity types. Therefore, if we are able to identify the relevant attributes, we can use this information to infer the entity type of the target entity. To test our prediction, we used 125 queries which were randomly extracted from the original query sample and we tested our model, using the classifier described in 2.2[1]. The results in terms of precision, recall and F-measure[2] are summarized in Table 4. The overall performance of our approach is satisfactory and promising, even though the evaluation measures show some interesting differences between the categories. The best result was obtained for the entity type "Event", whereas we obtained the weakest result for the type "Artifact". These

---

[1] The 125 queries constituting our test set were not part of the sample which was used to calculate the Bayesian Relevance Measures reported in Table 3(a).

[2] *Precision* is defined as the number of queries correctly assigned to the entity type divided by the total number of queries assigned to that type; *recall* is defined as the number of queries correctly assigned to the entity type divided by the total number of queries which should have been assigned to it; *F-measure* is the harmonic mean of precision and recall.

Table 3. Bayesian Relevance

(a) Entity types

Entity Type (e)	Attribute type (a)	p(e|a)
Person	first name	0.85
	surname	0.84
	occupation	0.89
	middle name	0.69
	pseudonym	0.33
	area of interest/activity	0.21
Organization	organization type	0.88
	organization name	0.73
	area of interest/activity	0.54
	url extention	0.29
Event	event name	0.96
	event type	0.95
	date:month	0.83
	date:year	0.81
	date:day	0.75
Artifact	artifact type	0.98
	features	0.90
	model name	0.89
	artifact name	0.86
	historical period/epoch	0.56
	nationality	0.50
Location	location type	0.84
	location name	0.65

(b) Location

Entity Type (e)	Attribute type (a)	p(e|a)
Tourist location	location name	0.74
	location type	0.28
	organization name	0.18
City	administrative role	0.68
	building name	0.68
	state name	0.48
	municipality	0.48
	country name	0.46
	city name	0.30
Shop	shop name	0.91
	product type	0.90
	brand	0.85
	shop type	0.79
	address:street	0.33
Hotel	hotel name	0.93
	hotel type	0.61
	number of stars	0.48
	price range	0.42
Restaurant	restaurant name	0.92
	type of cousine	0.90
	restaurant type	0.61
	services	0.47
	neighbourhood	0.43

results can be explained considering that some entity types are more frequently used to specify the context in which a target entity is placed. For example, a product can be identified with reference to the company name, an artwork can be searched with reference to its author. Therefore, the presence of some attribute types, like "organization name" or "person name" in our example, may weaken the performance of our disambiguation method because they are not unique for specific entity types. This may lead to an increase in the false positive rate. This interpretation is consistent, for example, with a high level of recall and a weaker level of precision observed for Person. On the contrary, since events are rarely used to identify other entity types, the presence of event attributes strongly indicates that the query is about an event.

Once the general type has been identified, a second step is discriminating between entities inside the same high level entity type. This problem can be formulated as follows. Which are the attributes that are relevant for certain entity subtypes belonging to the same high-level type? To answer this question, we performed the same analysis, restricting the domain to the low-level entity types. In Table 3(b) we report an example of this second level of analysis (for space reasons we report only an extract of the results).

From an overall analysis of the results it turns out that for the majority of entity types "name" is the most relevant attribute used by people to identify the target of their request. This result confirms the centrality of proper names within the referential expressions (see for example [7]). However not all entities can be

Table 4. Test-set Evaluation

Measures	Person	Organization	Event	Artifact	Location
Precision	0.72	0.87	1	1	0.85
Recall	1	0.91	0.91	0.66	0.96
F-measure	0.84	0.89	0.95	0.80	0.90

Overall Precision	Overall Recall	Overall F-measure
0.86	0.89	0.88

identified by means of a name. For example, pieces of clothing, accidents, or governments are entity types identified preferentially by means of other attributes. A particular case is represented by the entity type "product". Our analysis shows that the majority of products are identified by the "model name" and not by the proper name of a specific entity (a type-token issue not uncommon in the context of products [10]). This result reveals another important aspect of the identification process: only a subset of entities are prototypically *namable* entities (e.g. person). Since users need also to identify non-namable things in their queries, the problem of Entity Type Disambiguation can not be entirely solved by the detection of the named entity in a query and the classification of it into predefined classes (an example of this approach can be found in [5]). Given a query like "guitar Jimi Hendrix 1967", the named entities are "Jimi Hendrix" and "1967", but the target entity of the query is an artifact (the guitar). The example shows that the simple classification of the named entities can be uneffective to detect the type of the target entity of the query and supports the idea that the disambiguation can be improved by including information from different kinds of attribute, such as "organization type" for organizations (e.g. non profit), temporal attributes for events (e.g. date), qualitative attributes (e.g. "color" or "material") for artifacts.

3.2   Distribution Trends: Attribute Position

The second research question of our study was about the distribution of attributes inside the queries. We aimed to highlight possible trends of attributes that recur during the formulation process and that reflect, it is argued, the strategies used by users to organize their information need. To this purpose, we focused on the distribution of attributes in terms of position. The aim was to explore whether there is a preferential order followed by subjects when they organize the attributes within the query so that an attribute type is more likely used in a specific position in the query. For example, is the name of the target entity always the first attribute specified? In this case the position of the attribute becomes extremely informative to understand the entity search process and should be included in an integrated model of attribute relevance. The results of our experiment give support to this hypothesis. An example is reported in Fig.1 that shows the probability distribution of the attribute types for Person and Organization. We note that "first name" and "organization name" are the attributes with the highest probability in first position, respectively for Person and Organization. Instead, "surname" and "middle name" (for Person) and "organization type" and "location" (for Organization) are the preferred attributes in second position.

Since the same attribute types can be used in queries with a different target, the position of the attribute in the query can be used to improve the disambiguation process. For example, the two queries "Steve Jobs Apple" and "Apple Steve Jobs" would look for the same entity type, by simply applying the relevance measures discussed in section 3.1. Instead, from our analysis it comes out that the attribute type order of the first query is more likely to be used

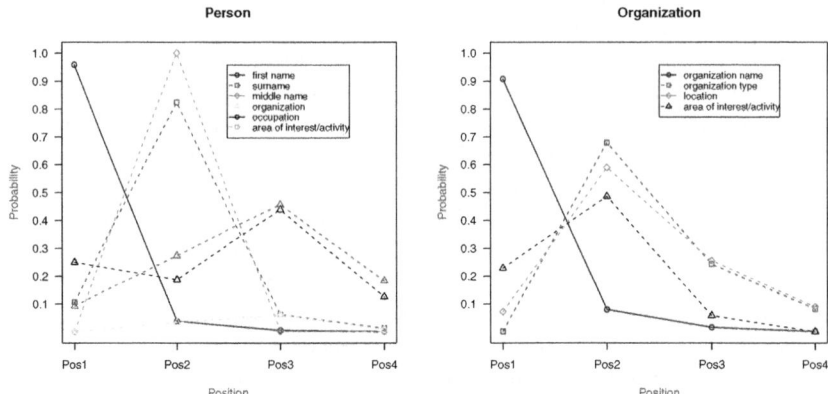

Fig. 1. Attribute Position

in queries about persons, whereas that of the second query is more frequent in queries about organizations.

3.3   An Extended Model of Attribute Relevance

Based on these results, we propose to extend the Bayesian model of attribute relevance presented in section 2.2 to incorporate position dependencies. This model assumes the positional independence for attribute types: the conditional probability of an attribute type given an entity type is independent from the position of the attribute in the query. To incorporate position dependencies, we suggest to weight the relevance measure described in section 2.2 by a position term Pos. For each entity type e and attribute type a, Pos_j is the probability of observing the attribute a in position j. The estimate of Pos_j is given as:

$$Pos_j = \frac{Npos_j}{N}$$

where $Npos_j$ is the number of queries of type e in which the attribute a occurs in position j and N is the number of queries of type e in which the attribute a occurs, regardless of the position in the query[3]. The results of the evaluation of the extended model on the same query sample used to evaluate the Naive Bayes Model are shown in table 5. The results show that the extended model may sensibly improve the disambiguation process compared with the original model.

An example of Entity Type Disambiguation. A sketch of a possible application of the results of our study can be the analysis of the queries Q_1:< George Washington> and Q_2:<Washington USA>. In both we have the same term <Washington> to be disambiguated. But since in Q_1 the term <Washington>

---

[3] When an attribute type is not observed in a given position, the relevance of it is multiplied by a small constant term to avoid the nullification of the relevance value.

Table 5. Test-set Evaluation: extended model

Measures	Person	Organization	Event	Artifact	Location
Precision	0.85	0.89	1	1	0.96
Recall	1	0.91	0.92	0.72	0.95
F-measure	0.92	0.90	0.95	0.84	0.96

Overall Precision		Overall Recall		Overall F-measure	
0.93		0.90		0.91	

is preceded by the term <George> that is likely to be a "first name", we can use the measures of relevance and position to infer that the type of attribute to assign to that term is more likely "surname" than "city name". Indeed, the attributes "name" and "surname" are the most relevant for the entity "Person" and "name" is more likely to be the first term specified in the query, whereas "surname" has the highest probability for the second position.

But how can we disambiguate that "Washington" in Q2 has a different meaning from "Washington" in Q1? From our data we know that when people search for a city, "city name" is one of the most relevant attribute, as well as the most likely attribute in the first position. Moreover, a location is frequently identified by means of another location. From that, we conclude that "Washington" in Q2 is more likely to be "city name" and the entity type of the query is "Location".

4 Conclusion

In this paper we have presented a cognitive study with the aim of investigating how people search for individual entities. This study is motivated out of specific needs that arise from ongoing work on an entity name system for the Semantic Web [3], where identifying a specific entity in a large entity repository based on a user query is one of the key problems, but also semantic search of information about individual entities is addressed. The conclusions we draw from our study are several. Firstly, we were able to confirm earlier findings from a different type of experiment which was performed to find out how people *describe* entities [2]. The combination of the results of both studies provide a community of ontology creators with a good background on how to model a certain set of entity types.

Second, we were able to extract certain patterns in the way people search for entities. In particular we studied the typical position of a certain type of attribute in a sequence of search terms. This result is relevant to tasks that have the objective of mapping keywords from a search query to a formal representation of a query that can be run against a system managing structured data (such as querying and RDF/OWL KB with SPARQL). The results can also play a significant role in the disambiguation of queries, e.g. for solving the problem of "George Washington" vs. "Washington USA" mentioned above. It is important to note that of course these findings are not limited to use cases of the Semantic Web, but we believe that especially the interdisciplinary view of an area between cognitive sciences, information retrieval and semantic systems on the Web can be a helpful contribution to future developments in the Semantic Web.

References

[1] Bazzanella, B., Palpanas, T., Stoermer, H.: Towards a general entity representation model. In: Proceedings of IRI 2009, the 10th IEEE Internationational Conference on Information Reuse and Integration, Las Vegas, USA, August 10-12, IEEE Computer Society, Los Alamitos (2009)

[2] Bazzanella, B., Stoermer, H., Bouquet, P.: Top Level Categories and Attributes for Entity Representation. Technical Report 1, University of Trento, Scienze della Cognizione e della Formazione (September 2008),
http://eprints.biblio.unitn.it/archive/00001467/

[3] Bouquet, P., Stoermer, H., Niederee, C., Mana, A.: Entity Name System: The Backbone of an Open and Scalable Web of Data. In: Proceedings of the IEEE International Conference on Semantic Computing, ICSC 2008, number CSS-ICSC 2008-4-28-25, pp. 554–561. IEEE Computer Society, Los Alamitos (August 2008)

[4] Broder, A.: A taxonomy of web search. SIGIR Forum 36(2), 3–10 (2002)

[5] Guo, J., Xu, G., Cheng, X., Li, H.: Named entity recognition in query. In: Proceedings of the 32nd International ACM SIGIR Conference on Research and Development in Information Retrieval SIGIR 2009 (2009)

[6] Jansen, B.J., Pooch, U.: A review of web searching studies and a framework for future research. Journal of the American Society for Information Science and Technology 52, 235–246 (2001)

[7] Kripke, S.: Naming and Necessity. Basil Blackwell, Oxford (1980)

[8] Rosh, E., Mervis, C.: Family resemblances: studies in the internal structure of categories. Cognitive Psychology 7, 573–605 (1975)

[9] Sartori, G., Lombardi, L.: Semantic relevance and semantic disorders. Journal of Cognitive Neuroscience 16, 439–452 (2004)

[10] Simons, P., Dement, C.W.: Aspects of the Mereology of Artifacts, pp. 255–276. Kluwer, Boston (1996)

[11] Taylor, R.S.: Process of asking questions. American Documentation 13, 391–396 (1962)

[12] Stoermer, H., Bouquet, P.: A Novel Approach for Entity Linkage. In: Proceedings of IRI 2009, the 10th IEEE Internationational Conference on Information Reuse and Integration, Las Vegas, USA, August 10-12. IEEE Computer Society, Los Alamitos (2009)

[13] Taylor, R.S.: Question-negotiation and information-seeking in libraries. College and Research Libraries 29, 178–194 (1968)

[14] Teevan, J., Alvarado, C., Ackerman, M.S., Karger, D.R.: The perfect search engine is not enough: a study of orienteering behavior in directed search. In: CHI 2004: Proceedings of the SIGCHI Conference on Human Factors in Computing Systems, pp. 415–422. ACM, New York (2004)

The Impact of Valence Shifters on Mining Implicit Economic Opinions

Claudiu Musat and Stefan Trausan-Matu

"Politehnica" University of Bucharest, 313 Splaiul Independentei,
Bucharest, 060032, Romania
{claudiu.musat,trausan}@cs.pub.ro

Abstract. We investigated the influence of valence shifters on sentiment analysis within a new model built to extract opinions from economic texts. The system relies on implicit convictions that emerge from the studied texts through co-occurrences of economic indicators and future state modifiers. The polarity of the modifiers can however easily be reversed using negations, diminishers or intensifiers. We compared the system results with and without counting the effect of negations and future state modifier strength and we found that results better than chance are rarely achieved in the second case. In the first case however we proved that the opinion polarity identification accuracy is similar or better than that of other similar tests. Furthermore we found that, when applied to economic indicators, diminishers have the effect of negations.

Keywords: Opinion Mining, Valence Shifters, Economic Predictions, Diminishers, Negations.

1 Introduction

With an ever growing number of economic information sources investors are faced with the impossibility of processing all the available public data before making an investment. Furthermore official predictions about the economic outlook are often countered or reinforced by opinions expressed by pundits on their personal weblogs, making the economic press jungle impregnable for the average investor. Thus a reduction of the amount of information that needs to be processed on average is desirable. One method of achieving that goal is grouping similar economic predictions together. We presented a method for extracting occult opinions regarding economic topics and within that model we highlighted the importance of valence shifters and the importance of giving the right interpretation to diminishers.

The paper continues with an introductory part in which we discussed how negations and, more generally, valence shifters are used in the art to increase the opinion extraction accuracy and then how opinion mining can be performed on economic texts. We then proposed a system to quantify the impact of various methods of employing valence shifters within the extraction of economic predictions, which is described in Section 2 and the results and conclusions follow in Sections 3 and 4 respectively.

D. Dicheva and D. Dochev (Eds.): AIMSA 2010, LNAI 6304, pp. 131–140, 2010.
© Springer-Verlag Berlin Heidelberg 2010

1.1 Opinion Mining in Economics

The idea of finding the overall opinion in a given text is one of the constants of opinion mining [1]. Even though many works have been concentrating on breaking the problem down [2,3], for classification purposes finding the prevalent polarity remains a priority. That is especially true in economics, where investors rely on their economic advisors' conclusions which they usually read in a hastily manner, eager to cut through the details and draw the conclusion. Thus finding economic opinions is an appealing task which has been receiving an increasing amount of attention in the last years. In a study regarding sentiment analysis of financial blogs, Ferguson et.al. [4] showed that using paragraph level annotations and a Bayesian classifier, more than 60% of the analyzed texts contained personal opinions. However their study was restrained to texts that explicitly referred to companies listed in the S&P 500 index. Mainstream media has been addressed by Ahmad *et al.* [5] in his studies on financial news streams. Furthermore opinions expressed in the press have been linked to relations between listed companies and to their stock prices [6].

1.2 Indirect Opinions

Generally the focus in opinion mining has been on analyzing occurrences of direct opinions in the studied texts [7]. Most methods above rely on the presence of explicit opinions in the analyzed texts, arguably a reason why opinion mining has been limited in many cases to extracting opinions about products such as films [2], where client feedback is direct.

However there are domains where this approach is less adequate, where we need to broaden this perspective in order to detect opinions that only derive from context and are not expressed directly. We limited the analysis to economic documents.

The lack of explicit opinions does not always imply the lack of an overall opinion. In their paper about user generated content management Amigo et al. [8] showed that the mere mention of a product in a review can express positive sentiment towards it. Twitter is a good example of highly polarized and informal environment where explicit expressions of sentiment are quite rare, given the message length constraints. Cruz et al. [9] found that when looking for public sentiment towards his company's clients very often the relevant results were factual mentions of their products.

Opinion expressed in the economic environment usually fall into the above category, as we rarely found an author ready to express his doubts about the outcome of a given process. The key to the interweaving of facts and opinions is empiricism. Certainty as the accumulation of positive examples has been named "conventional wisdom" by Galbraith [10].

1.3 Valence Shifters in Opinion Mining

Generally valence shifters are an important part of text mining but they play a crucial role in sentiment analysis. They include all the constructs that can alter the semantic orientation of the term they are referring to. There are many aspects that can modify the polarity of a certain term, depending on its role within a phrase [11]. From this pool of possible variations we focused on two types of valence shifters – negations and diminishers – and we tested their impact on the accuracy of an opinion extraction process designed for economic texts.

Negations are the ones that have until now received the greatest share of attention. By definition they reverse the polarity of their target with examples abounding in everyday language (*not, never, none*). Diminishers usually decrease the intensity of the polarity of the target but there are instances where their role is modified, as will be shown briefly.

Although there still are systems that do not specifically tackle the negations problem and rely on bag of words or statistical approaches, it has been shown [12] that including modifiers in the system significantly improves overall accuracy. The importance of negations in correctly extracting opinions from texts has been understood from the first works in the fields [13] and there have also been studies [12] that quantified the negations' effects on the employed system accuracies. Extraction systems such as OPINE [14] search for negations to improve the search for highly precise opinions, where high precision means the creation of (object, feature, opinion) tuples. Also, Stavrianou and Chauchat use a bag of words approach in [15], but before doing so they replace a word preceded by a negation with their WordNet [16] antonym.

2 Proposed System

The method presented below quantifies the impact of valence shifters in economic opinion extraction. Moreover, it provides a description of the way their interpretation affects the overall accuracy of the opinion identification task.

Its first part is the preprocessing, in which we determined the relevant terms and construct the corpora. Then we defined the necessary constraints that will apply to the key term combinations encountered in a text. An extraction task follows, in which we found valid combinations of different term classes and draw an overall document level polarity conclusion. We defined four different test cases and we repeated the polarity extraction phase for each valence modifier model.

2.1 Preprocessing

We searched for indirect opinions expressed in economic articles that were presented as facts and surfaced as certain combinations of economic indicators such as *"sovereign risk"*, *"unemployment"*, *"economy"* or *"bonds"* and phrases that suggested a future evolution, which we called modifiers.

Economic indicators are divided into two subsets – one that contains indices that rise when the economy is growing and one with indices that have a positive evolution during a period of economic decline. We called the first lot of indicators "positive", and their opposites "negative".

In economics variance is more important than absolute values. Thus, in order to assess the opinion orientation within an economic text, we searched for the suggested future state of the indicators encountered. The terms that indicate the growth or decrease of the indicator they are referring to were generally named "modifiers". These too were split into positive and negative, with the positive modifiers signaling a quantitative increase and the negative ones a decrease of their assigned indicator.

2.1.1 Indicators and Modifiers

The task of defining a relevant lexicon for an opinion mining system has been studied extensively and an excellent summary of the art is presented by Ding [17]. But in our case the above task must be divided into two subtasks – one of finding economic terms and yet another of finding their corresponding modifiers. We chose a supervised learning method that starts with two pools of known terms which are expanded afterwards. The decision derives from the very specific manner in which economic predictions are described within the system [18]. The system must know in advance which indicators have an inherent negative polarity and similarly which modifiers signal an increase or a decrease.

The economic terms and phrases were obtained by expanding the EconomyWatch glossary [19] and the modifiers by expanding an initial handpicked set of words signaling an increase or decrease. The first expansion relies on WordNet synsets and WordNet terms whose glosses contain occurrences of terms in the initial set. The method is conceptually similar to that of Hu [14], in which a word's context independent semantic orientation is found based on its WordNet vicinity.

The second key term extraction phase employs a Contextual Network Graph [20] that retrieves terms that tend to appear together with ones in the initial set in context. We divided the corpus presented below into two equal parts, one of which being used in the training phase. The method creates a term-document bipartite graph that represents the entire collection. The graph is then queried by distributing energy from the nodes that represent terms in the initial glossary and the resulting new relevant terms are those corresponding to nodes that receive energy above a predefined threshold.

We chose the combination above over unsupervised statistical methods such as Latent Dirichlet Allocation (LDA) [21] because of the latter's lack of specificity. For instance in the case of LDA the initial choice of the number of classes (k) affects the probabilities of a given term to appear in a given topic. Furthermore there is no method of specifying relationships that must be considered and relationships that must not – a constrained LDA. We therefore opted for the supervised learning techniques above and for very limited Part of Speech Tagging based combinations of relevant terms.

2.1.2 Valence Shifters

Detecting indicators and modifiers that were used with a reversed meaning was important for the accuracy of the system as the use of negated terms was often more prevalent than their use without a negation. Also we needed to elaborate on how the meaning of economic terms was reversed. We analyzed two different classes of valence shifters that co-occurred with economic indicators and modifiers – negations and diminishers. We chose these over intensifiers because of their ability to reverse the polarity of a construct and thus alter the polarity of the whole document.

Normally the only ones that match that criterion are the negations, with the most common being words like *no, not, never, less, without, barely, hardly* and *rarely* and the most common phrases being : *no longer, no more, no way, nowhere, by no means, at no time, not anymore*. But, for the more specific task of extracting economic predictions, we found that we obtained better results if we also considered that diminishers (such as *less* or *meager*) not only alter the polarity of their target but reverse it completely.

We used the negations and diminishers from the General Inquirer (GI) [22] categories. The negations there are presented separately while diminishers are presented as understatements. Also, in economics, strength is a positive attribute for any economic indicator, thus we included terms that imply weakness among the understatements.

We enhanced the GI sets of valence shifters with their WordNet synset members, as shown above in the indicators and modifiers section.

2.2 Constraints

Not all of the indicator-modifier pairs encountered are expressions of implicit opinions. The phrase "historically rising sovereign risk usually proves harmful for the economic recovery" does not signal a prediction about the evolution of the indicator. While bearing in mind that oversimplification could significantly reduce the opinion extraction system's recall, we used two heuristics to avoid uncertainties similar to the mentioned one.

The first imposed limitation was that the indicator should be either a noun (or, if the indicator is not a unigram, an n-gram containing a noun) or an adjective attached to a neutral noun. Neutral associate nouns are very frequent in the encountered economic texts, forming constructs such as "economic data" or "unemployment figures" or "mortality numbers".

The modifier is limited to the associated verb or, when the verb suggests a continuous state of fact, an adverb following that verb, as in "the recovery remains sluggish". For simplicity we referred to the entire family of verbs related to the one in previous example as *continuers*. In order to apply the heuristic adjustments above, a part-of-speech tagger was used.

2.3 Modifier and Valence Shifter Scope

We defined how indicators were associated with modifiers and how both indicators and modifiers were associated with their neighboring negations or diminishers by limiting each term's scope and finding their intersections. The words or n-grams that represent the considered key terms do not have to be right beside each other in the sentence. To determine which neighboring pairs are valid we applied a set of rules to filter out unwanted cases.

We did not to allow mixtures between sentences and also did not include pairs where a delimiter or conditional word delimiter [12] was found between them. Delimiters are words that can trim down the analyzed term's possible scope, by removing words from it. Examples of delimiters are "unless", "until", "hence" while terms such as "which", "who", "for", "like" are conditional delimiters. Any pairs of key terms that contain a delimiter between them are excluded.

We parsed the texts using a part of speech tagger and set the initial term scope to include all words between its closest nouns to the left and right side. Then, following the restrictions above, we trimmed their scopes down and if at the end the scopes of any two key terms intersected we considered them associated.

2.4 Corpus and Labeling

In this experiment we used the publicly available articles from The Telegraph Finance section and its subsections. A total of 21106 articles published between 2007 and 2010 were processed and within the corpus we have found 499 articles that were labeled as having a negative bias and 216 as having a positive one. Since human interpretation of the article polarity is itself subjective, for the annotation task we relied on examiner agreement [23] for a better selection. Two examiners singled out biased texts from the Finance (root), Markets, Jobs and Comments sections and only those articles on which examiners agreed were considered for future processing. All words in the processed documents were stemmed according to the Porter [24] algorithm and stop words were eliminated.

2.5 Opinion Extraction

After defining the key terms and corpora, the following task was to find the polarity of the present economic forecast. We defined it conventionally as the sum of the polarities of present co-occurrences of economic indicators and future state modifiers, reversed if necessary by the presence of attached valence shifters. A good example is that of recent discussions about the rise of sovereign default risk. The term risk is itself a negative indicator. When the modifier attached to it is a positive one, such as "will grow" the overall prediction is negative. Likewise, if the modifier is negative, like "fell" the result is positive. A simple formula summarizes the above:

$$P(O) = P(I) * P(M) * P(VS) \tag{1}$$

where

— P is the polarity of the construct
— O the resulting opinion
— I is the indicator involved
— M the attached modifier
— And VS are the valence shifters referring to the modifiers and indicators.

A document in which the majority of (indicator – modifier – valence shifter) tuples indicated the growth of a positive indicator or the decrease of a negative one was labeled as having a positive polarity, whereas a document in which the majority of pairs indicated the growth of a negative indicator or the decrease of a positive one had a negative label.

3 Comparative Results

We compared the system results for different values of the valence shifter polarity factor. We first considered the case where the valence shifter polarity in the formula above was always one, corresponding to completely ignoring valence shifters:

$$P(VS) = 1 \tag{2}$$

Then we only factored in the effect of negations (N), while disregarding diminishers (D):

$$P(VS) = \begin{cases} -1, VS \in N \\ 1, VS \in D \end{cases} \quad (3)$$

Thirdly we considered the effect of diminishers as reducing the construct polarity to half of its previous value:

$$P(VS) = \begin{cases} -1, VS \in N \\ 1/2, VS \in D \end{cases} \quad (4)$$

And finally the case where diminishers were thought of as regular negations, thus the polarity of all valence shifters is the same:

$$P(VS) = -1 \quad (5)$$

The results of the opinion extraction test phase are shown in Table 1 while the precision and recall of the experiments are presented in Table 2. As previously said, the tests were conducted on the second half of the selected biased articles in the corpus. Note that there were articles in the root section of both the positive and negative batches, not only in their subsections.

In Table 1 the resulting overall article polarities are numbered from 1 to 4, each representing the outcome of the four situations presented above. P_i is the number of articles marked as positive in the ith experiment where $i \in \{1..4\}$, while N_i and U_i are the number of articles marked as negative or unknown. An article was marked as unknown if the sum of the polarity bearing constructs within is negligible(i.e. between -1 and 1).

In Table 2, Pr_i denotes the precision and R_i the recall in all experiments. Precision is defined as the number of correctly retrieved documents divided by the total number of documents retrieved, while recall is the ratio of relevant retrieved documents to the total number of relevant documents. The first row – "Total" – represents the values obtained jointly for the negative and positive batches.

Table 1. Opinion retrieval data for the four considered situations

Corpus	Total	P_1	P_2	P_3	P_4	N_1	N_2	N_3	N_4	U_1	U_2	U_3	U_4
Negative													
All	499	144	86	81	72	207	252	262	304	148	161	156	123
Jobs	54	8	1	1	1	32	35	38	42	14	18	15	11
Markets	253	85	44	38	35	107	128	141	161	61	81	74	57
Comments	153	41	32	36	34	62	73	78	94	50	48	39	25
Positive													
All	216	76	87	118	117	81	50	48	48	59	79	50	51
Jobs	14	5	4	5	6	7	3	3	3	2	7	6	5
Markets	152	61	71	98	100	53	31	29	29	38	50	25	23
Comments	34	7	11	11	9	16	9	9	9	11	14	14	16

Table 2. System precision and recall

Corpus	Pr_1	Pr_2	Pr_3	Pr_4	R_1	R_2	R_3	R_4
TOTAL	55.71	71.37	74.66	77.82	71.05	66.43	71.19	75.66
Negative								
All	58.97	74.56	76.38	80.85	70.34	67.74	68.74	75.35
Jobs	80.00	97.22	97.44	97.67	74.07	66.67	72.22	79.63
Markets	55.73	74.42	78.77	82.14	75.89	67.98	70.75	77.47
Comments	60.19	69.52	68.42	73.44	67.32	68.63	74.51	83.66
Positive								
All	48.41	63.50	71.08	70.91	72.69	63.43	76.85	76.39
Jobs	41.67	57.14	62.50	66.67	85.71	50.00	57.14	64.29
Markets	53.51	69.61	77.17	77.52	75.00	67.11	83.55	84.87
Comments	30.43	55.00	55.00	50.00	67.65	58.82	58.82	52.94

The evolution of the system precision is easily observable. The most obvious conclusion is that not counting the impact of valence shifters (the first case) lead to a lower accuracy in a consistent manner. An interesting aspect of the results above is the evolution of the experiment recall when sequentially adding and modifying the valence shifter component. Firstly, by only adding negations we obtained a lower recall in conjunction with a significantly higher precision. This suggests that a high number of articles shifted from the wrongly classified set to the "Unknown" set.

We also observed that the improvements are not consistent over all the subsections of the analyzed corpora, as there were subsections where the precision dropped from one experiment to the next, although the overall precision improved. Of course, the impact of using diminishers on the experiment accuracy was limited by the percentage of articles where diminishers were encountered. Thus, as there were numerous articles where not a single diminisher was found, their impact on those article categories was reduced.

Also, the system recall was linked to the size of the training corpus, the worst results being obtained in most cases for the Jobs and Comment sections of the positive batch, where the number of articles was the lowest. Also we note that the comment sections for both the positive and negative articles presented a lower accuracy compared to the Markets and Jobs sections because of the much larger number of subjects discussed, from personal finance to macroeconomics.

4   Conclusions

One of the most important conclusions is that economic predictions are being expressed in a very succinct and identifiable manner. This comes from the fact that by using a rather small number of key phrases, instead of all the words present in the text, we obtain a detection rate similar to what other systems [4] obtain by using the contribution of the whole set of words used in the texts. However these predictions have proved to be difficult to extract without considering the contributions of various types of valence shifters.

If the first tests showed that the method's precision and accuracy were hardly better than chance, by successively adding negations and diminishers to the model we succeeded in increasing the precision by little under 20%. Also, by treating diminishers as outright negations helped to increase the system precision by an additional 3%. We believe that, although further tests are needed, this experiment proves that there are aspects about opinions expressed in the economic environment that are different from those expressed in most other fields and that the proper use of valence shifters is one of those differences.

Acknowledgments. This work is supported by the European Union Grant POSDRU/6/1.5/S/19 7713.

References

1. Pang, B., Lee, L.: Opinion Mining and Sentiment Analysis. Foundation and Trends in Information Retrieval 2, 1–135 (2008)
2. Wiebe, J., Wilson, T., Cardie, C.: Annotating Expressions of Opinions and Emotions in Language. Language Resources and Evaluation 39(2-3), 165–210 (2005)
3. Mei, Q., Ling, X., Wondra, M., Su, H., Zhai, C.: Topic Sentiment Mixture: Modeling Facets and Opinions in Weblogs. In: Proceedings of the 16th International Conference on World Wide Web, pp. 171–180 (2007)
4. Ferguson, P., O'Hare, M., Bermingham, A.: Exploring the use of Paragraph-level Annotations for Sentiment Analysis of Financial Blogs. In: First Workshop on Opinion Mining and Sentiment Analysis, pp. 42–52 (2009)
5. Ahmad, K., Cheng, D., Almas, Y.: Multi-lingual sentiment analysis of financial news streams. In: First International Workshop on Grid Technology for Financial Modeling and Simulation, pp. 984–991 (2006)
6. Ku, L.W., Lee, Y., Wu, T.H., Chen, H.H.: Novel Relationship Discovery Using Opinions Mined from the Web. In: Proceedings of AAAI 2006, pp. 213–221 (2006)
7. Liu, B.: Opinion Mining. In: Proceedings of WWW 2008, Beijing (2008)
8. Amigo, E., Spina, D., Bernardino, B.: User Generated Content Monitoring System Evaluation. In: First Workshop on Opinion Mining and Sentiment Analysis, pp. 1–13 (2009)
9. Cruz, F., Troyano, A., Ortega, F., Enriquez, F.: Domain Oriented Opinion Extraction Metodology. In: First Workshop on Opinion Mining and Sentiment Analysis, pp. 52–62 (2009)
10. Galbraith, J.: The Affluent Society, ch. 2. Asia Publishing House (1958)
11. Kennedy, A., Inkpen, D.: Sentiment Classification of Movie Reviews Using Contextual Valence Shifters. Journal of Computational Intelligence 22, 110–125 (2006)
12. Jia, L., Yu, C., Meng, W.: The effect of negation on sentiment analysis and retrieval effectiveness. In: Proceeding of the 18th ACM Conference on Information and Knowledge Management, pp. 1827–1830 (2009)
13. Pang, B., Lee, L.: Thumbs up?: sentiment classification using machine learning techniques. In: Proceedings of the ACL 2002 Conference on Empirical Methods in Natural Language Processing, vol. 10, pp. 79–86 (2002)
14. Popescu, A.M., Etzioni, O.: Extracting product features and opinions from reviews. In: Proceedings of the Conference on Human Language Technology and Empirical Methods in Natural Language Processing, pp. 339–346 (2005)
15. Stavrianou, A., Chauchat, J.H.: Opinion Mining Issues and Agreement Identification in Forum Texts. In: FODOP 2008, pp. 51–58 (2008)
16. Felbaum, C.: WordNet: An Electronic Lexical Database. MIT Press, Cambridge (1998)

17. Ding, X., Liu, B., Yu, P.: A holistic lexicon-based approach to opinion mining. In: Proceedings of the International Conference on Web Search and Web Data Mining, pp. 231–240 (2008)
18. Musat, C., Trausan-Matu, S.: Opinion Mining on Economic Texts. In: First Workshop on Opinion Mining and Sentiment Analysis, pp. 62–72 (2009)
19. Economy Watch, http://www.economywatch.com/
20. Ceglowski, M., Coburn, A., Cuadrado, J.: Semantic Search of Unstructured Data using Contextual Network Graphs (2003)
21. Blei, D., Ng, A., Jordan, M.: Latent Dirichlet Allocation. The Journal of Machine Learning Research 3, 993–1022 (2003)
22. Stone, P.J., Dunphy, D.C., Smith, M.S., Ogilvie, D.M.: The General Inquirer: A computer Approach to Content analysis. MIT Press, Cambridge (1966)
23. Balahur, A., Steinberger, R.: Rethinking Sentiment Analysis in the News. In: First workshop on Opinion Mining and Sentiment Analysis, pp. 79–89 (2009)
24. Porter, M.: An Algorithm for Suffix Stripping. In: New Models in Probabilistic Information Retrieval. British Library, London (1980)

Entity Popularity on the Web: Correlating ANSA News and AOL Search

Angela Fogarolli, George Giannakopoulos, and Heiko Stoermer

University of Trento,
Dept. of Information Science and Engineering,
Via Sommarive 14, 38123 Trento, Italy
{afogarol,ggianna,stoermer}@disi.unitn.it

Abstract. Annotated Web content, digital libraries, news and media portals, e-commerce web sites, online catalogs, RDF/OWL knowledge bases and online encyclopedias can be considered containers of *named entities* such as organizations, persons, locations. Entities are mostly implicitly mentioned in texts or multi-media content, but increasingly explicit in structured annotations such as the ones provided by the Semantic Web. Today, as a result of different research projects and commercial initiatives, systems deal with massive amounts of data that are either explicitly or implicitly related to entities, which have to managed in an efficient way. This paper contributes to Web Science by attempting to measure and interpret trends of entity popularity on the WWW, taking into consideration the occurrence of named entities in a large news corpus, and correlating these findings with analysis results on how entities are searched for, based on a large search engine query log. The study shows that entity popularity follows well-known trends, which can be of interest for several aspects in the development of services and applications on the WWW that deal with larger amounts of data about (named) entities.

1 Introduction

Entities – as opposed to documents – are gaining importance in the information space of the Web, due to the many activities that attempt to elevate the Web from a document-centric information space that is very hard to process for machine algorithms, towards something more fact-centric, powered by machine readable annotations. At the core lie (real-world) entities such as persons, locations, organizations, which often represent the first-class objects that facts are *about*.

It is entities mentioned on the web that this paper is concentrating on. Information systems that do not operate on closed, controlled environments, but which deal with information about entities on the web, often need to implement methods that rely on a measure of popularity to take operational decisions. These can relate to cache management (data about which entity do I maintain in the cache), data lifecycle management (are my data still up-to-date), improvement

D. Dicheva and D. Dochev (Eds.): AIMSA 2010, LNAI 6304, pp. 141–150, 2010.

of search results (in doubt, return the more popular entity as a higher-ranked result), and other use cases that we cannot yet foresee.

Measuring popularity of entities mentioned on the web however is a challenge. For an enormous information space like the Web, standard approaches like simple statistical measures seem hardly practical. Running the required Named Entity Extraction (NER) and entity consolidation methods on the web itself can – if at all – only be performed by a chosen few, which leads to the situation that to the best of our knowledge, no study about the behavior of entity popularity on the Web is available today.

To tackle the challenges posed by the sheer size of the Web, this paper attempts to derive facts from the analysis of a large, yet manageably sized corpus of news items, develop a hypothesis, and confirm this hypothesis by correlating it to the analysis of a large search engine query log. Following this approach, we hope to provide results that allow for a certain level of generalization, and thus create insights about the behavior of entity popularity on the Web itself.

The rest of this article is arranged as follows. First, in Sect. 2 we provide a working definition of what we consider to be an "entity". Then we give a short overview over related work in section 3. Section 4 presents the analysis performed using the Italian news agency (ANSA) Annotated Corpus, and develops and confirms a hypothesis with the help of the AOL search engine query log [16]. Sect. 5 discusses the work performed, and concludes the article, offering use-cases in which the results presented can create a benefit, and illustrating future work.

2   Definition of Entity

While the Semantic Web works with the notion of a "resource", which is basically anything that is identified by a URI, concepts, objects and properties alike, our interest is more similar to what traditional Description Logics knowledge bases call "individuals". A straight-forward and trivial definition of the notion of "entity" in our sense might be the idea of a "real-world object", but already the example of a geographical location such as a city leads to first complications (is a city an object?). And indeed, a definition that describes our intent and at the same time stands up acceptably well to philosophical requirements is surprisingly difficult to find. Several important aspects have to be considered, such as physical vs. abstract, spatial vs. temporal, individual vs. collection, behavior in time, agentivity, and more. Bazzanella has come up with working solution that covers our area of interest, being the *set of all particular and concrete entities (events included)* [1]. This means that classes, properties and abstract concepts do not count as entity in our sense, but people, organizations, and geographical locations do.

3   Related Work

Extracting entities from the news is a current trend as stated by the the International Press Telecommunication Council(IPTC) [8]:

Increasingly, news organizations are using entity extraction engines to find "things" mentioned in news objects. The results of these automated processes may be checked and refined by journalists. The goal is to classify news as richly as possible and to identify people, organizations, places and other entities before sending it to customers, in order to increase its value and usefulness.

The question which *trends* entity popularity follows – especially on the Web – has not been widely analyzed so far. There is related work from the areas of focused retrieval [5,14,15,9] or question answering [4,17], which mainly describes techniques to measure entity popularity on a given corpus to achieve certain goals. First studies on social networks on the web [12,11] deal with inferring a ranking of entities based on their status in social networks. None of these works however intend to generalize and create insights about the Web itself.

4   A Study on Entity Popularity

4.1   Entities in ANSA News: The Italian News Agency Corpus

When selecting a set of documents that would allow to make observations suitable for a generalization to the Web as a whole, several aspects need to be covered. First, the dynamic character of the Web should be reflected, meaning that for example a collection of literature texts would not be an optimal choice. Secondly, the corpus should cover a rather broad area of domains and not focus on a single topic. Third, it should reflect the fact that the Web often covers many topics of broad interest, especially related to *entities* according to our definition.

The ANSA[1] Annotated Corpus is very useful in this respect, as it fulfills these criteria: it contains almost 1.2 Million news articles from a period of 3 years, about everything that is new and noteworthy, including persons, organizations, and locations.

The annotation of the ANSA corpus is made by analyzing the news items using Name Entity Recognition (NER). NER is composed by an Entity Detection phase which parses the text and determine the entities, and an Entity Normalization phase which associates a detected entity with its identifier. The NER aims at the detection of people, organizations, locations and events inside the news articles. While Entity Detection is a stand-alone functionality completely developed by Expert System[2], for Entity Normalization (to provide entities with their ids) the Entity Name System (ENS) [3] is queried to retrieve unique identifiers for each detected entity.

---

[1] "ANSA (Agenzia Nazionale Stampa Associata), is the leading wire service in Italy, and one of the leaders among world news agencies. ANSA is a not-for-profit cooperative, whose mission is the distribution of fair and objective news reporting. ANSA covers national and international events through its 22 offices in Italy, and its presence in more than 80 cities in 74 countries in the world." (cf. http://en.wikipedia.org/wiki/Agenzia_Nazionale_Stampa_Associata)

[2] http://www.expertsystem.net/

Table 1. Number of distinct and total entities in the ANSA corpus, by entity type

Entity type	Abbr.	Distinct E.	Total E.
Location	LOC	33409	12474210
Organization	ORG	77888	847889
Person	PER	138143	4616958
Total		249440	17939057

The size of the corpus and metadata ensure that the trends and numbers we describe are statistically reliable. Table 1 gives more details about the number of entities contained. The entities in the corpus are classified into three categories: person(PER), organization(ORG) and location(LOC).

Formally, for our later analysis, we model an entity e as follows: $e = \langle l, t \rangle$ with string label l and type t, where $t = \{LOC, PER, ORG, AUT\}$

An article A is modeled as: $A = \langle E, d, id, C \rangle$ with $E = \{e\}$, the set of entities mentioned in A, the publication date d, the unique article ID (URL) id, and the set of classifiers C from several formal taxonomies.

Entity Lifespan. An important aspect we are interested in to gain a better understanding of entity behavior is the "lifespan". We define lifespan as the number of months between the first and last occurrence of an entity. Table 2 shows the lifespan quantiles for each entity type.

Table 2. Quantiles of entity lifespan in the ANSA corpus in months

Entity type	Min	1st Q.	Median	Mean	3rd Q.	Max
LOC	2.00	17.00	29.00	25.12	35.00	36.00
ORG	2.00	6.00	14.00	16.06	25.00	36.00
PER	2.00	8.00	18.00	18.17	28.00	36.00
Overall	2.00	9.00	20.00	19.46	30.00	36.00

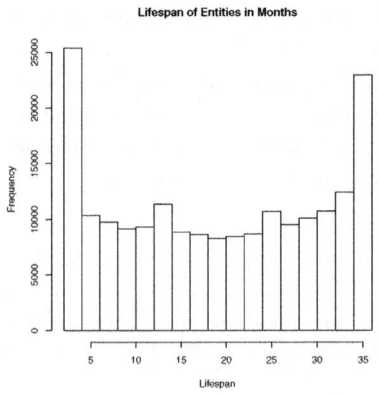

Fig. 1. The NE lifespan histogram

We call "NE lifespan" the average lifespan of the entities for the general named entity types: person, organization and location. The NE lifespan average for the entities in the corpus is 17.89 months. In Figure 1 we illustrate the histogram of the NE lifespan.

Entity Re-use. Looking closer at the data, it is possible to understand trends and behavioral patterns of entity popularity. This analysis is focused on learning how many times individual entities are mentioned in the corpus, i.e. how many occurrences inside the corpus exist for the same entity. From a different perspective, multiple occurrences of the same entity can be considered "re-use" of the entity. We analyze how many of times an entity is mentioned per month. In table 3 we summarize the average values of re-use for each entity type. The re-use is calculated dividing the total occurrences by the distinct ones for each entity type. The overall re-use is the average re-use rate on the three years of news covered by the corpus.

Table 3. Simple overall average re-use rate per entity type

Entity type	Re-use rate
LOC	373.38
ORG	53.66
PER	33.42

4.2   Hypothesis and Evaluation

In this section we evaluate our hypothesis about entity popularity and its trends. The section is divided in two parts, the first one correlates annotations about entities in the news corpus with searched entities in a search engine query log, while the second describes the analysis made to confirm that entity popularity follows a power-law distribution.

Entity Popularity. It can be argued that the popularity of an entity on the Web can not be measured by means of the number of times it is mentioned in a news article, but the popularity is more related to how many times Internet users search for that particular entity. In order to show that the entities mentioned in the news can however indeed give an idea of popularity of entities in the WWW, in the following experiment we correlate a news corpus with a search engine log, with the aim of confirming findings from the one with analysis of the other.

Due to the fact that search log files are not widely available, we restricted this evaluation to a the query log of the AOL (American On Line) search engine, which covers a period of three months. We compare the entities in the AOL logs with the ANSA entities mentioned in the same period. For the sake of the evaluation, to prove the generality of the "popularity" aspect, we contextualize this analysis on the entity type *person*.

Table 4. Correlation test results. Statistically significance indicated in **bold** letters

Test	Correlation Value	P-value
Pearson	**0.42**	0.0
Spearman	**0.53**	0.0
Kendall	**0.41**	0.0

The AOL logs cover the period from March 1st, 2006 to May 31st, 2006. In order to correlate the entities in the logs with the ones in the news article metadata, we processed the AOL log using a Lucene[3] index. Next, we used Lucene index functionalities such as specific filters[4] to count the occurrences of the ANSA entities in the search logs.

The hypothesis tested is that *the popularity of an entity in the news behaves similarly to the popularity of an entity in search queries*. We expect to find that people increase the number of searches for a particular entity when the entity appears more often in the news. The hypothesis can be written like this:

> The number of occurrences of an entity in the news is correlated to occurrences of an entity in search queries over the same period.

We ran three different correlation experiments on our datasets using: Pearson, Spearman and Kendall correlation. In all three correlation tests, the p-value is used to indicate the statistical significance of the observation. The lower the p-value, the more "significant" the result, in the sense of statistical significance. As in most statistical analyses, a p-value of 0.05 or less indicates high statistical significance. The person entities appearing in both the ANSA (news) and AOL (query) datasets are 2745.

The Pearson's correlation indicates whether there exists a linear correlation between two random variables and measures the strength of this correlation. A Pearson correlation value close to 0 indicates that there is not linear correlation between the random variables. A value near -1 or 1, indicates full negative or positive linear correlation, i.e., the value of one variable is a linear function of the other. In correlating the AOL dataset with the ANSA subset we learned that *there is a very powerful and statistically significant linear correlation between the variables*.

Spearman's and Kendall's rank correlation indicate whether two random variables tend to be modified in a correlated manner, without taking into account whether the correlation is linear. In other words, if the value of one random variable is increased when the other is increased, the rank correlations will show

---

[3] Lucene: http://lucene.apache.org/

[4] For entity identification we apply different Lucene techniques which include the use of a PhraseQuery for matching the query string in the logs containing a particular sequence of terms. PhraseQuery uses positional information of the term that is stored in an index. The distance between the terms that are considered to be matched is called slop. With used a slop set to three to ensure that all name and last name of the entity in the ANSA is present in the log.

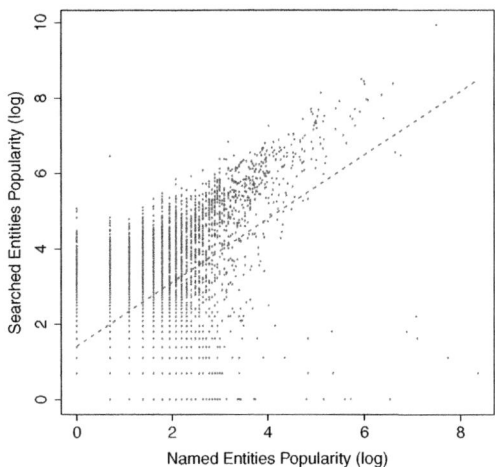

Fig. 2. Correlation between AOL searched entities and ANSA named entities (log-scale)

a value away from 0. The return value of the function describes the strength and direction (negative or positive) of the relationship between two variables. The results are between 1 and −1. The result of the rank coefficients indicate with a very high probability (since the p-value is almost 0.0) that there *is* a statistically significant correlation between entities in the AOL search logs and in the ANSA corpus. The intensity of the correlation is average.

The values of correlation mean that, if an entity is popular (i.e., appears often) in the news (ANSA corpus), it is expected to also be popular in the searches the users make and vice versa. In other words, each measure is a good predictor of the other over that period of three months.

Based on these tests, whose results are summarized in table 4, we have validated our hypothesis about entity popularity. In figure 2 we show the correlation graph between entities in the AOL logs (labeled "searched entities") and the "named entities" inside the ANSA corpus. Often, when an entity is popular in the ANSA, it is also popular in the AOL news.

Entity Popularity Distribution. The analysis described in this section aims to find a behavioral trend in the popularity of entities. The first part of the analysis plots – for each entity type (Person, Organization, Location) – the relationship between the type and its occurrences. What is evident from Fig. 3 is that entities generally follow a common trend, with few outliers. We note that, in order for the diagram to be readable we display the logarithm of the popularity. The graph suggests the hypothesis that the ANSA person entity occurrences follow a power-law distribution. The power-law distribution implies that entities with few occurrences are very common, while entities with many occurrences are rare. For this reason, we tested if the entity occurrences were distributed

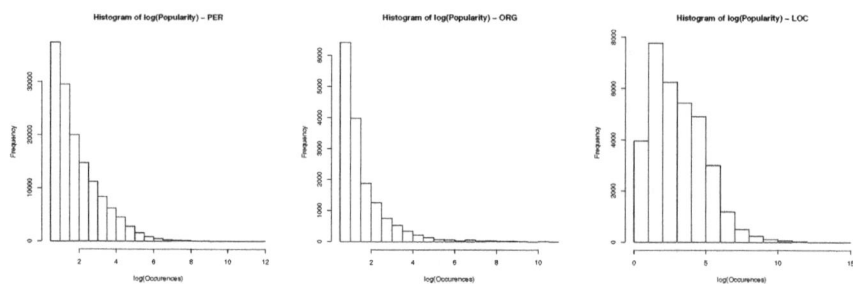

Fig. 3. Entity occurence distribution by category (logarithm): PER (left), ORG (middle), LOC (right)

according to a power-law distribution. In the past, scientists have discovered that many phenomena in the real world such as earthquakes or city growth are following a power-law pattern [19]. Detecting a true power-law distribution is however complicated [13] and requires proper tests. In the WWW, power law distributions have been used for predicting behavior on web site visits and links to pages, or for studying network links in social [18].

For calculating the power-law fitting, we follow the suggestions of Clauset et al. [6] – which are implemented in the Plfit R library[5] – and estimate the goodness of the fitting by using the D statistic of the Kolmogorov-Smirnoff test [7] (K-S test), provided by tge Plfit function. The test follows these steps:

1. First, we use the plfit library to estimate the possible parameters xmin and α where x is a vector of observations of some quantity to which we wish to fit the power-law distribution $p(x) \sim x^{-\alpha}$ for $x >= xmin$. This is done through a Maximum Likelihood Estimation process. The library returns an xmin that shows the most plausible x value, above which the power law stands for the data.

2. We use the D statistic of the K-S test to determine the p-value of the fit. If the p-value is over 0.05, we *cannot reject* the fact that the samples come from a power-law distribution. In this case, we will consider the power law a plausible model for tha data. Due to the fact that we do not look for the optimal model, we do not compare with other models (e.g., log-normal), but simply check the power model plausibility.

In the following paragraph we report the power-law estimation process on a sample of the ANSA dataset. The sample used for the analysis is the three month period of popularity: the same period where the experiments correlating searches to entity appearances, namely March 1st, 2006 to May 31st, 2006.

In a power-law we have $p(x) = x^{-\alpha}$. We are now able to extract the correct exponent $\alpha = 2.14$ and $xmin = 8$.

---

[5] R Plfit function created by Laurent Dubroca.

It appeared that, the power law distribution *is a good fit for the ANSA data*. In other words, focusing on the extracted sample data we found that a distribution $p(x) \sim x^{-2.14}, x \geq 8$ describes plausibly the empirical distribution of popularity for ANSA entities.

5   Conclusion

In this paper we have presented an approach for gaining insights into the behavior of entity popularity on the Web. We have analyzed a large corpus of news articles and their metadata, and made observations about lifespan, re-use, and popularity behavior. The latter aspect was investigated more deeply, as a first analysis had shown that popularity seems to follow a well-known distribution. To test this hypothesis, and to ensure that a generalization of the findings to the Web is viable, we have performed an additional experiment in which we correlate the entities in the news corpus with a large search engine query log. The results of the analyses are twofold. From one hand we defined entity popularity as the positive relationship between how many times an entity is mentioned and how many times is searched. On the other hand we supported this hypothesis by fitting this trend into a power-law distribution.

The results presented in this paper – in addition to their obvious contribution to "Web Science" as promoted by Tim Berners Lee et al. [2,10] – can play a role in the design of systems that deal with information about entities on the Web in various disciplines.

Acknowledgments

This work is partially supported by the by the FP7 EU Large-scale Integrating Project **OKKAM – Enabling a Web of Entities** (contract no. ICT-215032). For more details, visit http://www.okkam.org

References

1. Bazzanella, B., Stoermer, H., Bouquet, P.: Top Level Categories and Attributes for Entity Representation. Technical Report 1, University of Trento, Scienze della Cognizione e della Formazione (September 2008),
   http://eprints.biblio.unitn.it/archive/00001467/
2. Berners-Lee, T., Hall, W., Hendler, J., Shadbolt, N., Weitzner, D.J.: Creating a Science of the Web. Science 313(5788), 769–771 (2006)
3. Bouquet, P., Stoermer, H., Niederee, C., Mana, A.: Entity Name System: The Backbone of an Open and Scalable Web of Data. In: Proceedings of the IEEE International Conference on Semantic Computing, ICSC 2008, number CSS-ICSC 2008-4-28-25 in CSS-ICSC, pp. 554–561. IEEE Computer Society, Los Alamitos (August 2008)
4. Cheng, G., Ge, W., Qu, Y.: Falcons: searching and browsing entities on the semantic web. In: WWW 2008: Proceeding of the 17th International Conference on World Wide Web, pp. 1101–1102. ACM, New York (2008)

5. Cheng, T., Yan, X., Chang, K.C.-C.: Entityrank: searching entities directly and holistically. In: VLDB 2007: Proceedings of the 33rd International Conference on Very Large Data Bases. VLDB Endowment, pp. 387–398 (2007)
6. Clauset, A., Shalizi, C.R., Newman, M.E.J.: Power-law distributions in empirical data (2007)
7. Conover, W.J.: A Kolmogorov Goodness-of-Fit Test for Discontinuous Distributions. Journal of the American Statistical Association 67(339), 591–596 (1972)
8. International Press Telecommunications Council. Guide for implementers. Document revision 1, International Press Telecommunications Council (2009)
9. Demartini, G., Firan, C.S., Iofciu, T., Krestel, R., Nejdl, W.: A Model for Ranking Entities and Its Application to Wikipedia. In: Latin American Web Conference, LA-WEB 2008, pp. 28–38 (2008)
10. Hendler, J., Shadbolt, N., Hall, W., Berners-Lee, T., Weitzner, D.: Web science: an interdisciplinary approach to understanding the web. ACM Commun. 51(7), 60–69 (2008)
11. Jin, Y., Matsuo, Y., Ishizuka, M.: Ranking companies on the web using social network mining. In: Ting, H., Wu, H.-J. (eds.) Web Mining Applications in E-commerce and E-services, ch. 8, pp. 137–152. Springer, Heidelberg (2008)
12. Jin, Y., Matsuo, Y., Ishizuka, M.: Ranking entities on the web using social network mining and ranking learning. In: WWW 2008 Workshop on Social Web Search and Mining (2008)
13. Newman, M.E.J.: Power laws, pareto distributions and zipf's law. Contemporary Physics 46(5), 323–351 (2005)
14. Nie, Z., Ma, Y., Shi, S., Wen, J.-R., Ma, W.-Y.: Web object retrieval. In: WWW 2007: Proceedings of the 16th International Conference on World Wide Web, pp. 81–90. ACM, New York (2007)
15. Nie, Z., Wen, J.-R., Ma, W.-Y.: Object-level vertical search. In: CIDR, pp. 235–246 (2007), www.crdrdb.org
16. Pass, G., Chowdhury, A., Torgeson, C.: A picture of search. In: InfoScale 2006: Proceedings of the 1st International Conference on Scalable Information Systems. ACM Press, New York (2006)
17. Popov, B., Kitchukov, I., Angelova, K., Kiryakov, A.: Co-occurrence and Ranking of Entities. Ontotext Technology White Paper (May 2006)
18. Schnegg, M.: Reciprocity and the emergence of power laws in social networks. International Journal of Modern Physics 17(8) (August 2006)
19. Shiode, N., Batty, M.: Power law distributions in real and virtual worlds (2000)

Using Machine Learning to Prescribe Warfarin

Brent Martin, Marina Filipovic, Lara Rennie, and David Shaw

Computational Intelligence Research Lab
University of Canterbury, Christchurch New Zealand
brent.martin@canterbury.ac.nz,
marina@projectexperts.co.nz,
lara.rennie@gmail.com,
dpshaw@ihug.co.nz

Abstract. Predicting the effects of the blood thinner warfarin is very difficult because of its long half-life, interaction with drugs and food, and because every patient has a unique response to a given dose. Previous attempts to use machine learning have shown that no individual learner can accurately predict the drug's effect for all patients. In this paper we present our exploration of this problem using ensemble methods. The resulting system utilizes multiple ML algorithms and input parameters to make multiple predictions, which are then scrutinized by the doctor. Our results indicate that this approach may be a workable solution to the problem of automated warfarin prescription.

1   Introduction

Predicting the effect of the anticoagulant drug warfarin is a difficult and risky task. In this problem the history of dosages for cardiac and vascular patients is known, along with the corresponding International Normalised Ratio (INR), which measures the time the blood takes to clot and can be compared both between patients and across different countries. Noise is inherent in the data-set as it is impossible to control a patient's life-style, so confounding factors, including non-compliance (i.e. taking the wrong amount), are extremely likely. Other errors or missing data arise from the fact that the data have been obtained from hand-written doctor records of a patient. Errors when performing data input are also possible. It is therefore ideally suited as a data set to examine time-series data with machine learning. It is hoped some machine learning algorithm will be able to predict either the effect of the next dosage on the INR, or the optimal next dosage for a particular INR reading, with some degree of accuracy.

In this paper we report on studies into both of these aspects. Section 2 introduces the problem of warfarin prediction in more detail. In Section 3 we report on investigations into the ability of machine learning to predict the outcome of a warfarin dose. Based on these results we developed a prototype decision support system that presents predictions to a Doctor for a range of doses; this work is described in Section 4, which also reports a preliminary evaluation. We conclude and indicate future directions in Section 5.

D. Dicheva and D. Dochev (Eds.): AIMSA 2010, LNAI 6304, pp. 151–160, 2010.

2 Warfarin Prediction

Determining the appropriate dosage of warfarin is extremely difficult for many reasons. One of the most important issues is the way in which the drug effect is measured. Warfarin has a half-life of 36 hours [1], so it takes approximately 4-6 days to achieve new steady-state plasma concentrations after dose adjustments. As a result the maximum response to a dose is not visible for at least one or two days after ingestion, making it difficult to accurately adjust the last dosage given after a worrying INR reading. There is a large variance in individual responses to the drug due to its complex pharmacology [3], which is affected by many factors, including an individual's age, weight and gender, lifestyle habits such as alcohol and tobacco consumption, and even environmental factors such as diet. Patient compliance is also an issue [2]. The general health of the patient can also affect one's response to warfarin [1].

The simplest mechanism for prescribing warfarin is a "loading table" of rules specifying what dosage is needed following a given INR reading. This, however, does not acknowledge individual differences in warfarin response. Current advice is to adjust the dosage by 5-20% of the total weekly dose, depending on the current INR, the previous dose, and any reasons identified that might explain the undesirable current INR reading [4]. Graphical "nomograms" have also been developed to help with dose adjustment, such as that by Dalere, which is based on a model of warfarin activity [5]. In general success rates for physicians using nomograms, dosage adjustment tables or their own experience have not been particularly high. Schaufele, Marciello & Burke [6] demonstrated this by analysing 181 patients receiving warfarin treatment over a four-month period through a rehabilitation centre. Only 38% of all INR readings were found to be within the target range. Most physicians, however, achieve a 50-75% success rate for a particular patient [1]. This is still relatively low, especially when combined with the fact that between 1.1% and 2.7% of patients managed by anti-coagulant clinics suffered major bleeding [1].

There have been some attempts to utilise the machine learning capabilities of computers in warfarin treatment. The worth of machine learning as a potential approach has been demonstrated in the prescription of other drugs, as shown in a study by Floares et al [7], where neural networks were used successfully to compute an optimal dosage regimen for chemotherapy patients. These produced better results than the other conventional or artificial intelligence approaches reported in the study. Narayanan & Lucas [8] also attempted a machine learning solution to predicting INR levels after a given dosage, by using a genetic algorithm to select variables with which to train a neural network. However, no comparisons were offered to other solutions, examining only the benefits of the genetic algorithm in addition to the existing neural network. Neural networks have been also investigated by Byrne et al [9] and found to be twice as accurate as physicians at predicting the result of a given dosage. The benefits of extracting rules for warfarin dosage from ensemble learning have also been researched [10].

3 Applying ML to Warfarin

In our first study we explored the feasibility of using machine learning by trying to predict the effects of a given dose. Two options immediately presented themselves when considering the form that the output from the machine learning algorithms should take. The first of these was a numerical value of the expected INR reading after a given warfarin dosage. However, this severely restricted the number of machine learning algorithms that can be used, because most produce a nominal classification as their output. Since the actual value of INR is less important than its relative position to the therapeutic index, it was decided that dosage result could be usefully classified as either "low", "in range" or "high". We similarly binned the input INR values and added this as an extra set of attributes.

Several approaches were proposed as possible solutions, within which different algorithms and attributes could be used. These approaches principally differ in the source of data used for training the system as well as whether some form of ensemble learning is to be used. The simplest solution predicts a patient's response to warfarin solely by examining their history of interaction with the drug. This has the advantage that any model of the patient built by the machine learning algorithm would be individualised as much as possible. However, this solution would obviously not be ideal if the patient did not have a long history on which the algorithm could be trained. Alternatively, data from all patients could be used to train an algorithm, from which predictions for an individual patient could be made. In such a system all data points are used by the algorithm, no matter to which patient they belong.

We also trialled a "two layer" ensemble approach. Ideas from two popular ensemble learning techniques, "bagging" [11] and "stacking" [12], were combined. Each patient's history, including that of the patient currently under study, was modeled separately using just their own data and selecting the ML algorithm that best predicted their INR. Each patient model was then given the same data point, and their predictions of the resultant INR level for this data point fed into the "second-layer" algorithm, which was trained to use these predictions as a means for predicting the final INR value for this patient. The second-layer algorithm was varied to try to maximise performance.

Many different combinations of attributes were trialled to represent the patient's history for a given data-point. When learning from only one patient's data, up to three previous dosages and INR values were provided as input. More data was provided when working with a training set based on multiple patients, with anywhere from one to twelve previous dosages used; it was hypothesised that increasing the number of attributes would help the algorithms to select only relevant data points from other patients' histories.

The patient data was provided by volunteers and collated by the fourth author (Mr David Shaw, a cardiothoracic surgeon). Although data for over 70 patients were initially provided, not all of these could be used. First, patients could be divided into two groups: initial-state patients (within two years of their heart valve operation) or steady-state patients. The problem of predicting their response to warfarin differs significantly between groups, and because of time constraints it was decided to restrict it to steady-state patients only. Some patients also had to be excluded because their history was of insufficient length. Furthermore, for systems based on data from more than one patient, all patients studied had to have the same therapeutic range. Ten patients were selected at random on which to perform experiments. Dosages recorded are those given over a week. The performance of the various systems was compared to base accuracies provided by Naïve Bayes and the graphical nomogram detailed in

Dalere's study, the latter approximating what a physician might have prescribed [5]. For each prediction all history up to (but excluding) the dose in question was used.

Results when learning from individual histories showed different algorithms and attributes suited different patient histories. Accuracies achieved varied from patient to patient, and a significant improvement could be noted over time for patients with long histories. For example, the accuracy for patient 31 reached a maximum of 53% over the whole data-set, but when only the accuracy over the last 12 data-points was considered, 67% was observed. The best performing algorithm and attributes was not necessarily the same for these two situations, however. When learning from multiple patient histories the most predictive attributes and most accurate algorithm again varied between patients. However, there was less variance here than for individual histories. The success rate of the best solution varied between 61% and 90%. Most patients, however, needed a larger history than when trained on only their own data. There was also more consensus on the most successful algorithms, with Ridor [13], NNge [14] and a Bayesian Network (created by genetic search) the most successful algorithms on more than one occasion each. This is potentially a more useful solution, in that theoretically a patient does not need to have an extensive history to make predictions. Finally, we evaluate the 'two-layer' approach. Results were not as promising as initially hoped. Accuracies of the best second-layer algorithm ranged from 44% to 85% depending on the patient. Many machine learning algorithms for the second layer machine were trialled, yet the best result that could be achieved for a particular patient was usually achieved by many different algorithms. This suggests that the choice of algorithm for this second machine may not be important.

Comparisons were made between different ways of predicting the effect of a given warfarin dosage by examining the percentage of correct predictions made over the datapoints. However, predictions are only made from at least the 7th data-point, and on some occasions slightly later than this, if the attributes required dictate this. This allows sufficient history to be used to make predictions. Table 1 shows a comparison of the best machine learning solutions with graphical nomograms, the accuracy of the physician and a Naïve Bayesian prediction. The machine learning solutions compared are the best algorithm combination when learning on the patient's own history, the best when learning on multiple histories, and the best two-layer result obtained. The "base accuracy" to which each solution is compared is the percentage of the most popular class over the patient's history, or the accuracy that would be achieved if the most common result was predicted in every case.

One-way ANOVA was performed on the raw percentage success for each patient of the five different prediction methods. A significant difference was found ($F=7.198$; $p<0.001$). Post-hoc analysis using paired t-tests with a Bonferroni correction was hence applied. This showed a weakly significant difference between Naïve Bayes and the machine learning solutions, apart from the case of the two-layer approach from which Naïve Bayes was not significantly different. There was a weakly significant difference in accuracy between the two-layer approach and the nomogram (66.5% versus 52.5% respectively), $t=4.5$; $p<005$. More crucially, strongly significant differences were observed between the machine learning solution using only the individual histories and the nomogram (66.5% versus 52.5%), $t=8.2$, $p<0.001$. There was also a significant difference between the machine learning solutions using the patient's own

Table 1. Comparison of the success in predicting the effect of warfarin dosages of the best solution for each approach (percentage improvement over the base accuracy)

Patient ID	Individual patient history	Multiple histories	Two-layer	Nomogram	Doctor	Naïve Bayes
2	10	**17**	0	-18	0	4
5	0	8	0	-19	-64	**9**
31	5	**23**	-4	-8	0	-4
39	23	**32**	31	7	0	15
40	0	**3**	0	-20	-60	-13
44	13	**19**	5	3	-9	12
48	22	**36**	17	4	0	17

history versus learning on multiple patients, which achieved a mean accuracy of 79%, p<0.003. When learning on the history of multiple patients, the best machine learning solution was significantly superior to both the nomogram (t=11.2, p<0.0001) and the doctors' accuracy (41%), t=4.9; p<0.003.

4   A Decision Support Approach

The first study suggested that ML can potentially predict warfarin's effect. However, this is a more tightly constrained problem than the one we are trying to solve; in general we want to produce a model that indicates the *best* dose. Machine learning is not sufficiently accurate to do this directly owing to the lack of information about other factors (such as lifestyle) and the strong interaction between attributes, but it might prove useful in giving a doctor some guidance on the likely effect of a range of doses, such that the doctor could make a more informed decision. The main goal of this second study was to build on what was learned in the first study and develop a proto-type for a web-based predictor that provides precisely this information. This prototype allows the doctor to have enough flexibility to create and combine different attributes in order to create a model that will predict a dose's effect on the patient with the greatest accuracy, by allowing them to select/create the input attributes and combine them in order to make the best predictive model on a per-patient basis.

When creating an attribute there are a number of different functions that can be applied to any numeric data field, such as sum, average and delta. Each function can be applied to different length time periods. Calculated attributes can be repeated up to four times per example (i.e. for the current and previous three dosage intervals). General attributes (e.g. gender) can also be added or removed from the final group of attributes. Further, each patient's model can be created using their history only or by utilizing that of all patients in the database. Finally, there are potentially many different ML algorithms that can be used to create a model; the prototype uses the WEKA code base [13] to provide ML algorithms, and the doctor can select which ones they wish to use for a given patient.

The output of the system is a graph that plots the predicted effect of all doses from 0 to 15mg in 0.5mg increments. Two kinds of output are possible: nominal or numeric (as previously described). For numeric outputs the graph for each model is presented, while for nominal attributes we also add a simple ensemble learner: each model's output is translated into a number (low=-1; in-range=0; high=1) and we sum these values to get the combined prediction; our expectation was that this combined output would prove superior to any individual model and overcome the problem previously identified whereby every patient requires an individual model. For the purpose of this evaluation we used J48 (a variant of C4.5 [15]), NNGE [14] and IBk [16] for nominal predictions and IBk and SMO regression for numeric prediction, for all patients.

The patient data used in this research came from the same source as the first study. Among initial state patients there are some that show a steady reaction to warfarin from the early days (months); these patients should be relatively easy to predict. Others have an inconsistent history even after a year of taking the drug, possibly because they have made frequent lifestyle changes. Patients that have only recently started taking warfarin present the biggest challenge. To explore the feasibility of creating predictive models for these three groups of patients (initial, steady-state, varying), three patients were examined: one with a long and steady history (over 2 years), one with a moderate-length, inconsistent history and one with almost no history at all. The latter was in fact a copy of a patient that we do have history for, with most of the historic data excluded during training.

After some initial regression testing we concentrated on the sum, average and delta functions when calculating attribute values since they appeared to have some influence over INR readings. It has been recognized that a patient's lifestyle data would be of great value in predicting the right dose of warfarin, but the only data we have available at the moment is the gender (which is known for only some of the patients). This is specified as being female, male or missing, and was used as an attribute when learning on multiple patient histories to see if this improved accuracy. Regarding the interval over which to sum/average doses, it should be noted that the points in a patient's data set do not have the same temporal gap. However, almost all readings are 1-5 weeks apart, and the warfarin from a specific dose is processed within a week. We therefore used one week as the period for our testing purposes.

To evaluate the flexibility of the approach we developed five test conditions covering a range of attribute options. Test 1 used a single attribute only, consisting of the warfarin dose summed over the last seven days. Test 2 added the average and delta doses, and test 3 added gender. Test 4 investigated the efficacy of adding previous historic data by applying the same attributes as test 2 (sum, average and delta doses) for each of the *last three* dosage periods. Finally, test 5 used all of the attributes of test 3 plus the average INR and dose over the entire history period. Note that these tests do not necessarily capture the best way to predict for each of these patients. Rather, they illustrate how much variation can be quickly achieved using the system and show how it facilitates building useful models.

Fig. 1. shows some of the output graphs produced, representing tests 1, 4 and 5 for patient 2 (long history, varying response) using multiple patient models, for the output from the combined models and J48. We chose J48 for comparison because the results appeared reasonably useful (this was not true for the other algorithms). Test 1 is a

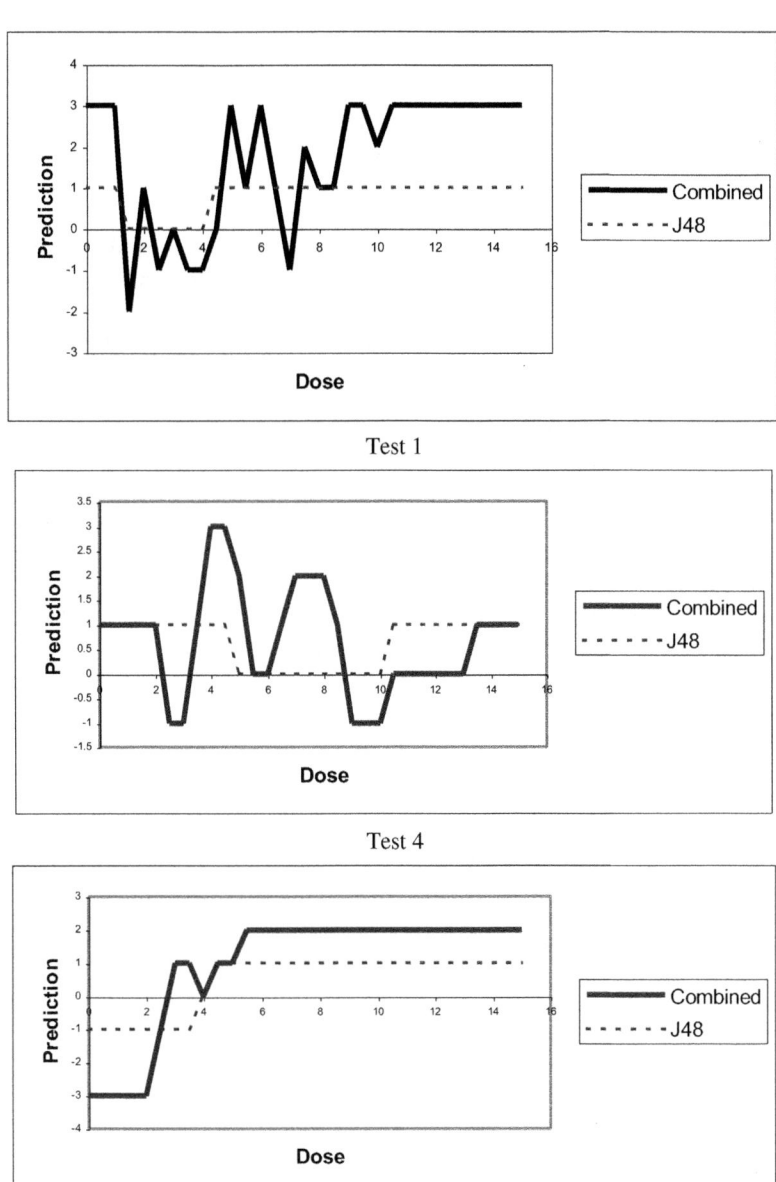

Test 1

Test 4

Test 5

Fig. 1. Sample outputs for patient 2 (long history, varying response)

baseline, which we would not necessarily expect to perform very well, and this is clearly the case; both the combined result and that of J48 are of little help for this patient. Adding further history (test 4) is of itself not very helpful because, as described previously, the other patients' response to warfarin may not match that of this patient, so the additional data is still of limited use. In contrast, adding the average

INR and dose for each patient improves the models considerably; for test 5 J48 produces a very clean model that predicts the correct dose to be 4.0, which agrees with this patient's history. The combined model, whilst messier, predicts the same dose if the mid-point between the −3 and +2 plateaus is used. In contrast there is no significant effect from adding this information for patient 1 (none of the models were very useful), whilst patient 3 is easily predicted from single patient data owing to the steady nature of their response.

Overall the models in test 1 were of limited use. Patient 1 (short history) has only a couple of days worth of history and therefore the only prediction possible would be by using other patient's history. The results obtained for each model showed that INRs are spread over the range of 2-8.5mg, which is of no use. For patient 2 (long history, varying response to warfarin) the single patient model learned by J48 gave a clean prediction of 4.5-5, while the combined model was noisier but nonetheless predicted the same value; this matches the patient's data. The multiple-patients model however was very noisy and did not produce any useful result. Patient3 (steady history) also produced good models for single patient data, with J48 predicting a dose of 8.5-10, and the combined model suggesting a dose of 8.5-9, both of which appear to be a good match to the data. Again the multiple patient model was too noisy to be of use. Test 2 (sum, average and delta over the last 7 days) showed similar results to the test 1 but with little bit more consistency. In test 3 the gender was added, but this didn't result in significant improvements to any of the models.

Test 4 also used the sum, average and delta values for the last 7 days, however it then repeated this calculation for the *previous two doses*. The purpose of this test was to illustrate how including additional history (as used in the first study) affects the predictions produced. Patient 1 (short history) was not included in this test since the history was not long enough and therefore the calculations on the previous intervals could not be performed. For patient 2 (varying response) this test produced more consistent results between the algorithms used. However, the suggested dose by all 3 nominal algorithms increased from 4-5 to to 5.5mg, which is not supported by the data. This is probably the result of using the delta information; all of our tests suggested this attribute is not useful. The accuracy of the model decreased for patient 3 (steady state) for both single and multiple patient models, even though this patient was the easiest to predict in previous 3 tests. This suggests additional dose history by itself is not useful for such patients.

Finally, test 5 added average totals of INR and dose. These attributes might be expected to have two potential benefits. First, for single patient models they would smooth out temporary fluctuations in response. Second, for multiple patient models they might eliminate patients that do not share a similar warfarin response. For all three patients the multiple-patient models improved. In the case of patient 1, this was the only set of parameters that produced a usable model; the predicted in-range dose was 3-4 OR 6-10. The latter is spurious and suggests the model is still not sufficiently selective. However, if we assume the doctor has a rough idea of the ideal dose then they could usefully use the 3-4 value as a guide. For the other two patients the multiple-patient model improves dramatically with both J48 and the combined models giving a clean response at around the dose expected (4.0 and 8.5-10 for patients 2 and 3 respectively). This suggests that with the right selection of attributes a multiple-patient model can be learned that produces useful results, which bodes well for dealing with patients whose output response is highly variable or whose history is fairly short.

5 Conclusion

The goal of this research is to develop an online warfarin prediction application that will initially be used by doctors, but which may in time become suitable for patient self-prescription. In this study we first investigated how well various machine learning approaches compare to more traditional ways of predicting the effect of a dose of warfarin in heart valve patients. A new "two-layer" approach was tried to test the hypothesis that the warfarin problem consists of multiple, potentially related data sets. Many different attribute combinations were attempted to provide the best representation of the data and any temporal patterns observed that could help with prediction. When tested on the data of heart valve patients it was found that the effect of a warfarin dosage could be predicted with the most accuracy by machine learning algorithms learning on the history of multiple patients. However, the best performing algorithm and attributes differed from patient to patient, making a one-fits-all solution unlikely.

Despite the shortcomings identified, in the second study we investigated whether machine learning is sufficiently accurate to be useful in guiding a doctor to make an informed prescription. We developed a prototype that outputs a prediction for a range of doses, from which the correct dose might be inferred. Overall we were able to produce plausible models for three test patients: one with very little history, one with a long history of varying warfarin response, and one with a steady response. For patients with little or varying history a multiple-patient model that included their average INR reading and dose performed best, while for the steady-state patient an excellent model was created based only on their most recent dose. We also investigated using a model that reported the combined results of several algorithms. In general this model outperformed any single ML algorithm for all three patients, suggesting an ensemble approach may overcome some of the problems of the highly individual responses to warfarin observed. The biggest challenge is predicting warfarin response accurately for new patients: such patients can only be predicted using the multiple-patient model, and its accuracy is directly proportional to the number of patients for which we have history. For the purpose of this research we had a small number of histories available (19) but still we were able to predict the right dose for our test Patient 1 to a reasonable degree of accuracy, which is a promising achievement.

Overall the results are sufficiently positive to encourage us to continue to explore this approach. We will further investigate combinations of attributes and parameters to determine the extent to which we can narrow the search. In particular, this study made very little use of previous INR results, which are likely to be useful. We will also investigate more sophisticated ensemble methods to try to increase the reliability of the combined model. Finally, we will add more data and trial the system with doctors who prescribe warfarin to evaluate its utility.

The prescription of warfarin remains a difficult problem with life-and-death implications. This research is a promising step towards better administration of this important therapy.

References

1. Gallus, A., Baker, R., Chong, B., Ockelford, P., Street, A.: Consensus guidelines for warfarin therapy. Medical Journal of Australia 172, 600–605 (2000)
2. Roland, M., Tozer, T.: Clinical Pharmacokinetics: Concepts and Applications. Williams and Wilkins, Philadelphia (1995)
3. Gage, B.F., Fihn, S.D., White, R.H.: Management and dosing of warfarin therapy. The American Journal of Medicine 5, 211–228 (2000)
4. Jaffer, A., Bragg, L.: Practical tips for warfarin dosing and monitoring. Cleveland Clinic Journal of Medicine 70, 361–371 (2003)
5. Dalere, G.: A graphical nomogram for warfarin dosage adjustment. Pharmacology 19, 461–467 (1999)
6. Schaufele, M.K., Marciello, M.A., Burke, D.T.: Dosing practices of physicians for anticoagulation with warfarin during inpatient rehabilitation. American Journal of Physical Medication and Rehabilitation 79, 69–74 (2000)
7. Floares, A.G., Floares, C., Cucu, M., Marian, M., Lazar, L.: Optimal drug dosage regimens in cancer chemotherapy with neural networks. Journal of Clinical Oncology 22, 2134 (2004)
8. Narayanan, M., Lucas, S.: A genetic algorithm to improve a neural network to predict a patient's response to warfarin. Methods of Information in Medicine 32, 55–58 (1993)
9. Byrne, S., Cunningham, P., Barry, A., Graham, I., Delaney, T., Corrigan, O.I.: Using neural nets for decision support in prescription and outcome prediction in anticoagulation drug therapy. In: Lavrac, N., Miksch, S. (eds.) Fifth Workshop on Intelligent Data Analysis in Medicine and Pharmacology, Berlin (2000)
10. Wall, R., Cunningham, P., Walsh, P., Byrne, S.: Explaining the output of ensembles in mdeical decision support on a case by case basis. Artificial Intelligence in Medicine 28, 191–206 (2003)
11. Breiman, L.: Bagging predictors. Machine Learning 26, 123–140 (1996)
12. Wolpert, D.H.: Stacked generalisation. Neural Networks 5, 241–259 (1992)
13. Witten, I.H., Frank, E.: Data mining. Morgan Kaufman, San Francisco (2000)
14. Martin, B.: Instance-based learning: nearest neighbour with generalisation. In: Computer Science. University of Waikato, Hamilton (1995)
15. Quinlan, R.: C4.5: Programs for Machine Learning. Morgan Kaufmann, San Mateo (1993)
16. Aha, D.W., Kibler, D., Albert, M.K.: Instance-based learning algorithms. Machine Learning 6, 37–66 (1991)

Single-Stacking Conformity Approach to Reliable Classification

Evgueni Smirnov[1], Nikolay Nikolaev[2], and Georgi Nalbantov[1]

[1] Department of Knowledge Engineering, Maastricht University,
P.O.BOX 616, 6200 MD Maastricht, The Netherlands
{smirnov,g.nalbantov}@maastrichtuniversity.nl
[2] Department of Computing, Goldsmiths College, University of London,
London SE14 6NW, United Kingdom
n.nikolaev@gold.ac.uk

Abstract. This paper considers the problem of constructing classifiers for road side assistance capable of providing reliability values for classifications of individual instances. In this context we analyze the existing approaches to reliable classification based on the conformity framework [16,18,19,27]. As a result we propose an approach that allows the framework to be applied to any type of classifiers so that the classification-reliability values can be computed for each class. The experiments show that the approach outperforms the existing approaches to reliable classification for road side assistance.

1 Introduction

In the last ten years machine-learning classifiers have been applied for many classification problems [12]. Nevertheless, only few classifiers have been employed in critical-domain applications [3,16,19]. This is due to the difficulty to determine whether a classification assigned to a particular instance is reliable or not.

The importance of the reliable-classification problem can be demonstrated in the context of the out-going EU project MYCAREVENT[1]. One of the goals of the project is to implement *a road side assistance decision support system* capable of providing manufacturer-specific car-repair information according to the problems identified by cars's Off-/On-Board-Diagnosis systems. One of the core parts of the system is a machine-learning classifier for road side assistance. The classifier has to predict the status of a car that experiences problems. Due to substantial costs involved, in addition to good generalization performance, the classifier has to provide reliability values for each possible classification of a car problem. In this way a system user can estimate the risks of actions s/he takes for each possible classification (car-problem status).

In this paper we present our first attempt to construct machine-learning classifiers for road side assistance capable of providing classification-reliability values. We start

---

[1] MYCAREVENT (Mobility and Collaborative Work in European Vehicle Emergency Networks) was a 3-year research project financed by the IST (Information Society Technology) program of the European Commission within the Sixth Framework Program. For more information, please visit http://www.mycarevent.com/

D. Dicheva and D. Dochev (Eds.): AIMSA 2010, LNAI 6304, pp. 161–170, 2010.
© Springer-Verlag Berlin Heidelberg 2010

in Section 2 where we formalize the classification task in the context of reliable classification. In Section 3 we briefly describe the existing approaches to reliable classification that we find useful for our problem, namely the conformity framework [13,15,27] and the meta-classifier conformity approach [18,22][2]. After analyzing both approaches we propose in Section 4 a novel approach to reliable classification that we call single-stacking conformity approach. The performance of the three approaches is compared in the context of our problem in Section 5. The comparison shows that the single-stacking conformity approach outperforms the conformity framework and the meta-classifier conformity approach for road side assistance. Finally, Section 6 concludes the paper.

2   Classification Task

Let $\mathbf{X}$ be an instance space and Y be a class set. Training data D is a set $\{(\mathbf{x}_1, y_1), (\mathbf{x}_2, y_2), \ldots, (\mathbf{x}_n, y_n)\}$ of n labelled instances $(\mathbf{x}_i, y_i)$ where instance $\mathbf{x}_i$ is in $\mathbf{X}$ and class y_i is in Y. Given a space H of classifiers h ($h : \mathbf{X} \to Y$), the classification task is to find classifier $h \in H$ that correctly predicts future, unseen instances.

If a classifier h has to be used for reliable classification, we need to solve additional two tasks [1,3,5,8,9,11,13,15,16,19,23,24,25,27]:

– to obtain reliability values for classification of individual instances; and
– to obtain a threshold on these values.

If the reliability value for a classification of an instance provided by the classifier h is greater than the threshold, the classification is considered to be reliable. Otherwise, it is unreliable and the instance is left unclassified.

3   Existing Approaches to Reliable Classification

When we have to construct a classifier capable of providing reliability values for individual instance classifications, we can employ either the conformity framework [16,27] or the meta-classifier conformity approach [18,19]. Both approaches are briefly described in this section.

3.1   The Conformity Framework

The conformity framework was proposed in [13,15,16,27] for constructing classifiers for reliable classification. The framework assumes that the data D with n training instances and an instance $\mathbf{x}_{n+1}$ to be classified are drawn from the same unknown distribution. Given a classifier h, the framework computes p-value $p(y_{n+1})$ for each class $y_{n+1} \in Y$ that the instance $\mathbf{x}_{n+1}$ can receive by h. The p-value is determined as follows:

$$p(y_{n+1}) = \frac{\#\{i : \alpha_i \geq \alpha_{n+1}\}}{n+1} \tag{1}$$

---

[2] Although version spaces are a successful approach to reliable classification [17,20,21], they are not considered in the paper since they do not provide classification-reliability values.

where α_i is the non-conformity value of instance $(\mathbf{x}_i, y_i)$ in the bag $\{(\mathbf{x}_1, y_1), ..., (\mathbf{x}_{i-1}, y_{i-1})\ (\mathbf{x}_{i+1}, y_{i+1}), ..., (\mathbf{x}_{n+1}, y_{n+1})\}$.

The class y_{n+1} with the largest p-value is the classification of $\mathbf{x}_{n+1}$. The credibility of this classification is the largest p-value and the confidence is one minus the second largest p-value.

To compute non-conformity values α_i for each instance $(\mathbf{x}_i, y_i)$ we need to construct a non-conformity function α for the classifier h used. If we have access to the internal structure of classifiers, then non-conformity functions can be constructed for nearest neighbor classifiers [13], decision trees, neural networks, support vector machines [15,26], and the naive Bayes classifier [16]. To provide an example, let us consider the non-conformity function for nearest neighbor classifiers [13]. Given an instance $\mathbf{x}_i$ labeled to belong to class y_i, the non-conformity function returns for $\mathbf{x}_i$ a non-conformity value α_i equal to $\dfrac{D_k^{y_i}}{D_k^{-y_i}}$, where $D_k^{y_i}$ is the sum of distances between $\mathbf{x}_i$ and k nearest neighbors of $\mathbf{x}_i$ that belong to class y_i, and $D_k^{-y_i}$ is the sum of distances between $\mathbf{x}_i$ and k nearest neighbors of $\mathbf{x}_i$ that do not belong to class y_i.

Although non-conformity functions were proposed for some basic types of classifiers (see above), there is no approach how to design these functions in general. Moreover, if we do not have access to the internal structure of classifiers (e.g., the classifier is a human expert), we cannot design non-conformity functions at all. Therefore, we may conclude that the applicability of the conformity framework is restricted.

3.2   Meta-classifier Conformity Approach

We proposed the meta-classifier conformity approach in [10,18,19,22] to allow the conformity framework to be applied for any type of classifiers. The approach was inspired by [2] and can be explained as follows. Assume that we have a classification problem that requires computing p-values for instance classifications but our best classifier B is not capable of providing such values. Then, if we have any conformity-based classifier M, the approach is to train M as a meta classifier that predicts the correctness of each instance classification of B. In this way, the p-values of the meta predictions can be viewed as estimates of the p-values of the instance classifications of the base classifier B. More precisely, if B assigns class y to instance $\mathbf{x}$, then the p-value of class y is set equal to the p-value p_0 of the meta class "correct classification" and the sum of p-values of all the remaining classes $Y \setminus \{y\}$ is set equal to the p-value p_1 of the meta class "incorrect classification". Thus, using the ensemble of B and M, denoted by $MCT(B:M)$, we can estimate the p-value (conformity) and confidence of each classification y provided by B.

An obvious drawback of the meta-classifier conformity approach is that when the base classifier B assigns a class y to the instance $\mathbf{x}$ we cannot estimate the p-value of each class in $Y \setminus \{y\}$. Thus, if $p_0 < p_1$, the classification with the highest p-value among classes $Y \setminus \{y\}$ cannot be identified. To overcome this drawback we propose in the next section a single-stacking conformity approach.

4   Single-Stacking Conformity Approach

This paper proposes the single-stacking conformity approach to allow the conformity framework to be applied for any type of classifiers so that the p-values can be computed

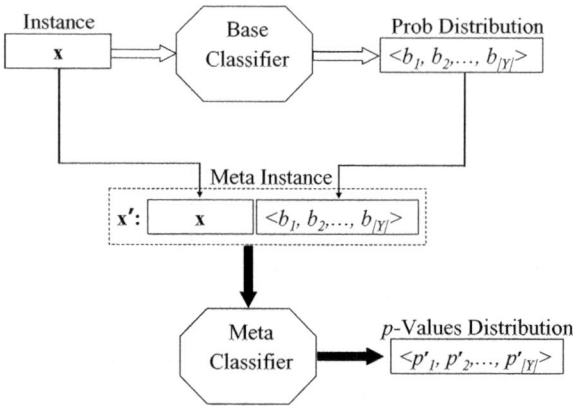

Fig. 1. Single-Stacking Conformity Approach

for each class involved in classification. As the name suggest the key idea behind the approach is to employ a stacking ensemble [28] consisting of one base classifier B and one meta classifier M (see Figure 1). In this way, if the classifier M is based on the conformity framework , the p-values of the meta predictions can be considered as the p-values of the instance classifications of the base classifier B.

The meta classifier M plays a central role in a single-stacking ensemble (see Figure 1). For each instance to be classified it has to receive classification information from the base classifier B in terms of the class probability distribution and then to compute the desired p-value for each class $y \in Y$. Therefore,

- meta instance space X' is defined by the attributes of the original instance space X plus $|Y|$ attributes representing the class probability distribution that the base classifier B computes for any instance in X;
- the meta class set Y' coincides with the class set Y.

The meta data D' are formed in $X' \times Y'$ using internal k-fold cross validation as follows: a labelled meta instance $(\mathbf{x}'_i, y_i) \in D'$ is formed from the labelled instance $(\mathbf{x}_i, y_i) \in D$ s.t. $\mathbf{x}'_i$ is a union of $\mathbf{x}_i$ and the class probability distribution computed by B for $\mathbf{x}_i$. Once meta data D' have been formed, the meta classifier M is trained on these data.

The single-stacking ensemble of the base classifier B and the meta classifier M is denoted by *SST(B:M)*. *SST(B:M)* is used for classification of an instance $\mathbf{x}$ as follows (see Figure 1). First, B classifies $\mathbf{x}$ by providing the class probability distribution $\langle b_1, b_2, ..., b_{|Y|} \rangle$ for $\mathbf{x}$. Then, the instance $\mathbf{x}$ and the distribution $\langle b_1, b_2, ..., b_{|Y|} \rangle$ are concatenated to form the meta instance $\mathbf{x}'$. The meta classifier M classifies the meta instance $\mathbf{x}'$ by providing the class probability distribution for $\mathbf{x}'$. If M is based on the conformity framework, the class probability distribution of M is a class p-values distribution $\langle p'_1, p'_2, ..., p'_{|Y|} \rangle$ consisting of p-values p'_i for the classes in Y. In this case, the single-stacking conformity approach approximates for the base classifier B the p-value

p_i for each class $y_i \in Y$ using the p-value p'_i of the meta classifier M. This approximation rule guarantees that the classification rule of the meta classifier M is preserved. Since the meta classifier M does contain information about misclassifications of B, the approximation rule can cause eventually correcting the classifications of B.

5   Road Side Assistance Classifiers

This section provides our experiments in constructing road side assistance classifiers using the conformity framework, the meta-classifier conformity approach, and the single-stacking conformity approach. In subsection 5.1 we describe the classification task of road side assistance. Then, in subsection 5.2 we specify and experiment with the classifiers based on the three approaches to reliable classification described in the paper.

5.1   Road Side Assistance Classification Problem

In the context of the MYCAREVENT project the road side assistance classifier had to be trained on historical patrol-car data of previously diagnosed faults and their symptoms provided by RAC (a UK-based motoring organisation originated from the Royal Automobile Club). The data were derived from the Call-Center-Operator dialogue.

The data are described using four discrete attributes *Brand* (40 discrete values), *Model* (229 discrete values), *Primary Fault* (35 discrete values), *Secondary Fault* (80 discrete values), and class attribute *Status*. The class attribute have three values (classes):

1. **Fixed:** The problem is solved by road side assistance and the car can continue its journey safely (3366 instances).
2. **Required tow:** The car needs to be towed to the workshop (1077 instances).
3. **Other:** Some parts of the problem cannot be solved by road side assistance but the car is able to get to the workshop by its own (1477 instances).

5.2   Experiments

To construct road side assistance classifiers we employed three standard classifiers: the C4.5 decision tree learner (*C4.5*) [14], the k-nearest neighbor classifier (*NN*) [5], and naive Bayes classifier (*NB*) [4] as well as one conformity-based nearest neighbor classifier (*TCMNN*)[3] [13]. *C4.5*, *NN*, and *NB* were used as independent classifiers and as base classifiers in conformity ensembles. *TCMNN* was used as an independent conformity-based classifier and as a meta classifier in conformity ensembles. The meta-classifier conformity approach was presented by three ensembles *MCT(C45:TCMNN)*, *MCT(NN:TCMNN)*, and *MCT(NB:TCMNN)*. The single-stacking conformity approach was presented by three ensembles *SST(C45:TCMNN)*, *SST(NN:TCMNN)*, and *SST(NB: TCMNN)*. The settings of the classifiers were chosen experimentally so that the classification performance was maximized. For the classification reliability values of *C4.5*, *NN*,

---

[3] The *TCMNN* non-conformity function is given in section 3.1. *TCMNN* computes p-values according to formula 1.

and *NB* we used the classification probabilities of these classifiers. For the classification reliability values of *TCMNN* and the conformity ensembles we used classification p-values these classifiers can generate.

We experimented with the classifiers by varying the reliability threshold r in the interval $[0, 1]$. If the reliability value for a classification of an instance was greater than r, the classification was considered as reliable; otherwise, the classification was considered as unreliable and the instance was left unclassified. For each value of the reliability threshold r we evaluated using the 10-fold cross validation:

- rejection rate: proportion of the *unclassified* instances;
- accuracy rate on the *classified* instances;
- rejection rate per class: proportion of the *unclassified* instances per class;
- true positive rate per class *TPr* [6] on the *classified* instances.

The results of our experiments are presented as accuracy/rejection and *TPr*/rejection graphs[4] [7]. They are provided in Figures 2, 3, and 4. To facilitate the comparison between the classifiers we extracted from the graphs the rejection rates for the accuracy rate of 1.0 and the *TPr* rate of 1.0 per class. These rates are presented in Table 1.

Table 1. Rejection rates for the accuracy rate of 1.0 and the *TPr* rate of 1.0. R is the rejection rate for the accuracy rate of 1.0. R_F is the rejection rate for the *TPr* rate of 1.0 for class "Fixed". R_R is the rejection rate for the *TPr* rate of 1.0 for class "Required tow". R_O is the rejection rate for the *TPr* rate of 1.0 for class "Other". The undefined rejection rates are denoted by "–".

Classifiers	R	R_F	R_R	R_O
NB	0.93	**0.56**	–	0.51
NN	0.98	0.90	–	0.97
C4.5	0.98	0.95	–	0.78
TCMNN	0.72	0.69	0.93	0.52
MCT(NB:TCMNN)	0.72	0.71	0.96	0.37
MCT(NN:TCMNN)	0.71	0.69	0.92	0.48
MCT(C4.5:TCMNN)	0.71	0.62	–	0.38
SST(NB:TCMNN)	0.68	0.67	0.88	**0.34**
SST(NN:TCMNN)	0.71	0.71	0.95	0.36
SST(C4.5:TCMNN)	**0.64**	0.68	**0.84**	**0.34**

An analysis of the results shows that for most of the cases the accuracy/rejection and TPr/rejection graphs of *TCMNN*, the *SST* ensembles, and the *MCT* ensembles dominate those of *C4.5*, *NN*, and *NB*. This implies the following two conclusions:

- the class probabilities of *C4.5*, *NN*, and *NB* are pure estimates of classification-reliability values. They can be used only for the majority class "Fixed"and fail bitterly for the minority classes "Required tow" and "Other"(see the last two columns of Table 1).

---

[4] We note that the accuracy/rejection (*TPr*/rejection) graph of the "always-right" classifier is determined by the segment $\langle (0, 1), (1, 1) \rangle$. If two classifiers have the same accuracy rate, we prefer the classifier with lower rejection rate.

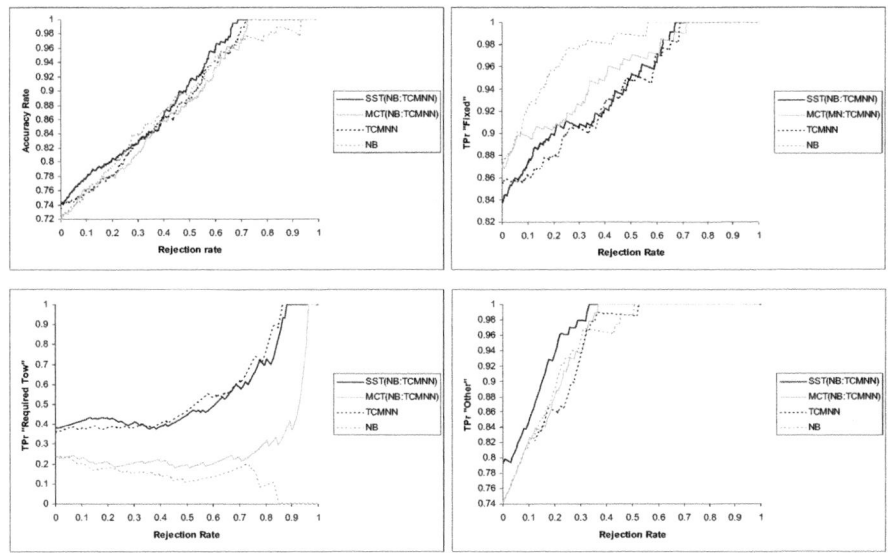

Fig. 2. Accuracy/rejection graph and TPr/rejection graphs for *NB*, *TCMNN*, *SST(NB:TCMNN)*, and *MCT(NB:TCMNN)*

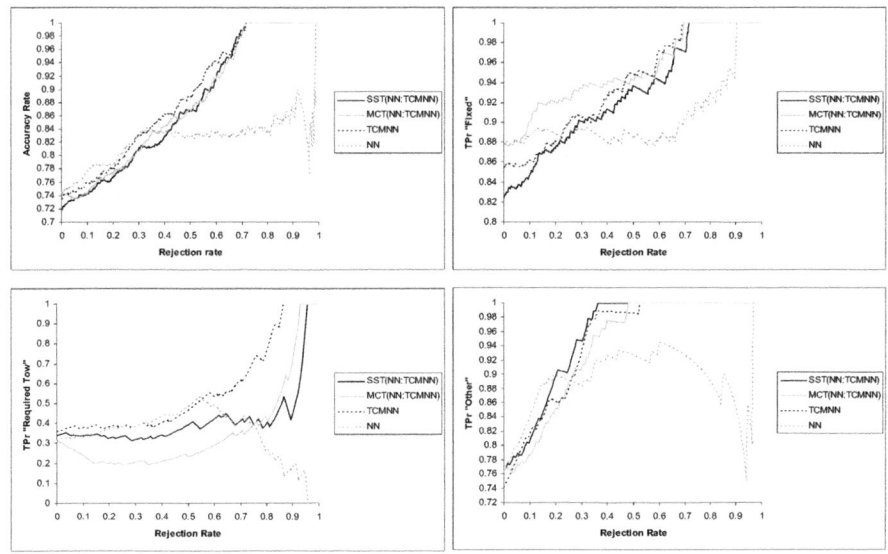

Fig. 3. Accuracy/rejection graph and TPr/rejection graphs for *NN*, *TCMNN*, *SST(NN:TCMNN)*, and *MCT(NN:TCMNN)*

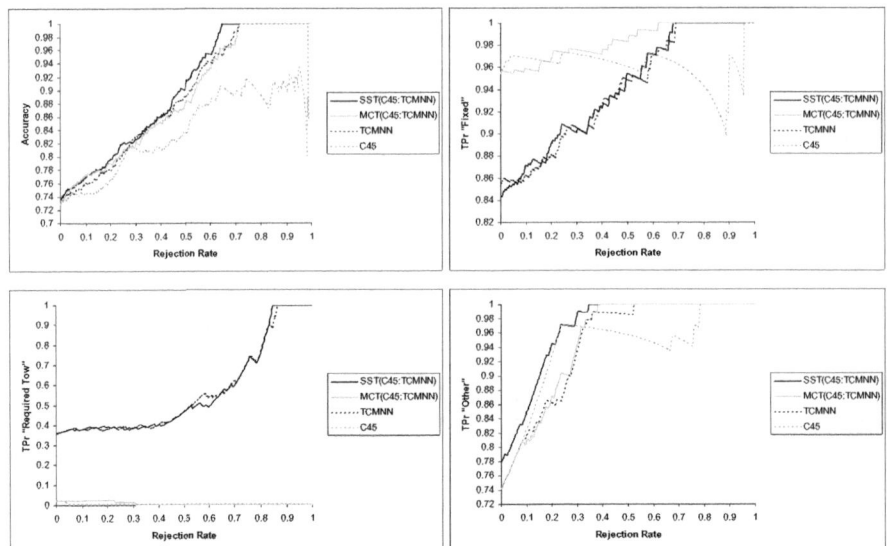

Fig. 4. Accuracy/rejection graph and TPr/rejection graphs for *C4.5*, *TCMNN*, *SST(C4.5:TCMNN)*, and *MCT(C4.5:TCMNN)*

- *SST* ensembles, *MCT* ensembles and *TCMNN* provide good classification-reliability values. They can be used for the majority class "Fixed" as well as for the minority classes "Required tow" and "Other".

A further analysis shows that *SST* ensembles outperform *TCMNN* and *MCT* ensembles on accuracy/rejection graphs (Table 1). For the majority class "Fixed" one of the *MCT* ensembles outperforms *SST* ensembles and *TCMNN*. For the minority classes "Required tow" and "Other" two of the *SST* ensembles outperform *MCT* ensembles and *TCMNN* .

The best classifier is the *SST(C4.5:TCMNN)* ensemble. Its rejection rate for the accuracy rate of 1.0 outperforms those of *MCT* ensembles and *TCMNN* with values 0.07 and 0.08, respectively (see Table 1). Although *SST(C4.5:TCMNN)* is outperformed by *MCT(C4.5:TCMNN)* on the majority class "Fixed", it provides the lowest rejection rates for the minority classes "Required tow" and "Other".

6   Conclusion and Future Research

In this paper we considered the problem of constructing machine-learning classifiers for road side assistance that are capable of providing classification-reliability values. We analyzed the conformity framework and the meta-classifier conformity approach in the context of this problem. As a result we proposed the single-stacking conformity approach that allows the conformity framework to be applied to any classifier so that the classification-reliability values can be computed for each class.

The experiments show that the single-stacking conformity approach allows constructing ensembles that are capable of outperforming a standard conformity-based classifier (*TCMNN*) and meta-classifier conformity ensembles for road side assistance. We explain this result by correcting mechanism that the single-stacking conformity approach employs. Another important result that follows from the experiments is that standard classifiers such as decision trees, nearest neighbor classifiers, and naive Bayes classifiers are not capable of providing good classification-reliability values. Thus, we may conclude that the conformity framework and its accompanying meta-conformity approaches are useful when classification-reliability values have to be plausibly estimated for practical problems like road side assistance.

Future research will focus on analysis of the single-stacking conformity approach, especially on conditions when the approach can be successfully applied. The analysis can be used for extending the approach to a multi-stacking mechanism for assigning classification-reliability values.

References

1. Baskiotis, N., Sebag, M.: C4.5 competence map: a phase transition-inspired approach. In: Proceedings of the 21th International Conference on Machine Learning (ICML 2004), pp. 73–80 (2004)
2. Bay, S., Pazzani, M.: Characterizing model errors and differences. In: Proceedings of the 17th International Conference on Machine Learning (ICML 2000), pp. 196–201 (2000)
3. Bosnic, Z., Kononenko, I.: An overview of advances in reliability estimation of individual predictions in machine learning. Intell. Data Anal. 13(2), 385–401 (2009)
4. Domingos, P., Pazzani, M.: Beyond independence: conditions for the optimality of the simple bayesian classifier. In: Proceedings of the Thirteenth International Conference on Machine Learning (ICML 1996), pp. 105–112. Morgan Kaufmann, San Francisco (1996)
5. Duda, R., Hart, P., Stork, D.: Pattern Classification, 2nd edn. Wiley, Chichester (2000)
6. Fawcett, T.: Roc graphs: Notes and practical considerations for researchers. Tech. Rep. HPL-2003-4, HP Laboratories (2003),
   http://citeseer.ist.psu.edu/fawcett04roc.html
7. Ferri, C., Hernndez-Orallo, J.: Cautious classifiers. In: Proceedings of the 1st International Workshop on ROC Analysis in Artificial Intelligence (ROCAI 2004), pp. 27–36 (2004)
8. Friedel, C., Ruckert, U., Kramer, S.: Cost curves for abstaining classifiers. In: Lachiche, N., Ferri, C., Macskassy, S. (eds.) Proceedings of the ICML 2006 Workshop on ROC Analysis (ROCML 2006), Pittsburgh, USA (June 29, 2006)
9. Hullermeier, E.: Instance-based prediction with guaranteed confidence. In: Proceedings of the 16th European Conference on Artificial Intelligence (ECAI 2004), pp. 97–101 (2004)
10. Kaptein, A.: Meta-Classifier Approach to Reliable Text Classification. Master's thesis, Maastricht University, The Netherlands (2005)
11. Kukar, M.: Quality assessment of individual classifications in machine learning and data mining. Knowl. Inf. Syst. 9(3), 364–384 (2006)
12. Perner, P.: Proceedings of 10th Industrial Conference on Data Mining (ICDM 2010), Berlin, Germany. July 12-14, IBaI Report (2010)
13. Proedru, K., Nouretdinov, I., Vovk, V., Gammerman, A.: Transductive confidence machines for pattern recognition. In: Elomaa, T., Mannila, H., Toivonen, H. (eds.) ECML 2002. LNCS (LNAI), vol. 2430, pp. 381–390. Springer, Heidelberg (2002)
14. Quinlan, J.R.: C4.5: programs for machine learning. Morgan Kaufmann, San Mateo (1993)
15. Saunders, C., Gammerman, A., Vovk, V.: Transduction with confidence and credibility. In: Proceedings of the 16th International Joint Conference on Artificial Intelligence (IJCAI 1999), pp. 722–726 (1999)

16. Shafer, G., Vovk, V.: A tutorial on conformal prediction. Journal of Machine Learning Research 9, 371–421 (2008)
17. Smirnov, E., Nalbantov, G., Nikolaev, N.: k-Version-space multi-class classification based on k-consistency tests. In: Proceedings of the European Conference on Machine Learning and Knowledge Discovery in Databases, ECML PKDD 2010 (2010) (accepted)
18. Smirnov, E., Kaptein, A.: Theoretical and experimental study of a meta-typicalness approach for reliable classification. In: Proceedings of the IEEE International Workshop on Reliability Issues in Knowledge Discovery (RIKD 2006), pp. 739–743. IEEE Computer Society, Los Alamitos (2006)
19. Smirnov, E., Nalbantov, G., Kaptein, A.: Meta-conformity approach to reliable classification. Intell. Data Anal. 13(6), 901–915 (2009)
20. Smirnov, E., Sprinkhuizen-Kuyper, I., Nalbantov, G., Vanderlooy, S.: Version space support vector machines. In: Proceedings of the 17th European Conference on Artificial Intelligence (ECAI 2006), pp. 809–810 (2006)
21. Smirnov, E., Sprinkhuizen-Kuyper, I., Nikolaev, N.: Generalizing version space support vector machines for non-separable data. In: Proceedings of the IEEE International Workshop on Reliability Issues in Knowledge Discovery (RIKD 2006), pp. 744–748. IEEE Computer Society, Los Alamitos (2006)
22. Smirnov, E., Vanderlooy, S., Sprinkhuizen-Kuyper, I.: Meta-typicalness approach to reliable classification. In: Proceedings of the 17th European Conference on Artificial Intelligence (ECAI 2006), pp. 811–812 (2006)
23. Tzikas, D., Kukar, M., Likas, A.: Transductive reliability estimation for kernel based classifiers. In: R. Berthold, M., Shawe-Taylor, J., Lavrač, N. (eds.) IDA 2007. LNCS, vol. 4723, pp. 37–47. Springer, Heidelberg (2007)
24. Vanderlooy, S., Sprinkhuizen-Kuyper, I., Smirnov, E.: An analysis of reliable classifiers through ROC isometrics. In: Lachiche, N., Ferri, C., Macskassy, S. (eds.) Proceedings of the ICML 2006 Workshop on ROC Analysis (ROCML 2006), Pittsburgh, USA, June 29, pp. 55–62 (2006)
25. Vanderlooy, S., Sprinkhuizen-Kuyper, I., Smirnov, E.: Reliable classifiers in ROC space. In: Saeys, Y., Tsiporkova, E., Baets, B.D., de Peer, Y.V. (eds.) Proceedings of the 15th Annual Machine Learning Conference of Belgium and the Netherlands (BENELEARN 2006), Ghent, Belgium, May 11-22, pp. 113–120 (2006)
26. Vapnik, V.: Statistical Learning Theory. John Wiley, NY (1998)
27. Vovk, V., Gammerman, A., Shafer, G.: Algorithmic Learning in a Random World. Springer, Heidelberg (2005)
28. Wolpert, D.: Stacked generalization. Neural Networks 5, 241–260 (1992)

A Relative Word-Frequency Based Method for Relevance Feedback

Zilong Chen[1] and Yang Lu[2]

[1] State Key Lab. of Software Development Environment, BeiHang University
HaiDian District, Beijing, P.R. China
chenzl@nlsde.buaa.edu.cn
[2] School of Software and Microelectronics, Peking University
HaiDian District, Beijing, P.R. China
sheep@pub.ss.pku.edu.cn

Abstract. Traditional relevance feedback methods, which usually use the most frequent terms in the relevant documents as expansion terms to enrich the user's initial query, could help improve retrieval performance. However, in reality, many expansion terms identified in traditional approaches are indeed unrelated to the query and even harmful to the retrieval. This paper introduces a new method based on the relative word-frequency to select good expansion terms. The relative word-frequency defined in this paper is a new feature and can help discriminate relevant documents from irrelevant ones. The new approach selects good expansion terms according to the relative word-frequency and uses them to reformulate the initial query. We compare a set of existing relevance feedback methods with our proposed approach, including the representative vector space models and language models. Our experiments on several TREC collections demonstrate that retrieval effectiveness can be much improved when the proposed approach is used. Experimental results show that the improvement of our proposed approach is more than 30% over traditional relevance feedback techniques.

Keywords: relevance feedback, expansion term, relative word-frequency, language model, vector space model.

1 Introduction

In modern Information Retrieval (IR), it is possible for the search to distinguish the relevant documents and the irrelevant ones for the query if user could identify all of the terms that describe the subject of interest. However, most users find it difficult to formulate queries which are well designed for retrieval purposes. In fact, a user's original query is usually too short to describe the information need accurately. Many important terms are missing from the query, leading to a poor coverage of the relevant documents.

Traditional relevance feedback technique, which adds additional terms extracted from the feedback documents, is an effective method to improve the retrieval

D. Dicheva and D. Dochev (Eds.): AIMSA 2010, LNAI 6304, pp. 171–180, 2010.

effectiveness. In general, the expansion terms are extracted according to the term distributions in the feedback documents.

However, the main disadvantage of most existing traditional methods is adding some irrelevant terms appearing in the relevant documents to the new query, when they use the terms in the relevant documents to enrich the user's initial query. A term with high frequency in relevant documents cannot be a feature of the relevant documents because it maybe appears in other documents frequently. Therefore some traditional expansion terms maybe have little impact on the retrieval effectiveness and even are harmful to the retrieval. Peat and Willett [2] found that the high frequency terms tended to be poor at discriminating between relevant and irrelevant documents. Cao et. al. also showed that many expansion terms identified in traditional approaches were indeed unrelated to the query and even harmful to the retrieval [3].

It is believed that the usefulness of the terms that appear in the feedback documents is different. The good expansion terms could improve the effectiveness; the bad expansion terms could hurt it and neutral expansion terms have little impact on the retrieval effectiveness. Therefore, it is ideal to use only good expansion terms to expand queries. There have been some works on this, Cao et. al. proposed a method to integrate a term classification process to predict the usefulness of expansion terms. Multiple additional features can be integrated in the process [3]. Tan even proposed a term-feedback technique based on the terms clustering to form some subtopic clusters that could be examined by users [4].

In this paper, we try to find most representative word for the relevant documents and irrelevant ones, and then a new relevance feedback method based on the relative word-frequency has been proposed. It is known that every word has a certain frequency in the natural language; the frequency of a word in one category is different from that in the other. For example, a word has a normal frequency in the entire documents; it will be the representative feature of the category whether it has a high or low frequency in this category. It will be useful to discriminate this category from other categories. Therefore, we consider that the relative word-frequency (defined in section 3) could be a useful feature to discriminate the relevant documents and irrelevant ones, and we can use it to select good expansion terms to enrich query.

The rest of the paper is organized as follows. In Section 2, we present a brief overview of related work. In Section 3, we describe the details of the relative word-frequency based relevance feedback algorithm. The evaluation results are shown in Section 4. At the end, Section 5 concludes with a summary.

2 Related Work

Relevance feedback has been widely used in IR. It has been implemented in different retrieval models, such as vector space model (VSM), probabilistic model, language model (LM), and so on. Since this work adopts both VSM and LM, the related studies in these models are specifically reviewed.

In VSM, the Rocchio method is a classical relevance feedback method [5] in VSM. The main idea is to add the terms appearing in the relevant documents to the initial query, and remove the terms appearing in the irrelevant documents.

Standard_Rocchio:

$$\overrightarrow{q_m} = \alpha\overrightarrow{q} + \frac{\beta}{|D_r|}\sum_{\forall \overrightarrow{d_j} \in D_r} \overrightarrow{d_j} - \frac{\gamma}{|D_n|}\sum_{\forall \overrightarrow{d_j} \notin D_n} \overrightarrow{d_j} \qquad (1)$$

Usually, the information contained in the relevant documents is more important than the information provided by the irrelevant documents [6]. This suggests making the constant γ smaller than the constant β. Experiments demonstrate that the performance is best when $\beta=0.75$ and $\gamma=0.25$.

In LM, the query model describes the user's information needs and is denoted by $P(\omega|\theta_o)$; this model is estimated with MLE without smoothing. In general, this query model has a poor coverage of the relevant terms, especially for short queries. Many terms related to the query's topic are missing (or has a zero probability) in this model. Relevance feedback is often used to improve the query model.

The relevance model [8] is the representative approach. It assumes that a query term is generated by a relevance model $P(\omega|\theta_R)$. However, it is impossible to define the relevance model without any relevance information. Therefore it exploits the top-ranked feedback documents by assuming them to be samples from the relevance model. The relevance model is estimated as follows:

$$P(\omega|\theta_R) \approx \sum_{D \in F} P(\omega|D)P(D|\theta_R) \qquad (2)$$

where F denotes the set of feedback documents. On the right side, the relevance model θ_R is approximated by the original query Q. By applying Bayesian rule and making some simplifications, we get the following formula:

$$P(\omega|\theta_R) \approx \sum_{D \in F} \frac{P(\omega|D)P(Q|D)P(D)}{P(Q)} = \sum_{D \in F} P(\omega|D)P(Q|D) \qquad (3)$$

That is, the probability of a term ω in the relevance model is determined by its probability in the feedback documents (i.e. $P(\omega|D)$) as well as the correspondence of the latter to the query (i.e. $P(Q|D)$). The above relevance model is used to enhance the original query model by the following interpolation:

$$P(\omega|\theta_q) = (1-\lambda)P(\omega|\theta_o) + \lambda P(\omega|\theta_R) \qquad (4)$$

where λ is the interpolation weight (set to be 0.5 in our experiments).

The relevant model assumes that the most frequent terms contained in the feedback documents (i.e. with a high $P(\omega|D)$) are related to the query and are useful to improve the retrieval effectiveness.

Several additional criteria have been used to select terms related to the query. For example, Amo [7] argued that treating the top N documents of the initial retrieval results as same will lose some useful information. Some rank index of the initial results will not be used. Robertson presented the concepts that the selected terms should have a higher probability in the relevant documents than in the irrelevant documents

[9]. Smeaton and Van Rijsbergen examined the impact of determining expansion terms using minimal spanning tree and some simple linguistic analysis [1].

3   Relative Word-Frequency Based Method

3.1   Definition

We use the following notations in the rest of the paper: C is the set of entire documents; D_r is the set of positive feedback documents; D_n is the set of negative feedback documents; $|C|$, $|D_r|$, $|D_n|$ are the document number of the set C, D_r, D_n; $c(\omega, d_j)$ is the count of word ω in document d_j; $c(\omega, C)$ is the count of word ω in C; $n_r(\omega)$ is the count of documents which contain the word ω in D_r; $n_n(\omega)$ is the count of documents which contain the word ω in D_n; $|d_j|$ is the length of document d_j (the count of words contained in the document d_j); $avdl$ is the average document length.

Then, we define the relevant relative word-frequency $RF_R(\omega)$ and the irrelevant relative word-frequency $RF_{NR}(\omega)$ as following:

$$RF_R(\omega) = \frac{\dfrac{1}{|D_r|} \cdot \displaystyle\sum_{\forall d_j \in D_r} \dfrac{c(\omega, d_j)}{|d_j|}}{\dfrac{c(\omega, C)}{|C| \cdot avdl}} ; \quad RF_{NR}(\omega) = \frac{\dfrac{1}{|D_n|} \cdot \displaystyle\sum_{\forall d_j \in D_n} \dfrac{c(\omega, d_j)}{|d_j|}}{\dfrac{c(\omega, C)}{|C| \cdot avdl}} \quad (5)$$

Presented this way, the relative word-frequency $RF_R(\omega)$ (or $RF_{NR}(\omega)$) represents the ratio of the average frequency of the word in positive feedback documents D_r (or in negative feedback documents D_n) and the frequency of the word in the entire collection C.

Base on the relative word-frequency, the relevant word importance information $I_R(\omega)$ and the irrelevant word importance information $I_{NR}(\omega)$ could be defined for each term as following:

$$I_R(\omega) = | \lg RF_R(\omega) | \cdot \frac{n_r(\omega)}{|D_r|} ; \quad I_{NR}(\omega) = | \lg RF_{NR}(\omega) | \cdot \frac{n_n(\omega)}{|D_n|} \quad (6)$$

According to this definition, the term will be more important in the following situations: 1) it has a much different word-frequency in relevant documents (or irrelevant documents) and in all documents; 2) it appears in more feedback documents. In situation (1), if the word frequency of a term in relevant documents is either greater or less than that in all documents, it is thought as with a more different word frequency and will be selected to enrich the initial query.

3.2   General Strategies for Relative Word-Frequency Based Method

Usually, we can get both relevant documents and irrelevant ones from the set of feedback documents. Sometimes it is not natural to mix these two kinds of information as in the case of using generative models for feedback [11], extension has to be made to make them work for negative feedback [10]. The query modification strategy aims to mix both positive and negative information together in a single query model. A more flexible alternative strategy is to maintain a positive query representation and a negative query representation separately, and combine the scores of a document with both representations.

With this strategy, negative examples can be used to learn a negative query representation which can then be used to score a document based on the likelihood that the document is a distracting irrelevant document; such a score can then be used to adjust the positive relevance score between the original query and the corresponding document. Intuitively, a document with higher relevance score to the negative query representation can be assumed to be less relevant, thus the final score of this document can be computed as

$$S_{combined}(Q,D) = \alpha \times S(Q_{new},D) + \beta \times (1 - S(Q_{neg},D)) \qquad (7)$$

where Q_{new} is the new positive query representation based on the original query and the positive feedback, Q_{neg} is a negative query representation and α, β is a parameter to control the influence of positive feedback and negative feedback. When $\beta=0$, we do not perform negative feedback, and the ranking would be the similar as the original ranking according to new positive query Q_{new}. A larger value of β causes more penalization of documents similar to the negative query representation (we set $\alpha=0.75$, $\beta=0.25$, like Standard_Rocchio Equations).

The above equation shows that either a high score of $S(Q_{new},D)$ or a low score of $S(Q_{neg},D)$ would result in a high score of $S_{combined}(Q,D)$. According to the above definition, we can select the most representative terms to reformulate the new positive relevant query Q_{new} and the irrelevant query Q_{neg}. In next subsection, we will discuss some specific ways of implementing these general strategies in both VSM and LM.

3.3 Relative Word-Frequency Based Method in VSM

In VSM, our approach is different from the traditional Rocchio method. We only use good expansion terms which can represent relevant documents or irrelevant documents to modify the initial query.

As we know, query can be represent as a vector $q=(q_1, q_2, ..., q_n)$, n is the total number of terms in all documents. q_i is the weight of term k_i in the query. In our experiments, we use the follow expression to modify the query. We use $q(k_i)$ to denote the weight of term k_i in the original query. $d_j(k_i)$ is the weight of term k_i in the document d_j. The new modified relevant query q_{new} and irrelevant query q_{neg} can be calculated as follow,

$$q_{new}(k_i) = \begin{cases} q(k_i) & , \ I_R(k_i) < \delta \\ \alpha \times q(k_i) + \beta \times \lambda_j \times \sum_{\forall d_j \in D_r} d_j(k_i) & , \ I_R(k_i) \geq \delta \end{cases} \qquad (8)$$

$$q_{neg}(k_i) = \begin{cases} 1 - q(k_i) & , \ I_{NR}(k_i) < \delta \\ \dfrac{1}{|D_n|} \sum_{\forall d_j \in D_n} d_j(k_i) & , \ I_{NR}(k_i) \geq \delta \end{cases} \qquad (9)$$

It is known that in the initial retrieval result, documents are ranked by the similarity with query. The top ranked document is more relevant. However, this position information is absent in most existing feedback methods. In our approach, we calculate the linear interpolation parameter λ according to the document's position [12], the weight of document decreases when position index of the document increases and the sum weight of all the documents is 1. Therefore, the documents will be more important in the feedback when its position is in the front. We assume that j is the index of document in the positive feedback documents D_r; n is another representation of $|D_r|$. Then, λ_j, the weight of the relevant document d_j, is computed as

$$\lambda_j = \frac{1}{2n} + \frac{n+1-j}{n(n+1)} \qquad (10)$$

Then we calculate the relevant score between the q_{new}, q_{neg} and each document. Then the left documents will be re-ranked according to the new score $S_{combined}(Q,D)$. Experiments showed that our approach is more effective than the traditional Rocchio method. More details will be discussed in section 4.

3.4 Relative Word-Frequency Based Method in LM

In LM, we use the follow score function to calculate the similarity between documents and query:

$$Score(d,q) = \sum_{\omega \in V} P(\omega | \theta_q) \log P(\omega | \theta_d) \qquad (11)$$

where V is the vocabulary of the whole collection C, θ_q and θ_d are respectively the query model and the document model. The document model needs to be smoothed to solve the zero-probability problem. A commonly used smoothing method is Dirichlet smoothing:

$$P(\omega | \theta_d) = \frac{tf(\omega,d) + uP(\omega | C)}{|d| + u} \qquad (12)$$

where $|d|$ is the length of the document d, $tf(\omega, d)$ is the term frequency of ω within d, $P(\omega|C)$ is the probability of ω in the whole collection C estimated with MLE (Maximum Likelihood Estimation), and u is the Dirichlet prior (set at 1,500 in our experiments).

To use only good expansion terms to modify the initial query model, we re-define $P(\omega|\theta_q)$ and $P(\omega|\theta_R)$ compared with traditional relevance model.

$$Score(Q_{new},D) = \sum_{\omega \in V} P(\omega | \theta_{q_{new}}) \log P(\omega | \theta_d) \qquad (13)$$

$$Score(Q_{neg},D) = \sum_{\omega \in V} P(\omega | \theta_{q_{neg}}) \log P(\omega | \theta_d) \qquad (14)$$

$$P(\omega | \theta_{q_{new}}) = \begin{cases} P(\omega | \theta_0) & , \quad I_R(\omega) < \delta \\ \alpha P(\omega | \theta_0) + \beta P(\omega | \theta_R) & , \quad I_R(\omega) \geq \delta \end{cases} \qquad (15)$$

$$P(\omega \,|\, \theta_{q_{neg}}) = \begin{cases} 1 - P(\omega \,|\, \theta_O) & , \quad I_{NR}(\omega) < \delta \\ P(\omega \,|\, \theta_{NR}) & , \quad I_{NR}(\omega) \geq \delta \end{cases} \qquad (16)$$

Let n be the number of relevant documents in the D_r. We use a linear interpolation technique [12] to estimate the likelihood of our relevance model:

$$P(\omega \,|\, \theta_R) = \sum_{i=1}^{n} \lambda_i P_R(\omega \,|\, d_i); \quad \sum_{i=1}^{n} \lambda_i = 1; \quad \lambda_i = \frac{1}{2n} + \frac{n+1-i}{n(n+1)} \qquad (17)$$

$P_R(\omega|d_i)$ is estimated by the maximum likelihood of the term w occurring in the document d_i, and λ is the interpolation weight used to turn the importance of each relevant document for the relevance model (similar as the definition in section 3.3).

If we assume n is the number of irrelevant documents in the D_n. $P(\omega|\theta_{NR})$ can be defined similar as follow. $P_{NR}(\omega|d_i)$ is estimated by the maximum likelihood of the term w occurring in the document d_i.

$$P(\omega \,|\, \theta_{NR}) = \frac{1}{n} \sum_{i=1}^{n} P_{NR}(\omega \,|\, d_i) \qquad (18)$$

4   Experiments

To evaluate the effectiveness of our proposed approach described in the previous section, we have constructed our test collections based on two representative TREC data sets: ROBUST track and Web track data sets.

4.1   Data Sets

Our first data set is from the ROBUST track of TREC 2004 including all documents on TREC disks 4&5. It has around 556,000 news articles and 311,000 documents marked whether relevant with each query or not. In average, each document has 467 terms. We use all the 249 queries as our base query set. The second data set is the GOV data set used in the Web track of TREC 2003 and 2004. It was about 18 GB in size and contained 1, 247, 753 Web pages crawled from the ".gov" domain in 2002. In average, each document had 1,094 terms. In total, we had 125 queries in our base set (50 from Web track 2003 and 75 from Web track 2004).

For both data sets, data pre-processing was standardized: terms were stemmed using the Porter Stemming and stop words were removed by using standard stop word list. For each dataset, we split the available topics into two parts: the feedback data and the test data. We used the top N documents in the first retrieval results as the feedback data. Every document in the feedback data was examined for relevance by user, we used these feedback documents to modify the query model and re-rank the left documents (remove the feedback documents from all documents). We named these left documents as test data. We would change the parameter N and evaluate the effectiveness of different feedback methods on the test data. At last, we averaged the results in these two data sets and showed them. More detail experimental results will be described in next subsection.

4.2 Comparison of Different Relevance Feedback Algorithms

In the experiments, the following methods were compared on the same TREC data set:

VSM: the VSM with the original queries; **ROCCHIO**: the VSM with Rocchio method; **VSM+RWF**: the VSM with our relative word-frequency based method; **LM**: the KL-divergence retrieval model with the original queries; **REL**: the relevance model; **LM+RWF**: the KL-divergence retrieval model with our relative word-frequency based method;

To measure the performance of our algorithms, we used two standard retrieval measures: (1) Mean Average Precision (MAP), which was calculated as the average of the precision after each relevant document is retrieved, reflected the overall re-trieval accuracy. (2) Precision at 20 documents (P@20), which was not well average and only gives us the precision for the first 20 documents. It reflected the utility per-ceived by a user who might only read up to the top 20 documents on the first page.

The results for the comparison methods on TREC collections are presented in Figure 1-2 and Table 1.

From the table and figures, we observed that both relevance model and Rocchio method could improve the retrieval effectiveness of LM and VSM significantly. This observation is consistent with previous studies. The MAP we obtained with these two methods represented the effectiveness of most existing traditional feedback techniques.

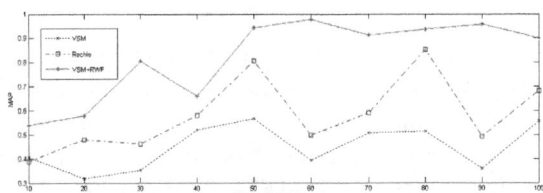

Fig. 1. MAP of VSM, Rocchio, VSM+RWF

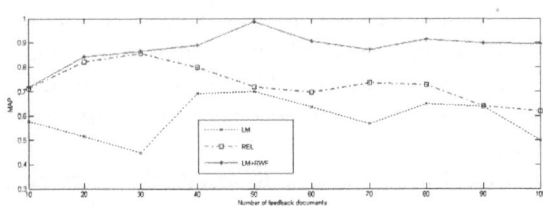

Fig. 2. MAP of LM, REL, LM+RWF

Table 1. P@20 of VSM, Rocchio, VSM+RW, LM, REL, LM+RWF

Feedback documents	P@20					
	VSM	Rocchio	VSM+RWF	LM	REL	LM+RWF
10	0.40	0.45	0.50	0.50	0.70	0.70
20	0.45	0.45	0.65	0.55	0.75	0.75
30	0.35	0.40	0.70	0.40	0.70	0.80
40	0.55	0.55	0.65	0.60	0.65	**0.85**
50	0.30	0.40	**0.70**	0.65	0.65	**0.95**
60	0.35	0.35	**0.80**	0.65	0.60	**0.90**
70	0.50	0.50	**0.85**	0.50	0.50	**0.90**
80	0.35	0.45	**0.75**	0.55	0.55	**0.95**
90	0.40	0.35	**0.85**	0.50	0.50	**0.90**
100	0.45	0.60	**0.85**	0.40	0.40	**0.90**

Comparing to the relevance model and the relative word-frequency based method, it showed that the latter performs better. The reason could be the relative word-frequency based method relies more on the difference between the feedback documents and the entire collection to select the expansion terms, than the relevance model. By doing this, it could filter out more bad or neutral expansion terms. By comparing the Rocchio method and the relative word-frequency based method, we found the result is very close to the above conclusion.

As expected, the relative word-frequency based method performed very well in our experiments. When this method was used together with information retrieval model (both VSM and LM), the effectiveness is always improved. The improvements were more than 30% and were statistically significant.

The relative word-frequency based method could help identify the useful expansion terms: although maybe not all the useful expansion terms were identified, those identified were highly related and useful. As the weight of these terms was increased, and the relative weight of the other terms was decreased, which made their weights in the final query model smaller. These experiments illustrated the reason that the relative word-frequency based method can improve the retrieval effectiveness.

5 Conclusion

Traditional relevance feedback, which adds additional terms extracted from the feedback documents, is an effective method to improve the query representation and the retrieval effectiveness. The basic assumption is that most strong terms in the feedback documents are useful for IR. In reality, the expansion terms determined in traditional ways are not always useful. Only a small proportion of the suggested expansion terms are useful, and many others are either useless or even harmful.

Motivated by these observations, we proposed to further classify expansion terms using relative word-frequency. The relative word-frequency defined in this paper is a new feature of the terms and can help discriminate relevant documents from irrelevant ones. Then we proposed a new feedback method based on this feature, and aim to select the representative expansion terms directly according to their possible impact on the retrieval effectiveness. This method is different from existing ones, which often rely on other criteria that do not always correlate with the retrieval effectiveness.

Our experiments, based on two TREC collections, showed that the selected expansion terms using our method were significantly better than the traditional expansion terms. This study shows the importance to examine the crucial problem of usefulness of expansion terms before the terms are used. We also provided a general framework to integrate our approaches with existing retrieval models.

Acknowledgements

We acknowledge support from State Key Lab. of Software Development Environment, BeiHang University's programs. These programs ID are 2005CB321901, 2005CB321903. We also appreciate valuable suggestions from anonymous reviewers.

References

[1] Smeaton, A.F., Van Rijsbergen, C.J.: The retrieval effects of query expansion on a feedback document retrieval system. Computer Journal 26(3), 239–246 (1983)
[2] Peat, H.J., Willett, P.: The limitations of term co-occurrence data for query expansion in document retrieval systems. Journal of the American Society for Information Science 42(5), 378–383 (1991)
[3] Cao, G.H., Nie, J.Y., Gao, J.F., Robertson, S.: Selecting Good Expansion Terms for Pseudo-Relevance Feedback. In: The Proceedings of SIGIR 2008, pp. 243–250 (2008)
[4] Bin, T., Velivellia, Hui, F., et al.: Term feedback for information retrieval with language models. In: The Proceedings of SIGIR 2007, pp. 263–270 (2007)
[5] Rocchio, J.J.: Relevance feedback in information retrieval. In: The SMART Retrieval System. Experiments in Automatic Document Processing, pp. 313–323 (1971)
[6] Salton, G., McGill, M.J.: Introduction to Modern Information Retrieval. McGraw-Hill, New York (1983)
[7] Amo, P., Ferreras, F.L., Cruz, F., et al.: Smoothing functions for automatic relevance feedback in information retrieval. In: Proc of the 11th International Workshop on Database and Expert Systems Applications, pp. 115–119 (2000)
[8] Lavrenko, V., Croft, B.: Relevance-based language models. In: The Proceedings of SIGIR 2001, pp. 120–128 (2001)
[9] Robertson, S.E.: On term selection for query expansion. Journal of Documentation 46(4), 359–364 (1990)
[10] Wang, X., Fang, H., Zhai, C.: Improve retrieval accuracy for difficult queries using negative feedback. In: CIKM, pp. 991–994 (2007)
[11] Zhai, C., Lafferty, J.: Model-based feedback in the language modeling approach to information retrieval. In: Proceedings of ACM CIKM 2001, pp. 403–410 (2001)
[12] Miller, D., Leek, T., Schwartz, R.: A hidden markov model information retrieval system. In: The Proceedings of SIGIR 1999, pp. 214–221 (1999)

Sequence Detection for Adaptive Feedback Generation in an Exploratory Environment for Mathematical Generalisation*

Sergio Gutierrez-Santos, Manolis Mavrikis, and George Magoulas

London Knowledge Lab
{sergut,m.mavrikis}@lkl.ac.uk, gmagoulas@dcs.bbk.ac.uk

Abstract. Detecting and automating repetitive patterns in users' actions has several applications. One of them, often overlooked, is supporting learning. This paper presents an approach for detecting repetitive actions in students who are interacting with an exploratory environment for mathematical generalisation. The approach is based on the use of two sliding windows to detect possible regularities, which are filtered at the last stage using task knowledge. The result of this process is used to generate adaptive feedback to students based on their own actions.

1 Introduction

Detecting repetitive actions in users' behaviour with interactive applications has several advantages. First, automating execution of regular tasks can increase users' productivity. This observation led to the creation of script languages for the automatic execution of repetitive tasks in commercial applications. Subsequently macros or programming by demonstration systems [1] emerged, and recent advantages in machine learning allowed the development of user interface learning agents (e.g., [2,3]) that attempt to take away the burden of explicit demonstration and by observing users' actions and identify repetitive patterns and provide suggestions to automate them (e.g., [2]). Also, identifying regularities in users' interactions can help system designers and developers understand better how a system is used, identify unnecessary actions, problems and design flaws [4].

A somehow overlooked area for the identification of regularities in actions is supporting learning in technology-enhanced learning scenarios. In particular, our interest in identifying patterns of actions stems from the requirement to support young learners while interacting with eXpresser, an Exploratory Learning Environment (ELE) for mathematical generalisation, in classroom conditions. This is a domain where repetition and commonality is the basis of hypothesis formation and subsequently generalisation [5].

ELEs are virtual spaces that provide learners with relatively open tasks and therefore have been shown to be useful tools for discovery learning [6,7]. However,

---

* This work is funded in part from the MiGen Project, grant TLRP RES-139-25-0381.
Thanks to the rest of the MiGen team for support and ideas.

D. Dicheva and D. Dochev (Eds.): AIMSA 2010, LNAI 6304, pp. 181–190, 2010.

as with other constructivist approaches (c.f. [8]), research in the learning sciences suggests that students require explicit support and guidance to learn from their experience [9]. For example, students often pursue unproductive learning paths, or just adopt a playful but unproductive interaction. Our work responds to the need to provide learners with the right information at the right time when interacting with the eXpresser.

In particular, the work presented in this paper focuses on automating the detection of latent repetitive actions while students are constructing models in an ELE. With the ability to detect repetitive actions, the system is able to provide personalised feedback that draws the attention of the learner to what they have been repeating (i.e. making the implicit structure explicit) and supporting them in their efforts to find the general rule.

2   Background

2.1   The eXpresser: An Exploratory Learning Environment

The eXpresser is a microworld (i.e. a special kind of ELE where students can create their own objects and construction to think with) that engages students in generating figural patterns. This is quite central to the British Mathematical curriculum, and is considered one of the major routes to algebra [10,5,11]. For a detailed description of the system, the pedagogical rationale behind its design and its potential in supporting student autonomy in their learning, and the difficulties that students face in mathematical generalisation, the reader is referred to [12]. In relation to the work presented in this paper, one of the objectives behind the design of the microworld is to support students to develop an orientation towards looking for the underlying structure of patterns by identifying their building blocks, a central 'habit of mind' in algebraic thinking [13].

As an example, consider one of the tasks students undertake in the system (see Figure 1). Through exploration of the environment, students are encouraged to 'see' the structure of the pattern in a multitude of different ways and make explicit the rule that pertains their construction using the metaphor of colouring the pattern. In order to do that, students can use building blocks of square tiles to make patterns.

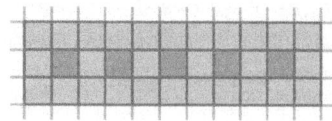

Fig. 1. The footpath task: How many green tiles do you need to surround n red tiles?

2.2   The Problem of Detecting Structure

Many learners in our target age (11–12 years old) find it challenging to express the general case of how many green tiles are needed to surround the red tiles (e.g. $n_g = 5 \cdot n_r + 3$). The crucial determinant of progress in this task is the ability to perceive a structure within the pattern and articulate it to themselves and subsequently to the system.

Although the system provides patterning and grouping facilities so that any structure can be expressed explicitly, many learners fail to be precise about the building block that they perceive in their patterns. They often revert to constructions that are specific to numerical cases. One pedagogic strategy used to support students in this situation capitalises on the 'rhythm' of their action [5], by drawing attention to repeated sequences of actions that highlight some recognition of structure by the student. More often than not, the perceived structure is implicit both to the system and to the learner in that they themselves often do not realise that they are working systematically. Findings from early studies with the system were in line with the literature in mathematics education [14,10,5], suggesting that prompting students to observe the rhythm in their actions has the potential to make the implicit structure evident, to enable them to identify the variants and invariants of a pattern and capture them in their constructions and expressions.

In the classroom, identifying and reinforcing the rhythm of actions is usually part of the teacher's role. However, the overall aim of our project, which is to facilitate integration of the ELE in classroom, introduced the technical challenge of delegating some of the responsibilities of the teacher to an intelligent system. One of the requirements therefore is to identify students' latent structure and provide personalised feedback that draws the learners' attention to what they have been repeating. In other words, by making the implicit structure explicit, the system can support students in their efforts to find the general rule.

3   Detection of Repetitive Actions

This section describes the algorithm designed to detect "rhythm" in the actions of the students, and how it is used to provide feedback to them.

3.1   General Description

Our approach consists of three stages (see Figure 2). First, sequences of events (i.e. in the case of eXpresser, tiles' positions) are analysed using two sliding windows. The distances between the windows are calculated, using a string metric that is appropriate for the problem according to a pedagogical analysis done in advance. These distances are stored in a table. Those windows that exhibit a higher similarity to other subsequent windows are selected as possible student rhythms, i.e. inherent structures in their minds. A final step makes an additional fine-filtering according to specific knowledge about the task to discard patterns that would not be adequate from a pedagogical point of view.

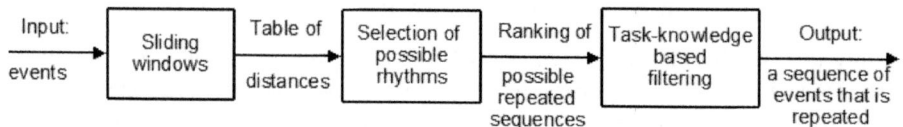

Fig. 2. Detection of repeated structures in users' actions in three stages

In the first stage the two sliding windows contain sub-sequences from the total sequence of events. The windows are moved to cover all the events and never overlap as it does not make sense in this case to look for similarities between a sequence of events and a particular copy of itself. For each pair of positions, the distance between both windows is calculated (a specific example is shown later in Table 1). Distance calculation is application-dependent, being determined by the different kinds of "noise" or "errors" that are commonly observed in the sequences of events for a particular application: e.g. when rhythm sequences are expected to differ because of insertion errors, a Levehnstein metric could be used. A survey that studies several distance functions between strings can be found in [15].

The movement of both sliding windows is described using pseudo-code in Figure 3. It should be noted that the window size is not fixed, but varies between two thresholds $T1$ and $T2$. These thresholds are also application-dependent, and must be set beforehand according to the characteristics of the task. It is important to strike a balance between low thresholds, which may result in too many spurious results, and high thresholds, which may result in no findings at all. One of the two windows (*main_window*) is moved from the beginning of the sequence of events to the end. The other (*auxiliary_window*) is moved from the last position of the main window to the end of the stream. The windows never overlap because it does not make sense in our case to look for similarities between a sequence of events and a partial copy of itself.

A simple example illustrates these concepts in Figure 4, where a sequence of events (represented as letters) is shown. The four pairs of lines below represent examples of sliding windows, the main one on the left. The string distance

```
FOR_EACH window size (between two thresholds T1 and T2)
   FOR_EACH position (from beginning to end of stream)
      get main_window
      FOR_EACH window size (between two thresholds)
         FOR_EACH position (from the end of the main window until end)
            get auxiliary_window
            calculate and store distance between both windows
         END_FOR
      END_FOR
   END_FOR
END_FOR
```

Fig. 3. Sliding windows

Fig. 4. Sliding windows. The distance between both windows depends on the distance used: e.g. if a Hamming distance is used, the distances are 0, N/A, 2, 5; if a Levehnstein distance is used, the distances are 0, 1, 1, 4.

between each pair of windows is calculated using both a Hamming and Levehnstein distance (cf. [15]).

3.2   Detection of Rhythm in eXpresser

This section describes the application of this process in the context of eXpresser and the footpath task. In this case, the events are generated by tiles placed on the canvas of eXpresser, and a rhythm –when detected– indicates some hidden structure that can be further used to provide adaptive feedback.

Preparation. The first step is choosing the two thresholds on which the sliding window algorithm operates. This usually involves some level of knowledge elicitation from the experts in the domain. In our case, this elicitation process produced a series of possible "rhythmical structures" or possible "building blocks" that can be used to construct the footpath pattern. These are depicted in Figure 5. The shortest ones have 3 tiles, while the longest ones have 5 tiles. Therefore, we chose $T1 = 3$ and $T2 = 6$, to allow insertion errors in the longer rhythms.

The next design decision is choosing which distance will be measured between windows. In our case, after observing several recorded sessions with students and discussing the most common typical errors on the part of the students with the pedagogical team, we came to the conclusion that most would be in the form of insertions/deletions, e.g. putting some tiles out of place in the sequence of positions. Therefore, we decided to use the Levehnstein distance in the first phase. Allowing the possibility of a certain Levehnstein distance between the windows permits the detection of rhythm in the presence of noise: e.g. in our case, tiles out of place.

In the next section, the process is executed step by step for a particular scenario.

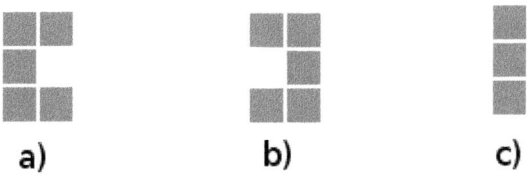

Fig. 5. Typical rhythmical structures in the footpath task: the "C", the "inverted C", and the "column"

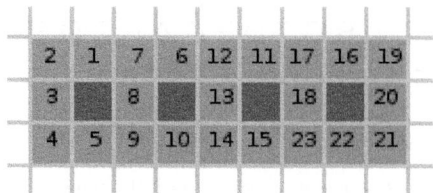

Fig. 6. Tiles position on a footpath-like task. Numbers show the order in which individual tiles have been placed on the canvas.

Rhythm of a student. In this section we illustrate all the steps in the process with an example. We start from Figure 6, that shows the order in which the tiles were placed in eXpresser. As it has been explained in Section 2, the goal of this task in eXpresser is to find how many green tiles are there for any given number of red tiles (i.e. $5 \cdot reds + 3$), and students are encouraged to construct the patterns themselves to see the relationships between both types of tiles. However, some students just put tiles on the eXpresser canvas "drawing" the pattern with colour tiles.

In this example, we see that the student has dropped tiles with high regularity until the end, where the characteristic "C" shape is abandoned to put first the tiles at the end. This and other similar behaviours have been observed by our team in pilot studies with the eXpresser as well as sessions in the context of a mathematical lesson in classrooms [12].

The output of the sliding window phase is stored in a table. Some sample rows of this table are shown on Table 1 to illustrate the result of this phase. In the table, p denotes the position of a window and s denotes its size.

The full table is analysed to find those windows that have a higher number of repetitions with lower distances. In this case, there are four windows that are repeated (i.e. reproduced with distance zero) four times. They are shown in Figure 7.

Table 1. Some distances for the example in Figure 6 (p denotes the position of a window and s denotes its size). Notes: † This corresponds to Figure 7c. ‡ This 3-tile-rhythm is discarded in favour of the longer rhythms (of size 5), in which it is embedded. ◊ This corresponds to a copy of the "C" (Figure 7a), and is ignored in favour of the first instance (window of size 5 on position 1).

p_{main}	s_{main}	p_{aux}	s_{aux}	$distance$	comments
1	5	6	5	0	
3	5	8	5	0	†
2	3	19	3	0	‡
1	5	7	5	2	
6	5	11	5	0	◊
1	5	16	5	4	

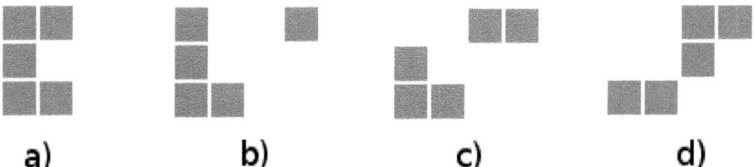

Fig. 7. Possible rhythms detected in the tiles position shown in Figure 6. All of them are repeated four times, i.e. there are four entries for each of them that show a distance of zero. The windows from which (a) and (c) are derived have been highlighted in Table 1.

It should be noted that the result of the former two stages might have been only option a) in Figure 7, if only the student's rhythm had been regular until the last tiles. In that case, the "C" structure would have been repeated five times and would have been the only output at this stage. However, this kind of small irregularities are very common among the students. This introduces the need for additional filtering using task knowledge. Although the four options are equivalent from a mathematical point of view (i.e. all of them have five green tiles), the expertise distilled from the pedagogical team in the project showed that for our target group building blocks with connected tiles facilitate children's reasoning process and realisation of structure. For this reason, we employ a filter that accepts only those rhythms in which all tiles are connected. This results in the rejection of options 7b, 7c, and 7d, and choosing 7a as the only possible rhythm in the actions of the student. The result is the same as a human teacher would have chosen.

Once the rhythm has been detected, feedback is provided to the student suggesting the use of such a building block to create the figure in a more structured way: repeating the building block to create a pattern. An example of such a form of feedback is shown in Figure 8. This feedback can be scaffolded in different ways in the architecture of the system (the interested reader is referred to [16]).

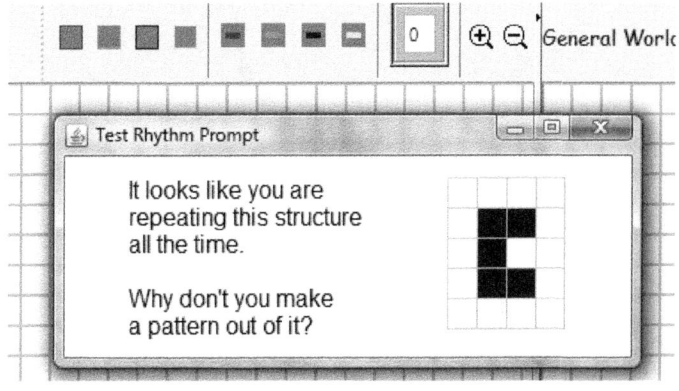

Fig. 8. Example of "rhythm" feedback

4   Evaluation

Detection accuracy. We have tested our algorithm with two different tasks in eXpresser, choosing from a a set of tasks that lend themselves to the appearance of "rhythm" in the action sequences of the student. Both tasks have been taken from the British National Curriculum, and are tasks that are commonly used in classrooms for the learning of mathematical generalisation. We have generated a series of scenarios for each task using five simulated students [17]. Each scenario corresponds to a student laying tiles on eXpresser, one by one. In five of these scenarios, simulated students show some rhythm (from those identified by the pedagogical team), but with some errors (e.g. spurious tiles among the ones that form the rhythm sequence); in the other cases, the laying of tiles did not follow these rhythmical patterns.

The algorithm was set up according to the characteristics of each task, i.e. the thresholds according to a pedagogical analysis as explained in Section 3.2, and the appropriate task filters are used. The algorithm was able to detect the rhythm in all test scenarios that included rhythm sequences. In the cases in which there was no rhythm in the actions of the artificial student, no rhythm was detected.

Computational cost and responsiveness. The highest computational cost of this approach comes from the first phase, where the algorithm's complexity is quadratic. In our application in eXpresser this is not an important handicap, since most of the students' interactions rarely involve more than a hundred tiles. Therefore, the response time of the algorithm is adequate for providing timely feedback when needed. As an extreme example, Figure 9 shows the response time for up to 300 single tiles generated by simulated students.

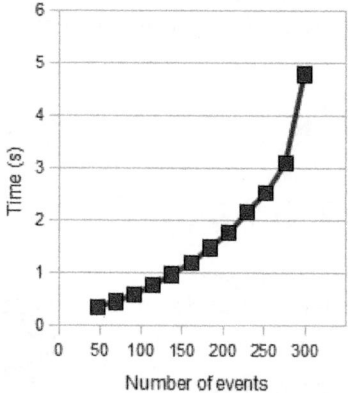

Fig. 9. Time used by our implementation to detect rhythm as a function of the number of tiles. For the scope of tasks in eXpresser (i.e. less than 100 tiles) the response time of the feedback is less than a second.

5   Related Work

There is a variety of sequential pattern mining algorithms for identifying repetitive patterns in sequential data (see [18]). Previous research suggested that an efficient approach to address our problem would be to mine for sub-sequences of sequential patterns over data stream of sliding windows (e.g. [19,20,21]).

Such approaches have been traditionally applied in online situations. However, challenges remain and they tend to cluster around the need to tolerate noise in the sequence [22] and to produce results with small amounts of data. The approach we presented here attempts to address these issues by combining two sliding windows with additional task knowledge filtering at the end, thus avoiding taking into account spurious effects.

6   Conclusions and Future Work

This paper has described a novel application of a method for detecting repetitive patterns in user actions on interactive systems. Detection of regularities in exploratory learning environments can be used to help students reflect on their actions. In our case, we use this to detect "rhythm" in the actions of students as they construct patterns in eXpresser tile by tile.

When detected, this personal "rhythm" is used to provide adaptive feedback to students that demonstrates to them the inherent structure in their actions. This feedback is crucial in helping them to articulate their implicit structure and derive a rule for the pattern they are constructing. Although any kind of related feedback would be useful, being able to detect the students' own inherent structure gives them a sense of ownership and contributes to helping them relate and internalise the feedback.

After the technical evaluation of the approach, we are in the process of validating the pedagogical relevance and appropriate timing of the feedback generated by this rhythm–detection component of our system. We are interacting with the pedagogical team comparing the feedback generated by the system and the interventions of the teacher related to rhythm in the actions of the students (i.e. gold–standard validation).

Currently, our approach relies on the task–specific window sizes. However, as more learners are interacting with the environment and data are collected from realistic interactions, our future efforts will concentrate on the design on an ensemble classifier for the selection of rhythms on the second stage. This would enable the automatic selection of window size.

References

1. Cypher, A., Halbert, D.C., Kurlander, D., Lieberman, H., Maulsby, D., Myers, B.A., Turransky, A. (eds.): Watch what I do: programming by demonstration. MIT Press, Cambridge (1993)
2. Caglayan, A., Snorrason, M., Jacoby, J., Mazzu, J., Jones, R., Kumar, K.: Learn sesame - a learning agent engine. Applied Artificial Intelligence 11, 393–412 (1997)

3. Ruvini, J., Dony, C.: Learning Users' Habits to Automate Repetitive Tasks, pp. 271–295. Morgan Kaufmann, San Francisco (2001)
4. Ramly, M.E., Stroulia, E., Sorenson, P.: Mining system-user interaction traces for use case models. In: IWPC 2002: Proceedings of the 10th International Workshop on Program Comprehension, p. 21. IEEE Computer Society, Los Alamitos (2002)
5. Radford, L., Bardini, C., Sabena, C.: Rhythm and the grasping of the general. In: Novotná, J., Moraová, H., Krátká, M., Stehlíková, N. (eds.) Proceedings of the 30th Conference of the International Group for the Psychology of Mathematics Education, vol. 4, pp. 393–400 (2006)
6. Noss, R., Hoyles, C.: Windows on Mathematical Meanings: Learning Cultures and Computers. Kluwer Academic Publishers, Dordrecht (June 1996)
7. van Jong, D.T., Joolingen, W.R.: Discovery learning with computer simulations of conceptual domains. Review of Educational Research 68, 179–201 (1998)
8. Mayer, R.E.: Should there be a three-strikes rule against pure discovery learning? - the case for guided methods of instruction. American Psychologist, 14–19 (2004)
9. Kirschner, P., Sweller, J., Clark, R.E.: Why minimal guidance during instruction does not work: An analysis of the failure of constructivist, discovery, problem-based, experiential and inquiry-based teaching. Educational Psychologist 41(2), 75–86 (2006)
10. Mason, J., Graham, A., Wilder, S.J.: Developing Thinking in Algebra. Paul Chapman Publishing, Boca Raton (2005)
11. Orton, A.: Pattern in the teaching and learning of mathematics. Cassell (1999)
12. Noss, R., Hoyles, C., Mavrikis, M., Geraniou, E., Gutierrez-Santos, S., Pearce, D.: Broadening the sense of 'dynamic': a microworld to support students' mathematical generalisation. Zentralblatt fur Didaktik der Mathematik 41, 493–503 (2009)
13. Cuoco, A.E., Goldenberg, E.P., Mark, J.: Habits of mind: An organizing principle for mathematics curriculum. Mathematical Behavior 15(4), 375–402 (1996)
14. Noss, R., Healy, L., Hoyles, C.: The construction of mathematical meanings: Connecting the visual with the symbolic. Educational Studies in Mathematics 33(2), 203–233 (1997)
15. Navarro, G.: A guided tour to approximate string matching. ACM Computing Survey 33(1), 31–88 (2001)
16. Gutierrez-Santos, S., Mavrikis, M., Magoulas, G.: Layered development and evaluation for intelligent support in exploratory environments: the case of microworlds. In: Intelligent Tutoring Systems, ITS 2010 (to appear 2010)
17. Vanlehn, K., Ohlsson, S., Nason, R.: Applications of simulated students: an exploration. J. Artif. Intell. Educ. 5(2), 135–175 (1994)
18. Han, J., Cheng, H., Xin, D., Yan, X.: Frequent pattern mining: current status and future directions. Data Mining & Knowledge Discovery 15(1), 55–86 (2007)
19. Chang, L., Wang, T., Yang, D., Luan, H.: Seqstream: Mining closed sequential patterns over stream sliding windows. In: IEEE International Conference on Data Mining, pp. 83–92 (2008)
20. Zhou, A., Cao, F., Qian, W., Jin, C.: Tracking clusters in evolving data streams over sliding windows. Knowledge and Information Systems 15, 181–214 (2008)
21. Gama, J., Rodrigues, P.P., Sebastian, R.: Evaluating algorithms that learn from data streams. In: SAC 2009: Proc. of the 2009 ACM Symposium on Applied Computing, pp. 1496–1500. ACM, New York (2009)
22. Wang, W., Yang, J., Yu, P.S.: Mining patterns in long sequential data with noise. SIGKDD Explor. Newsl. 2(2), 28–33 (2000)

A Deep Insight in Chat Analysis: Collaboration, Evolution and Evaluation, Summarization and Search

Mihai Dascalu[1,2], Traian Rebedea[1], and Stefan Trausan-Matu[1,3]

[1] University "Politehnica" of Bucharest,
313, Splaiul Indepentei, 060042 Bucharest, Romania
[2] S.C. CCT S.R.L.,
30, Gh Bratianu, 011413 Bucharest, Romania
[3] Romanian Academy Research Institute for Artificial Intelligence,
13, Calea 13 Septembrie, Bucharest, Romania
{mihai.dascalu,traian.rebedea,stefan.trausan}@cs.pub.ro

Abstract. Chat-like conversations are very popular and are often used by members of communities of practice to share ideas, propose solutions and to solve problems. With the advent and increasing popularity of Computer Supported Collaborative Learning (CSCL), chat conversations have also started to be used in a formal educational setting. However, the main problem when tutors tried to assess their students' chat conversations is that this process is very difficult and time consuming. Therefore, several efforts have been undertaken during the last years to develop a tool that supports the analysis of collaborative chat conversations used in an educational context. In this paper, we propose a solution based on Natural Language Processing (NLP), Social Network Analysis (SNA) and Latent Semantic Analysis (LSA) that can be used for several tasks: assessment of utterances and participants, important topics visualization, summarization and semantic search. Furthermore, a summary with the key findings from the first validation round of using this approach is presented.

Keywords: collaboration, chat assessment, extractive summarization, semantic search, Bakhtin's implications, Tagged LSA.

1 Introduction

Chat is probably the most practical and simple to use web communication technology at this point and thus justifies its popularity among users. Furthermore, during the last couple of years, various chat-like systems have appeared and have become very appealing: Twitter and Facebook status updates are just the most renowned examples. It seems that the dialog (which implies real-time, synchronous inter-change of utterances) using short textual messages fits naturally the needs of a large number of users. Although, most times chat-like technologies are used only for socialization and similar activities, they have also been adopted in education. In informal learning, discussions between the members of communities of practice often take place using chats. Moreover, CSCL advocates the use of chat as a supplement for standard teaching and learning strategies [1]. This way, chat has been introduced in formal education as well

D. Dicheva and D. Dochev (Eds.): AIMSA 2010, LNAI 6304, pp. 191–200, 2010.

and is used by students to solve problems and debate difficult topics in order to develop their knowledge about the domain and learn from their peers. However, this situation raised new problems for the tutors that need to assess and provide feedback to the students that participate in a chat conversation related to a course. It is considerably more difficult to assess a collaborative chat than a normal text written by an individual student.

Therefore it is important to build a tool that supports and provides preliminary feedback to both learners and tutors that use chat conversations. Recently, there have been several approaches for the analysis of chat-like conversations of learners and members of communities of practice, employing powerful classification mechanisms of the utterances [2], NLP processing focused on LSA [3, 4] and combining LSA with SNA and other techniques and heuristics [4, 5]. However, very few of these approaches have been used and validated in an educational context.

In this paper, we present an analysis technique for chats that relies primarily on LSA, but also employs SNA and NLP. This approach is used by the PolyCAFe chat and forum analysis environment that is currently under development. In the next sections we present the main functionalities of the analysis tools. Thus, section 2 describes the LSA training procedure for chats together with the important keywords visualization service. Section 3 explains how LSA and SNA can be used for assigning scores to each utterance. In the next section we provide a novel method for measuring collaboration within a chat, while sections 5 and 6 describe two other applications: chat summarization and semantic search. The paper ends with the key results of the first validation round and conclusions.

2   Tagged LSA and Semantic Space Visualization

Latent Semantic Analysis starts from the *vector-space model,* later on used also for evaluating similarity between terms and documents, now indirectly linked through concepts after projection is performed ([6], [8]).

The usage of LSA in our system is based on a *term-document matrix* build upon a corpus of chats with terms extracted from utterances after applying stop words elimination and spellchecking. Term Frequency – Inverse Document Frequency is applied and the final transformations are singular value decomposition and projection after the optimal considered empiric value for the number of dimensions k of 300, a value agreed by multiple sources [7].

Two important aspects must be addressed: *tagging* and *segmentation.* POS tagging is applied on all remaining words and a specific improvement is the reduction of existing verb forms by applying stemming only on verbs. According to [7] and [9], stemming applied on all words reduces overall performance because each form expresses and is related to different concepts. Therefore the terms included in the supervised learning process are made up of original words/stems plus their corresponding part of speech.

Segmentation divides chats into meaningful units by considering cohesion and unity between utterances. In the current implementation we use the utterances of the same participants because of their inner consistency and fixed non-overlapping windows for defining the maximum document length. A future improvement will include segmentation based on thread identification.

A small modification in the determination of the corresponding vector of each utterance is the use of the logarithmic function for smoothing the impact of a concept in a single utterance.

One immediate application of LSA is for *topic identification* which relies on the cosine similarity of each word with the whole document. The maximum value reflects the best correlation between the concept and all terms used in the discussion.

Another important aspect is the visualization of the resulted vector space. Therefore a network that is very similar to a social graph is generated by using the terms as nodes and the similarity between terms as the strength of the edge between vertices. This network is obtained by applying Breath First Search from the starting term (manually introduced by the user into a form with auto-complete options) and by inducing two heuristics maximum *k*-nearest neighbors selection and a minimal threshold for inter-word similarity for massively reducing weak links between terms. The actual visualization is performed using the Prefuse framework [10] offering 2 distinct models:

- a *physical driven model* similar to the planetary one where words have their own mass – *1+importance(word) / max(all importance)* and the importance is the sum of all similarities of the current word with all words represented in the vector space;
- a *radial model* for a central perspective – the resulted graph is focused on the central word and its neighbors; other included facilities are redesigning capabilities and searching for a word in the network.

3   The Grading Process of Chat Utterances

The first step in the grading process involves building the utterance graph in which two types of utterances can be identified: explicit ones, added manually by participants during their conversations by using a facility from the conversation environment – in our case, Concert Chat; the second is based on implicit links automatically identified by means of co-references, repetitions, lexical chains and inter-animation patters [5]. Each utterance is a node and the weights of edges are given by the similarity between the utterances multiplied by the trust assigned to each link. By default, each explicit link has a trust equal to 1 and each implicit link has an inner trust related to the factor used for identification scaled from 0 to 2 multiplied by an attenuation factor – in our current implementation 0.5 for allowing further fine-tuning of the grading process. The orientation of each edge follows the timeline of the chat and the evolution of the current discussion in time. The link is actually a transpose of the initial referencing from current utterance to the previous one to which it is logically linked explicitly or implicitly.

The actual grading of each utterance has *3 distinct components*: a *quantitative*, a *qualitative* and a *social* one. The *quantitative perspective* evaluates on the surface level the actual utterance. The assigned score, only from the quantitative view, considers the length in characters of each remaining word after stop words elimination, spellchecking and stemming are applied. For reducing the impact of unnecessary repetitions used only for artificially enhancing the grade, we use the logarithm of the number of actual occurrences of each word.

A more interesting dimension is the *qualitative* one which involves similarity computation between the current utterance and:

- the vector assigned to the entire chat for determining the correlation with the overall discussion;
- the vector of a specific set of keywords specified by the tutor or teacher as important topics for the discussions (these had to be addressed in each evaluated chat, but if free topics would be chosen, this factor would be 1);
- each later utterance linked with the current one for assessing the strength of the edge between them.

The social perspective implies an evaluation from the perspective of social networks analysis performed on the utterance graph. Here, the two previous perspectives are also present: the quantitative one reflected in the number of links and the qualitative one assessed by using LSA similarity. In the current implementation only two measures from graph theory are used: in-degree and out-degree, but other metrics specific to SNA (for example, betweenness) and minimal cuts will be considered. The centrality degree from social networks analysis is not very relevant because all the links follow the flow of the conversation and therefore they are all oriented in the same direction.

By combining all previous dimensions, the formula used for marking each utterance is:

$$\text{mark}(u) = \left(\sum_{remaining\ words} length(stem) \times (1 + \log(no\_occurences)) \right) \times$$

$$\times emphasis(u) \times social(u)$$

$$emphasis(u) = Sim(u, whole\_document) \times \qquad (1)$$

$$\times Sim(u, predefined\_keywords)$$

$$social(u) = \prod_{\substack{all\ social\ factors\ f \\ (quantitative\ and\ qualitative)}} (1 + \log(f(u)))$$

4   Collaboration Assessment in a Chat Conversation

Collaboration in a chat environment is assessed based on the following measures: social cohesion and collaboration, quantitative, mark based and gain based collaborations which will be thoroughly described in the next sub-sections.

4.1   Social Cohesion and Collaboration

Starting from the analysis of the social network performed both at the surface level and at the semantic one with participants as nodes of a graph, the following metric is derived for assessing equitability and cohesion for all participants. An generally accepted remark regarding a collaborative environment is that the more the participants have equal involvements (analyzed at surface level using only the number of interchanged utterances) and knowledge (relative to the topics taken into consideration and evaluated upon the marking process with LSA as support for semantic analysis) the more they collaborate being on the same level of interest and common/interchanged knowledge.

In order to have a normalized measure for the spread of each factor taken into consideration for social network analysis, the coefficient of variation is computed for each metric. The overall result represents 100% minus the mean value of all partial results because it is considered that the more the participants have similar involvements, the better the collaboration. A later improvement will involve using weighted influences for each factor in the final collaboration assessment (for example in-degree represents more interest in the current speaker, whereas out-degree expresses gregariousness in a certain manner; therefore their corresponding weights should have different values because of the nature of interactions expressed by these factors).

4.2 Quantitative Collaboration

The most straightforward approach for assessing the collaboration level of a chat is by using the number of explicit links and of the implicit ones with a trust coefficient assigned to them. Collaboration is essentially illustrated when certain ideas or viewpoints present within the current utterance are up-taken or are transferred to a latter one, issued by a different speaker. In other words, our opinion is that collaboration can be measured using linked utterances that have different speakers.

Starting from this idea, a quantitative collaboration score can be computed using the following formula:

$$\text{quantitati ve collaborat ion} = \frac{\sum_{\substack{\text{all links } 1 \\ \text{with different speakers}}} \text{attenuatio n(l) * trust(l)}}{\text{total number of links (implicit/ explicit)}} \tag{2}$$

where:

- attenuation(l) is used for assigning a different importance to the explicit links that are considered more valuable (score of 1) relatively to the implicit ones (score of 0.5);
- trust(l) is the assigned trust for an implicit link (for example, in the case of direct repetition this value is set to 2); for all explicit links, trust is set to 1.

By combining attenuation (values of 0.5 and 1.0) with trust (values in [0..2]), all explicit links have a score of 1 for their contribution, whereas implicit links can achieve the same maximum value only if their corresponding trust is maximum.

4.3 Gain Based Collaboration

The following two measures are computed starting from the assessment of the following formulas for any utterance u:

$$\text{gain(u)} = \text{personalgain(u)} + \text{collaborative gain(u)} \tag{3}$$

$$\text{personal gain(u)} = \sum_{\substack{\text{link l exists between u and v,} \\ \text{v is an earlier utterance and} \\ \text{u and v have same speaker}}} \begin{array}{l} ((\text{mark(v)} + \text{gain(v)}) * \text{similarity (u, v)} * \\ * \text{attenuatio n(l) * trust(l)}) \end{array} \tag{4}$$

$$\text{collaborative gain(u)} = \sum_{\substack{\text{link l exists between u and v,} \\ \text{v is an earlier utterance and} \\ \text{u and v have different speakers}}} \frac{((\text{mark(v)} + \text{gain(v)}) * \text{similarity}(u, v) *}{* \text{attenuation(l)} * \text{trust(l)})} \qquad (5)$$

The fundamental ideas are derived from information theory [13] in the sense that each utterance has an overall gain composed of both personal evolution (links to previous personal utterances) and collaborative building (ideas shared, future implications), with transfer of knowledge, ideas or concepts through the identified links. Therefore, for each utterance u we take into consideration all previous utterances v where a link (explicit or implicit) exists and we try to determine the amount of information that is transferred. Each utterance v has a score assigned after the marking process and a gain computed using its previous inter-linked utterances. Starting from these two components (cumulative build for gain and current importance using the corresponding mark), LSA is used for assessing the actual transfer, impact and correlation, in other words similarity between the two utterances (u and v).

The main difference between the two gains is that personal gain assumes that inner knowledge is built, whereas collaborative gain is expressed relative to others and expresses working together, sharing ideas and influencing the other participants of the discussion.

Our ideas are correlated with Bakhtin's dialogistic theory [11], in the sense that to achieve genuine collaboration, true polyphony, with an actual dense interlacing of voices is needed. The concept of voice at Bakhtin and in our approach has a larger extent than the usual acoustic one [5].

Links, both explicit and implicit, express the idea of inter-changing concepts or ideas, which are later processed with a new voice as result. In the end, a particular sense is given to each voice, taking into consideration echoes from previous voices related by the determined links. From this point, gain can be considered the echo of all previous inter-linked voices, attenuated corresponding with similarity and trust, and the mark is the actual implication, strength and importance of the current voice. Personal gain expresses and measures implicit inner dialog (individual voices), whereas the collaborative gain addresses explicit dialog (by definition between two participants), also highlighting external voices.

Using the previous formulas, the following two metrics are derived for assessing collaboration in a chat conversation:

- Formula (6) is used for estimating the percentage of overall utterances' importance/marks relatively to information build/transferred in a collaborative manner.

$$\text{mark based collab} = \frac{\sum_{\text{all utterances } u} \text{collaborative gain(u)}}{\sum_{\text{all utterances } u} \text{mark(u)}} \qquad (6)$$

- Formula (7) is used for assessing collaboration relatively to overall gain (practically excluding inner build).and

$$\text{gain based collab} = \frac{\sum_{\text{all utterances } u} \text{collaborative gain(u)}}{\sum_{\text{all utterances } u} \text{gain(u)}} \qquad (7)$$

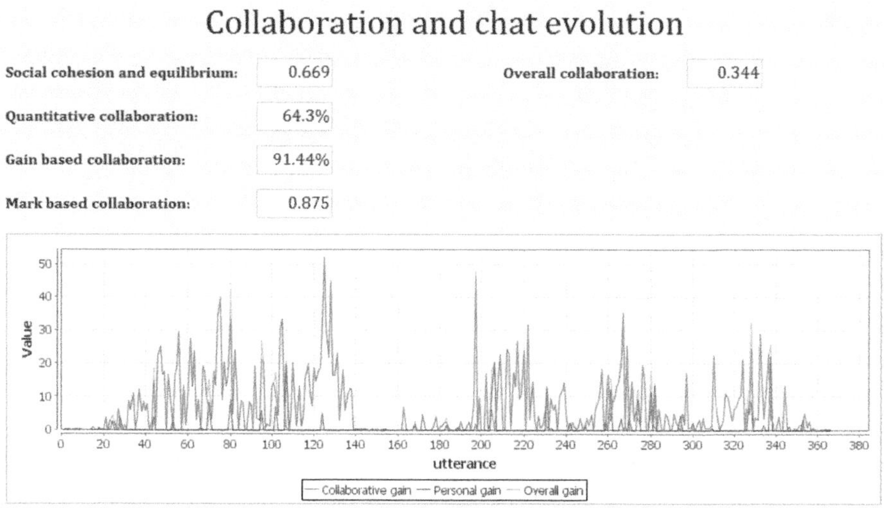

Collaboration and chat evolution

Social cohesion and equilibrium:	0.669	Overall collaboration:	0.344
Quantitative collaboration:	64.3%		
Gain based collaboration:	91.44%		
Mark based collaboration:	0.875		

Fig. 1. Collaboration assessment and chat evolution visualization

The overall collaboration mark is obtained by multiplying the four previously defined metrics:

$$\text{overall collaboration} = \text{social cohesion} \times \text{quantitative} \times \text{gain based} \times \text{mark based} \qquad (8)$$

5 Semantic Extractive Summarization

The most important aspect that can be identified regarding the summary generation is the tight link between obtaining a summary and evaluating chat participants. Therefore, in order to obtain a good extractive summary each utterance must be assessed – in our case, the marking process has been previously described and it provides the basis for identifying the key utterances.

A set of predefined percentages of extraction is provided, allowing the user to select the level of interest at which the overall view of the chat is desired. Starting from this percentage the number of utterances to be displayed is determined. After that, utterances are extracted in decreasing order of importance. A very important fact is that this process has a *gain based selection* (utterances are ordered based on the sum of corresponding marks and gains - both personal and collaborative ones), therefore combining collaboration and the marking process of each utterance. The approach of granting bonuses to the next level utterances in explicit threads and correspondingly weighted with trust for the implicit ones is reflected by the previously described gain which takes into account both implicit and explicit links, and fundaments its measure on semantic similarity.

Summary

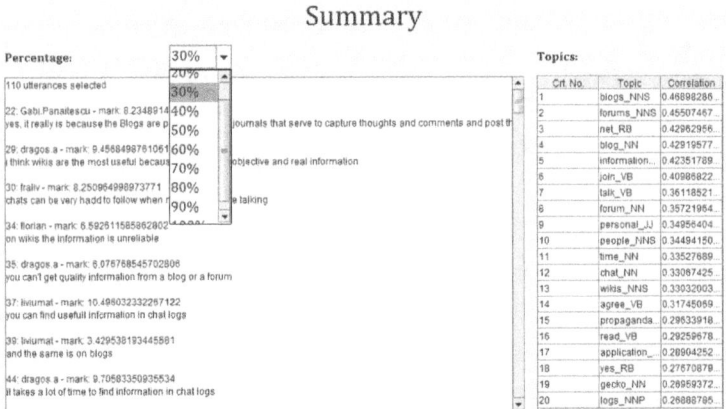

Fig. 2. Automatic summary and topics determination

The impact of using the automatic summary system was the reduction with more than 30% of the time spent by the tutor for the evaluation of a single chat.

6   Semantic Search

A distinctive and yet important component of the system addresses enhanced search capabilities within a conversation. Starting from a given query, two types of results are being assessed relatively to a single chat: a classification of participants with the criteria of best overall performance with regards to the given query and an utterance list in descending order of scores. In our case, these search facilities are centered on a scoring mechanism of each utterance.

In order to evaluate the relevance of each utterance to the given query, three steps are performed. First, the query is syntactically and semantically enriched by using synsets from WordNet and by selecting the most promising neighbors from the vector space model computed using LSA. For reducing the number of possible newly added items to the query, a threshold and maximum number of neighbors have been enforced. The final scope is to obtain a list of words and a query vector.

The next step addresses the syntactical level by evaluating the number of actual occurrences in each utterance. In this process different weights for original/enriched words and stems are also considered for reflecting the importance of each utterance.

The third step focuses on a semantic assessment by measuring the cosine similarity between the query vector and each utterance vector. The final score is obtained by multiplying the semantic score with the social one and with the sum of mark and gain determined in the previous analysis of the chat. These last two factors are important by taking into consideration the actual importance and cumulative gain of each utterance.

The results can be viewed in Fig. 3 in two different windows with regards to the searched factor – participants or utterances.

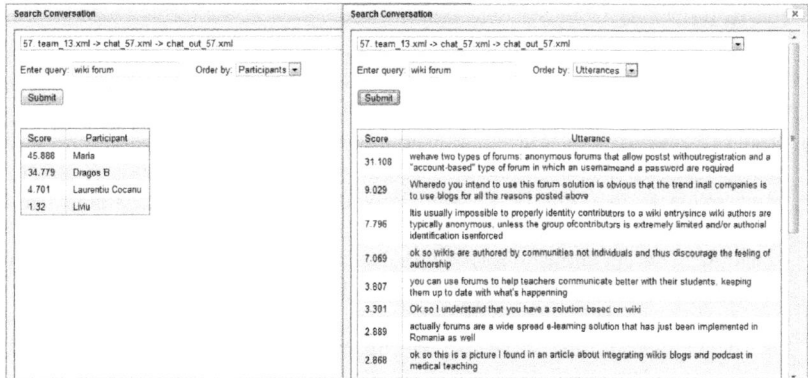

Fig. 3. Web widget displaying the results of the semantic search

7 Validation Results and Conclusions

All the functionalities described in the previous chapters have been wrapped up in the chat and forum analysis service called PolyCAFe. In order to measure the effectiveness and utility of the indicators computed by the system, a first validation round has been undertaken using a group of 5 tutors and 9 senior students that studied the Human-Computer Interaction course. The students were divided into two groups that had to debate which the best web collaboration technology is. After that, both the students and the tutors used the system to provide preliminary feedback about the chats. The 35 questions that were addressed to the tutors received average scores between 3.50-5.00 (where 5 is the maximum score and 1 the minimum). The most important results were that the tutors find the system relevant and useful for their activity, that the time for providing final feedback to the students is definitively reduced and the quality of the feedback is improved. On the other side, the students have validated 28 out of 32 questions, with average scores between 3.66-5.00.

In this paper, we have presented a novel analysis technique for collaborative chat conversations that combines various techniques from NLP and SNA in order to provide preliminary feedback to students and tutors. Although the system is still under development, it is available online using web widgets and services at http://ltfll-lin.code.ro/ltfll/wp5 and it has been successfully validated in a formal education context. However, several important issues have to be addressed, such as improving some of the indicators (such as the semantic search ranking) and facilitating the transferability of this approach to other domains and languages (besides HCI and English).

Acknowledgements

The research presented in this paper was partially supported by the FP7 EU STREP project LTfLL (www.ltfll-project.org) and the national CNCSIS grant K-Teams.

References

1. Stahl, G.: Group cognition: Computer support for building collaborative knowledge. MIT Press, Cambridge (2006)
2. Rose, C.P., Wang, Y.C., Cui, Y., Arguello, J., Stegmann, K., Weinberger, A., Fischer, F.: Analyzing Collaborative Learning Processes Automatically: Exploiting the Advances of Computational Linguistics in Computer-Supported Collaborative Learning. International Journal of Computer Supported Collaborative Learning (2007)
3. Dong, A.: Concept formation as knowledge accumulation: A computational linguistics study. Artif. Intell. Eng. Des. Anal. Manuf. 20(1), 35–53 (2006)
4. Dascalu, M., Chioasca, E.V., Trausan-Matu, S.: ASAP - An Advanced System for Assessing Chat Participants. In: Dochev, D., Pistore, M., Traverso, P. (eds.) AIMSA 2008. LNCS (LNAI), vol. 5253, pp. 58–68. Springer, Heidelberg (2009)
5. Trausan-Matu, S., Rebedea, T.: Polyphonic Inter-Animation of Voices in VMT. In: Stahl, G. (ed.) Studying Virtual Math Teams, pp. 451–473. Springer, Boston (2009)
6. Landauer, T.K., Foltz, P.W., Laham, D.: An Introduction to Latent Semantic Analysis. In: Discourse Processes, vol. 25, pp. 259–284 (1998)
7. Lemaire, B.: Limites de la lemmatization pour l'extracftion de significations. JADT 2009: 9^{es} Journees Internationals d'Analyse Statistique des Donnes Textuelles (2009)
8. Manning, C., Schutze, H.: Foundations of Statistical Natural Language Processing. MIT Press, Cambridge (1999)
9. Wiemer-Hastings, P., Zipitria, I.: Rules for Syntax, Vectors for Semantics. In: Proceedings of the Twenty-third Annual Conference of the Cognitive Science Society (2001)
10. The Prefuse Visualization Toolkit, http://prefuse.org/
11. Bakhtin, M.: Problems of Dostoevsky's poetics (C. Emerson, Trans.). University of Minnesota Press, Minneapolis (1984)
12. Shannon, C.: Prediction and entropy of printed English. The Bell System Technical Journal 30, 50–64 (1951)

Ontology-Based Authoring of Intelligent Model-Tracing Math Tutors

Dimitrios Sklavakis and Ioannis Refanidis

University of Macedonia, Department of Applied Informatics,
Egnatia 156, P.O. Box 1591, 540 06 Thessaloniki, Greece
{dsklavakis,yrefanid}@uom.gr

Abstract. This paper describes the MATHESIS Ontology, an OWL ontology developed within the MATHESIS project. The project aims at the development of an intelligent authoring environment for reusable model-tracing math tutors. The purpose of the ontology is to provide a semantic and therefore inspectable and re-usable representation of the declarative and procedural *authoring knowledge* necessary for the development of any model-tracing tutor, as well as of the declarative and procedural knowledge of the specific tutor under development. While the declarative knowledge is represented with the basic OWL components, i.e. classes, individuals and properties, the procedural knowledge is represented via the *process model* of the OWL-S web services description ontology. By using OWL-S, every authoring or tutoring task is represented as a composite process. Based on such an ontological representation, a suite of authoring tools will be developed at the final stage of the project.

Keywords: intelligent tutoring systems, model-tracing, cognitive tutors, web based tutors, authoring systems, ontologies, OWL-S.

1 Introduction

Intelligent tutoring systems (ITSs) and especially model-tracing tutors have been proven quite successful in the area of mathematics [1]. Despite their efficiency, these tutors are expensive to build both in time and human resources. It is currently estimated that one hour of tutoring takes 200-300 hours of development [2]. Moreover, their development requires teams of Ph.D. level scientists in education, cognitive science and artificial intelligence.

As a solution, authoring programs have been built having as their main purpose the reduction of development time as well as the lowering of the expertise level required to build a tutor. An extensive overview of these authoring tools can be found in [3].

Still, these tools suffer from a number of problems such as isolation, fragmentation, lack of communication, interoperability and re-usability of the tutors that they build. What seems important in the case of ITSs is the creation of repositories of interoperable components that can be automatically retrieved and used by ITS shells or authoring tools [4].

This last research goal is in direct relationship with the research being conducted in the field of the Semantic Web and especially Semantic Web Services, a framework

D. Dicheva and D. Dochev (Eds.): AIMSA 2010, LNAI 6304, pp. 201–210, 2010.
© Springer-Verlag Berlin Heidelberg 2010

for the (semi-)automatic discovery, composition and invocation of web services and, in the case of ITSs, learning services [5].

The main goal of the ongoing MATHESIS project is to develop authoring tools for model-tracing tutors in mathematics, with knowledge re-use as the primary characteristic for the authored tutors but also for the authoring knowledge used by the tools. The basic research hypothesis of the project is that this goal can be better served by the basic Semantic Web technologies, that is by developing an OWL ontology for the description of the declarative knowledge and using the OWL-S web service description language [6] to represent the procedural knowledge for both the authoring tasks and the tutoring tasks of the developed tutors. This paper describes the MATHESIS ontology that implements this representational scheme.

The MATHESIS ontology is an OWL ontology developed with the Protégé-OWL ontology editor [7]. Its development is the second stage of the MATHESIS project. Aiming at the development of real-world, fully functional model-tracing math tutors, the project is being developed in a bottom-up approach. In the first stage the MATHESIS Algebra Tutor was developed in the domain of expanding and factoring algebraic expressions [8]. The tutor is web-based, using HTML and JavaScript. The authoring of the tutor as well as the code of the tutor were used to develop the MATHESIS ontology in an opportunistic way, that is in a bottom-up and top-down way, as it will be described later.

The rest of the paper is structured as follows: Section 2 gives a brief presentation of the process model of OWL-S which is the key technology of the MATHESIS ontology. Section 3 describes the part of the ontology that represents the model-tracing tutor(s), while Section 4 describes the representation of the authoring knowledge. Section 5 concludes with a discussion about the ontology and further work to complete the MATHESIS project.

2 The OWL-S Process Model

OWL-S is a web service description ontology designed to enable the following tasks:

- Automatic discovery of Web services that can provide a particular class of service capabilities, while adhering to some client-specified constraints.
- Automatic Web service invocation by a computer program or agent, given only a declarative description of the service.
- Automatic Web service selection, composition and interoperation to perform some complex task, given a high-level description of an objective.

It is the last task that is of interest for the MATHESIS project and therefore we will focus on that. To support this task, OWL-S provides, among other things, a language for describing service compositions (see Fig.1).

Every service is viewed as a *process*. OWL-S defines Process as a subclass of ServiceModel. There are three subclasses of Process, namely the AtomicProcess, CompositeProcess and SimpleProcess. Atomic processes correspond to the actions a service can perform by engaging it in a single interaction; composite processes correspond to actions that require multi-step protocols; finally, simple processes provide an abstraction mechanism to provide multiple views of the same process.

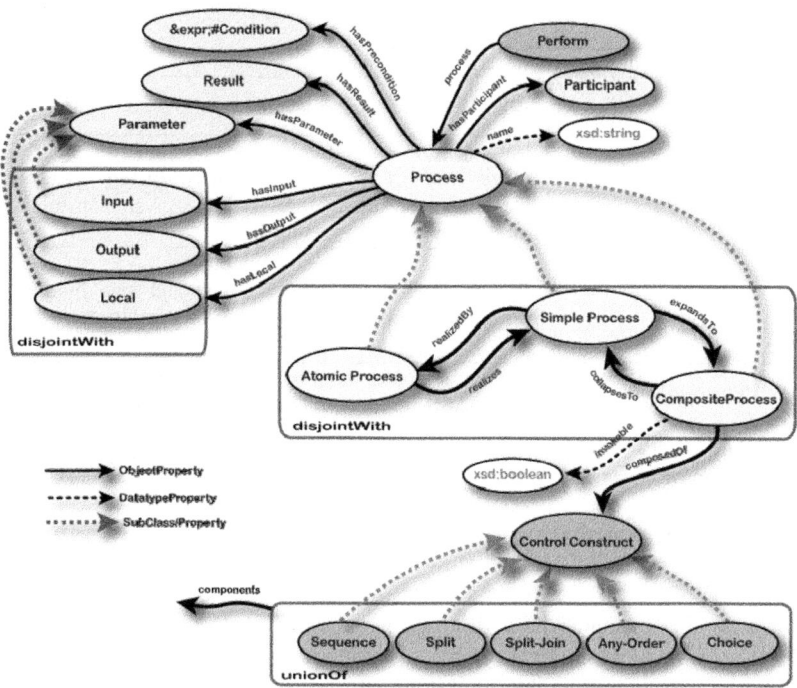

Fig. 1. Top level of OWL-S *process* ontology

Composite processes are decomposable into other composite or atomic processes. Their decomposition is specified by using control constructs such as Sequence and If-Then-Else.

A CompositeProcess has a composedOf property by which is indicated the control structure of the composite process, using a ControlConstruct. Each ControlConstruct, in turn, is associated with an additional property called components to indicate the nested control constructs from which it is composed and, in some cases, their ordering. For example, any instance of the control construct Sequence has a components property that ranges over a ControlConstructList (a list of control constructs). Table 1 shows the most common control constructs that OWL-S supports.

Any composite process can be considered as a tree whose non-terminal nodes are labeled with control constructs, each of which has children specified using the components property. The leaves of the tree are invocations of other processes, composite or atomic. These invocations are indicated as instances of the Perform control construct. The process property of a Perform indicates the process to be performed.

This tree-like representation of composite processes is the key characteristic of the OWL-S process model used in the MATHESIS ontology to represent both authoring and tutoring procedural knowledge, as it will be described in the subsequent sections.

Table 1. Common control constructs supported by OWL-S process model

Control Construct	Description
Sequence	A list of control constructs to be performed in order
Choice	Calls for the execution of a single construct from a given bag of control constructs (given by the components property). Any of the given constructs may be chosen for execution
If-Then-Else	It has properties ifCondition, then and else holding different aspects of the If-Then-Else construct
Repeat-While & Repeat-Until	The initiation, termination or maintenance condition is specified with a whileCondition or an untilCondition respectively. The operation of the constructs follows the familiar programming language conventions.

3   Tutor Representation in MATHESIS Ontology

The MATHESIS project has as its ultimate goal the development of authoring tools that will be able to guide the authoring of real-world, fully functional model-tracing math tutors. That means that during the authoring process and in the end, the result will be program code that implements the tutor. The ontology must be able to represent the program code. For this reason, in the first stage of the MATHESIS project the MATHESIS Algebra tutor was developed to be used as a prototype target tutor [8].

The MATHESIS Algebra tutor is a Web-based, model-tracing tutor that teaches expanding and factoring of algebraic expressions: monomial and polynomial operations, identities, factoring. It is implemented as a simple HTML page with JavaScript controlling the interface interaction with the user and implementing the tutoring, domain and student models. The user interface consists of common HTML objects as well as Design Science's WebEq Input Control applet, a scriptable editor for displaying and editing mathematical expressions that provides the same functionality as Equation Editor for Microsoft Word. Therefore, it is the representation of the HTML and JavaScript code that forms the low-level MATHESIS ontology of the tutor as described in the subsequent subsections.

3.1   Representation of the Tutor's HTML Code

The representation of the HTML code and the corresponding Document Object Model (DOM) of the user interface are shown in Figure 2. Each line of the HTML code is represented as an instance of the HTMLProgramLine class having three properties: the HTMLCode, hasNextLine and correspondingHTMLObject. The last one points to the HTMLObject defined by the code.

Each HTMLObject has the corresponding HTML properties as well as the hasFirstChild and hasNextSibling which implement the DOM tree. Therefore, there are two representations of the HTML code enabling a bottom-up creation of the ontology (from HTML code to DOM) and a top-down (from the DOM to HTML code). Given

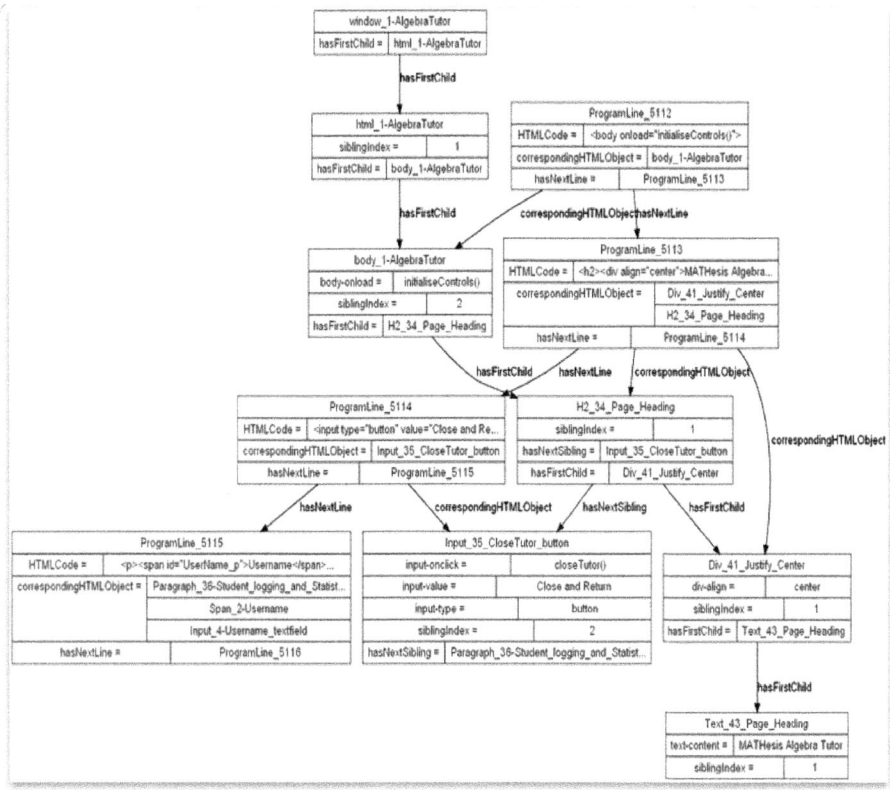

Fig. 2. The HTML code and the corresponding DOM representation

the appropriate parsing tools, one can build the interface in any Web-page authoring program and then generate from the HTML code the DOM ontology or, in the reverse order, the tutor authoring tools will create the DOM representation and then the HTML code.

3.2   Representation of the Tutor's JavaScript Code

The representation of the JavaScript code is shown in Figure 3. Each line of the Java-Script code is represented as an instance of the JavaScript_ProgramLine class having three properties: the javascriptCode, hasNextLine and hasJavaScriptStatement. The last one points to a JavaScript_Statement instance which is an AtomicProcess of OWL-S process model.

Once again, there are two representations of the JavaScript code enabling a bottom-up creation of the ontology (from JavaScript code to JavaScript_Statement atomic processes) and a top-down (from the JavaScript_Statement atomic processes to JavaScript code). Given the appropriate parsing tools, one can take JavaScript and then generate the corresponding JavaScript_Statement atomic processes or, in the reverse order, the tutor authoring tools will create the JavaScript_Statement atomic processes and then the JavaScript code.

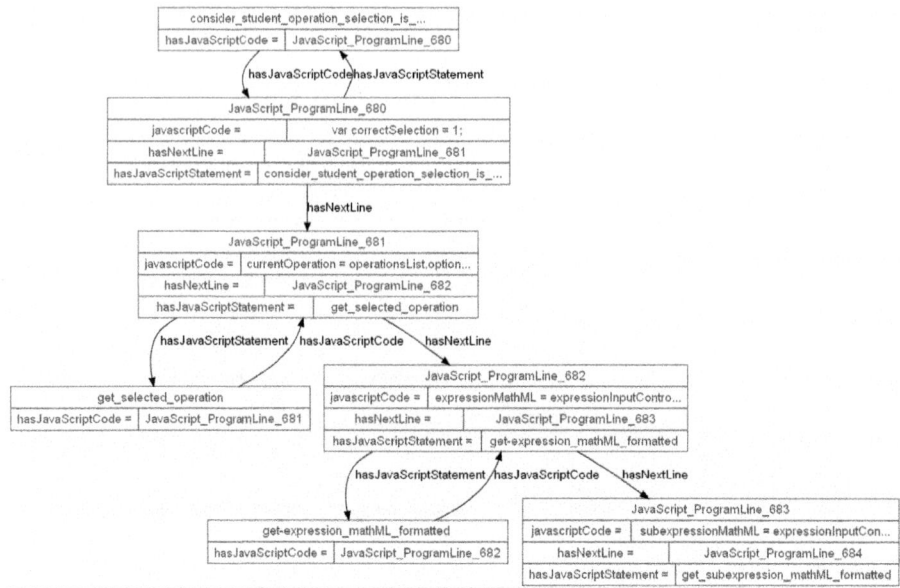

Fig. 3. The JavaScript code and the corresponding JavaScript_Statement Atomic processes

3.3   Representation of the Tutoring Model

Model-tracing tutors are named after their fixed tutoring model, the model-tracing algorithm. Being procedural knowledge, the model-tracing algorithm is represented in the MATHESIS ontology as a composite process named Model_Tracing_Algorithm (Figure 4). Each step of the algorithm is also a composite process (notice the small 'c' letter at the end of the names). That means that each step can be further elaborated by another algorithm (composite process). For example the Task_Execution_Expert_Process step can be described by an algorithm (not developed yet) that performs other composite processes like Step-Execution-Expert-Process, Task-Asked-Identification-Expert-Process, Task-Given-Identification-Expert-Process, Task-Sub-Task-Execution-Expert-Process (Figure 5).

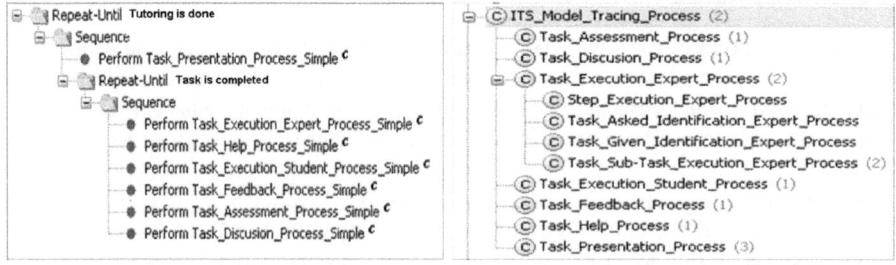

Fig. 4. The Model-Tracing-Algorithm process   **Fig. 5.** The model-tracing processes taxonomy

During the authoring of a specific tutor, the authoring tools will parse the tree of the Model-Tracing-Algorithm composite process and will invoke for each tutoring task a corresponding authoring task represented also as a composite process in order to implement the tutoring task for the specific tutor (see Section 4).

3.4   Representation of the Domain Expert Model

In a model-tracing tutor, the Domain Expert Model executes the next step of the problem and produces the correct solution(s) to compare them with the student's proposed solution. If the solution step is simple, then it is represented as an instance of the atomic process JavaScript_Statement (see Section 3.2). If the step is complex, then it is represented as a composite process. This analysis ends when the produced composite processes contain only atomic processes, that is JavaScript_Statement instances. An example for the monomial multiplication task from the MATHESIS Algebra tutor will

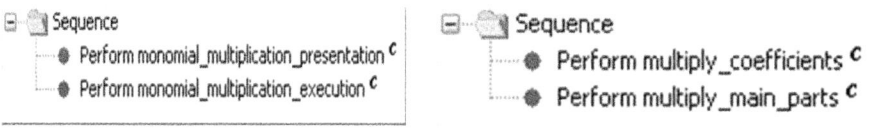

Fig. 6. Part of the monomial-multiplication model-tracing algorithm composite process

Fig. 7. The monomial-multiplication-execution composite process

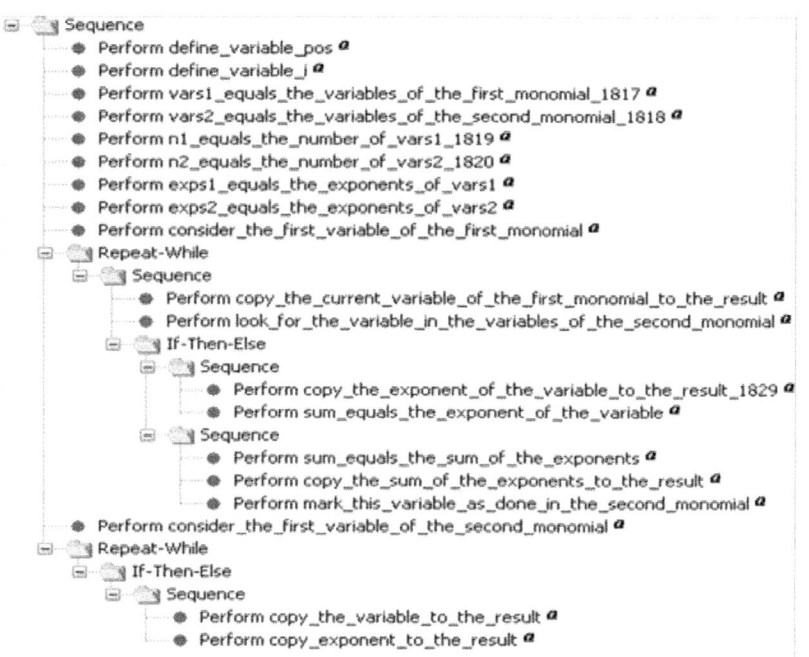

Fig. 8. The multiply-main-parts composite process

illustrate this technique. In Figure 6 a part of the model-tracing algorithm is presented as implemented for the monomial multiplication task according to Section 3.3. As the monomial_multiplication_execution step is a complex one, it is represented as a composite process with two other composite processes, multiply_coefficients and multiply_main_parts (Figure 7). Finally, in Figure 8 the multiply_main_parts process is described, containing only instances of the atomic process JavaScript_Statement (see Section 3.2).

4 Authoring Knowledge Representation in MATHESIS Ontology

As mentioned in Section 3.3, for each tutoring task of the model-tracing algorithm, there is a corresponding authoring task in the MATHESIS ontology, represented also as a composite process. The authoring_task_execute_task_by_expert (Figure 9) for example corresponds to the Task_Execution_Expert_Process_Simple tutoring task (see Figure 4). One of the composite processes that form this authoring task, the define_data_structures_for_knowledge_components is shown in Figure 10.

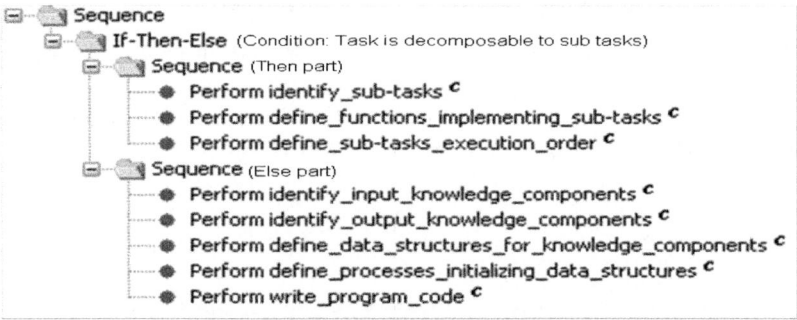

Fig. 9. The authoring process for the Execute-Task-By-Expert tutoring task

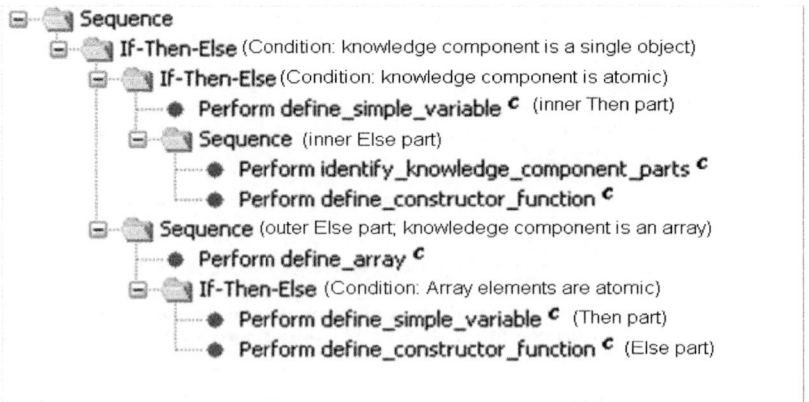

Fig. 10. The define-data-structures authoring process

Based on all the above representations, the overall authoring process will have as follows: The tools will parse the model-tracing algorithm (Section 3.3). For each step of the algorithm, the corresponding authoring process will be called and traced. This authoring process will guide the author in creating recursively the various parts of the tutor (Sections 3.1, 3.2, 3.4); as a consequence, any newly created tutor part becomes new knowledge in the ontology to be used later. In addition, the authoring process can present known issues, design principles, examples and existing implementations for the tutoring task to which it corresponds, maximizing that way knowledge re-use.

5 Discussion and Further Work

In an overview of intelligent tutoring system authoring tools [3], it is suggested that authoring tools should support:

- *Interoperability*: Tutors should be able to work with other systems providing *inspect-ability* (provide information to other applications), *record-ability* (broadcast data about their state) and *script-ability* (controlled by other applications)
- *Re-usability*: Tutors should be able to be used in a context different from their original one
- *Durability* and *Scalability*: Tutors should be able to cover big-sized domains and accommodate many simultaneous users
- *Accessibility*: Learner or content developer should locate and download material in a reasonable time.

Even in this preliminary form, the MATHESIS ontology provides a proof-of-concept that it can serve as the basis for the development of authoring tools and implemented tutors that will match these criteria. The main reason for this claim is that the ontology provides an open, modular and multi-level representation (ranging from conceptual design to program code) of both authoring and tutoring knowledge. In addition, being an OWL ontology with all procedural knowledge encoded as OWL-S processes opens a window towards semantic web services and Service Oriented Architectures [9]. Finally, the tutors, implemented in HTML and JavaScript, possess in a great extent the listed characteristics.

Of course, it is obvious that a lot of work has to be done: the ontology must be extended, refined and in a great extend formalized. This will be done by representing the whole MATHESIS Algebra tutor into the ontology. Because of the tremendous workload this task entails, an initial set of authoring tools are needed: parsers for HTML to/from MATHESIS interface model and for JavaScript to/from MATHESIS JavaScript_Statements/Program code translation; interpreters for the authoring and tutoring OWL-S processes; visualization tools for the authoring processes, the tutoring model (model-tracing algorithm) and the tutor parts being developed. These tools constitute our current research line.

References

1. Koedinger, K.R., Anderson, J.R., Hadley, W.H., Mark, M.A.: Intelligent Tutoring Goes to School in the Big City. International Journal of Artificial Intelligence in Education 8, 30–43 (1997)
2. Aleven, V., McLaren, B., Sewall, J., Koedinger, K.R.: The Cognitive Tutor Authoring Tools (CTAT): Preliminary Evaluation of Efficiency Gains. In: Ikeda, M., Ashley, K.D., Chan, T.-W. (eds.) ITS 2006. LNCS, vol. 4053, pp. 61–70. Springer, Heidelberg (2006)
3. Murray, T.: An Overview of Intelligent Tutoring System Authoring Tools: Updated Analysis of the State of the Art. In: Authoring Tools for Advanced Technology Learning Environments, pp. 491–544. Kluwer Academic Publishers, Netherlands (2003)
4. Davies, J., Studer, R., Waren, P. (eds.): Semantic Web Technologies: trends and research in ontology-based systems. John Wiley & Sons Ltd., England (2006)
5. Dicheva, D., Mizogichi, R., Capuano, N., Harrer, A. (eds.): Proceedings of the Fifth International Workshop on Ontologies and Semantic Web for E-Learning, SWEL 2007 (2007), http://compsci.wssu.edu/iis/Papers/SWEL07-Proceedings.pdf
6. Martin, D., Paolucci, M., McIlraith, S., Burstein, M., McDermott, D., McGuinness, D., Parsia, B., Payne, T., Sabou, M., Solanki, M., Srinivasan, N., Sycara, K.: Bringing Semantics to Web Services: The OWL-S Approach. In: Cardoso, J., Sheth, A.P. (eds.) SWSWPC 2004. LNCS, vol. 3387, pp. 26–42. Springer, Heidelberg (2005)
7. The Protégé ontology editor, http://protege.stanford.edu/overview/protege-owl.html
8. Sklavakis, D., Refanidis, I.: An Individualized Web-Based Algebra Tutor Based on Dynamic Deep Model-Tracing. In: Darzentas, J., Vouros, G.A., Vosinakis, S., Arnellos, A. (eds.) SETN 2008. LNCS (LNAI), vol. 5138, pp. 389–394. Springer, Heidelberg (2008)
9. Gutiérrez-Carreón, G., Daradoumis, T., Jorba, J.: Exploring Semantic Description and Matching Technologies for Enhancing the Automatic Composition of Grid-based Learning Services. In: Proceedings of the Fifth International Workshop on Ontologies and Semantic Web for E-Learning, SWEL 2007 (2007), http://compsci.wssu.edu/iis/Papers/SWEL07-Proceedings.pdf

Comparative Analysis of Distributed, Default, IC, and Fuzzy ARTMAP Neural Networks for Classification of Malignant and Benign Lesions

Anatoli Nachev

Business Information Systems, Cairnes School of Business & Economics, NUI Galway, Ireland
anatoli.nachev@nuigalway.ie

Abstract. Only one third of all breast cancer biopsies made today confirm the disease, which make these procedures inefficient and expensive. We address the problem by exploring and comparing characteristics of four neural networks used as predictors: fuzzy, distributed, default, and ic ARTMAP, all based on the adaptive resonance theory. The networks were trained using a dataset that contains a combination of 39 mammographic, sonographic, and other descriptors, which is novel for the field. We compared the model performances by using ROC analysis and metrics derived from it, such as max accuracy, full and partial area under the convex hull, and specificity at 98% sensitivity. Our findings show that the four models outperform the most popular MLP neural networks given that they are setup properly and used with appropriate selection of data variables. We also find that two of the models, distributed and ic, are too conservative in their predictions and do not provide sufficient sensitivity and specificity, but the default ARTMAP shows very good characteristics. It outperforms not only its counterparts, but also all other models used with the same data, even some radiologist practices. To the best of our knowledge, the ARTMAP neural networks have not been studied for the purpose of the task until now.

Keywords: data mining, neural networks, ARTMAP, breast cancer, ROC.

1 Introduction

Mammographic and sonographic screening procedures significantly reduce the mortality of breast cancer, however only 35% or less of women who undergo biopsy for histopathologic diagnosis of the disease are found to have malignancies [14]. A large percentage of those who undergo biopsy are unnecessarily subjected to discomfort, expense, and potential complications. In addition, the financial burden of these procedures (thousands of euros per biopsy) is significant in the present political and economic effort to reduce expenditures.

This study is focused on computer-aided diagnosis of breast lesions that had already been identified by radiologists as suspicious enough to warrant biopsy. In other words, these cases are generally considered indeterminate and more challenging, and any reduction in the number of benign biopsies represents an improvement over the status quo, provided high sensitivity is maintained.

D. Dicheva and D. Dochev (Eds.): AIMSA 2010, LNAI 6304, pp. 211–220, 2010.
© Springer-Verlag Berlin Heidelberg 2010

A considerable amount of research has been done in this area and variety of statistical and machine learning techniques have been used to build predictive models, such as linear discriminant analysis (LDA), logistic regression analysis (LRA), multilayer perceptrons (MLP), support vector machines (SVM), etc. [7, 12]. Various sources of medical information have been used to build predictive models, such as digitized screen-film mammograms, sonograms, magnetic resonance imaging (MRI) images, and gene expression profiles, etc. [7]. Current applications, however, tend to use only one information source, usually mammographic data in the form of data descriptors defined by the Breast Imaging Reporting and Data System (BI-RADS) lexicon [1]. Recently, Jesneck et al. [13] proposed a novel combination of BI-RADS mammographic and sonographic descriptors and some suggested by Stavros et al. [16], which in combination with MLP show promising results. The MLP have been largely applied for the task, but they have a drawback: the model assumes predefined network architecture, including connectivity and node activation functions, and training algorithm to learn to predict. The issue of designing a near optimal network architecture can be formulated as a search problem and still remains open. This study takes an alternative approach based on ARTMAP neural networks which feature well established architecture and fast one-pass online learning. We explore four members of the ARTMAP family as predictors and compare their performance and characteristics. To the best of our knowledge this has not been done before, particularly in a combination with both mammographic and sonographic data descriptors.

The paper is organized as follows: Section 2 provides a brief overview of the ATRMAP neural networks and discusses four members of the family: fuzzy, distributed, default, and ic; Section 3 introduces the dataset used in the study, its features, and the preprocessing steps; Section 4 presents and discusses results from experiments; and Section 5 gives the conclusions.

2   ARTMAP Neural Networks

The adaptive resonance theory (ART) introduced by Grossberg [10] led to the creation of a family of self-organizing neural networks for fast one-pass learning, pattern recognition, and prediction. In contrast, the most popular among practitioners neural nets multi-layer perceptrons (MLP) offer an off-line slow learning procedure that requires availability of all training patterns at once in order to avoid catastrophic forgetting in an open input environment. Some of the most remarkable ART neural nets are the unsupervised ART1, ART2, ART2-A, ART3, fuzzy ART, distributed ART and the supervised ARTMAP, instance counting ARTMAP, fuzzy ARTMAP, distributed ARTMAP, and default ARTMAP.

ARTMAP is a family of supervised neural networks that consists of two unsupervised ART modules, *ARTa* and *ARTb*, and an *inter-ART* module called map-field (Figure 1) [2]. An ART module has three layers of nodes: input layer *F0*, comparison layer *F1*, and recognition layer *F2*. A set of real-valued weights W_j is associated with the *F1-to-F2* layer connections between nodes. Each *F2* node represents a recognition category that learns a binary prototype vector w_j. The *F2* layer is connected through weighted associative links to a map field F^{ab}.

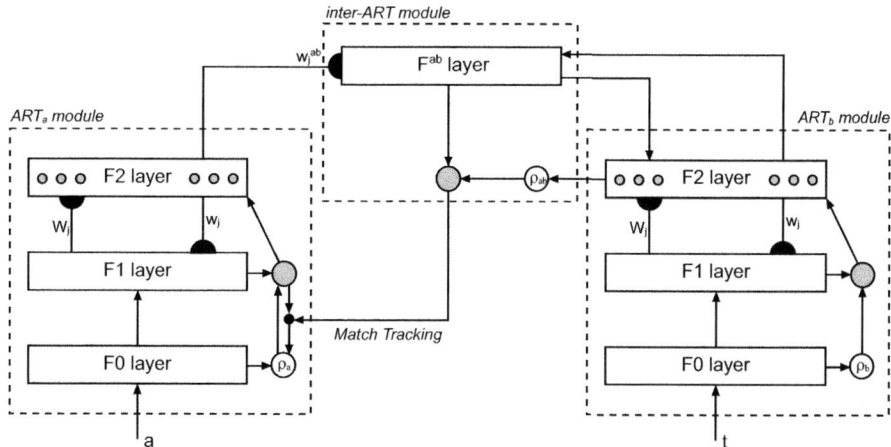

Fig. 1. Block diagram of an ARTMAP neural network adapted from [2]

The ARTMAP learning can be described by the following algorithm [2]:

1. *Initialization.* Initially, all the *F2* nodes are uncommitted, and all weight values are initialized. Values of network parameters are set.

2. *Input pattern coding.* When a training pattern is presented to the network as a pair of two components *(a, t)*, where *a* is the training sample and *t* is the class label, a process called complement coding takes place. It transforms the pattern into a form suited to the network. A network parameter called vigilance parameter (ρ) is set to its initial value. This parameter controls the network 'vigilance', that is, the level of details used by the system when it compares the input pattern with the memorized categories.

3. *Prototype selection.* The input pattern activates layer *F1* and propagates to layer *F2*, which produces a binary pattern of activity such that only the *F2* node with the greatest activation value remains active, that is, 'winner-takes-all'. The activated node propagates its signals back onto *F1*, where a vigilance test takes place. If the test is passed, then resonance is said to occur. Otherwise, the network inhibits the active *F2* node and searches for another node that passes the vigilance test. If such a node does not exist, an uncommitted *F2* node becomes active and undergoes learning.

4. *Class prediction.* The class label *t* activates the F^{ab} layer in which the most active node yields the class prediction. If that node constitutes an incorrect class prediction, then another search among *F2* nodes in Step 3 takes place. This search continues until an uncommitted *F2* node becomes active (and learning directly ensues in Step 5), or a node that has previously learned the correct class prediction becomes active.

5. *Learning.* The neural network gradually updates its adaptive weights towards the presented training patterns until a convergence occur. The learning dynamic can be described by a system of ordinary differential equations.

Fuzzy ARTMAP (FAM) was developed as a natural extension to the ARTMAP architecture. This is accomplished by using fuzzy ART modules instead of ART1, which replaces the crisp (binary) logic embedded in the ART1 module with a fuzzy one. In fact, the intersection operator ($\cap$) that describes the ART1 dynamics is replaced by the fuzzy AND operator ($\wedge$) from the fuzzy set theory $((p \wedge q)_i \equiv min(p_i, q_i))$ [3]. This allows the fuzzy ARTMAP to learn stable categories in response to either analog or binary patterns in contrast with the basic ARTMAP, which operates with binary patterns only.

ART1 modules in an ARTMAP net map the categories into *F2* nodes according to the winner-takes-all rule, as discussed above, but this way of functioning can cause category proliferation in a noisy input environment. An explanation of that is that the system adds more and more *F2* category nodes to meet the demands of predictive accuracy, believing that the noisy patterns are samples of new categories. To address this drawback, Carpenter [4] introduced a new distributed ART module, which features a number of innovations such as new distributed *instar* and *outstar* learning laws. If the ART1 module of the basic ARTMAP is replaced by a distributed ART module, the resulting network is called *distributed ARTMAP*. Some experiments show that a distributed ARTMAP retains the fuzzy ARTMAP accuracy while significantly reducing the network size [4].

Instance counting (ic) ARTMAP adds to the basic fuzzy ARTMAP system new capabilities designed to solve computational problems that frequently arise in prediction. One such problem is inconsistent cases, where identical input vectors correspond to cases with different outcomes. A small modification of the fuzzy ARTMAP match-tracking search algorithm allows the ic ARTMAP to encode inconsistent cases and make distributed probability estimates during testing even when training employs fast learning [5].

A comparative analysis of the ARTMAP modifications, including fuzzy ART-MAP, IC ARTMAP, and distributed ARTMAP, has led to the identification of the *default ARTMAP* network, which combines the winner-takes-all category node activation during training, distributed activation during testing, and a set of default network parameter values that define a ready-to-use, general-purpose neural network for supervised learning and recognition [6]. The default ARTMAP features simplicity of design and robust performance in many application domains.

3   The Dataset

This study uses data from physical examination of patients that contains findings from mammographic and sonographic examinations, family history of breast cancer, and personal history of breast malignancy, all collected from 2000 to 2005 at Duke University Medical Centre [13]. The data set contains 803 samples of which 296 malignant and 507 benign. Each sample contains 39 features, 13 of which are mammographic BI-RADS features, 13 are sonographic BI-RADS features, six are sonographic features suggested by Stavros et al. [16], four are other sonographic features, and three were patient history features [1, 13, 15]. The samples also contain a class label that indicates if the sample is malignant or benign.

The features are as follows: mass size, parenchyma density, mass margin, mass shape, mass density, calcification number of particles, calcification distribution, calcification description, architectural distortion, associated findings, special cases (as defined by the BI-RADS lexicon: asymmetric tubular structure, intramammary lymph node, global asymmetry, and focal asymmetry), comparison with findings at prior examination, and change in mass size. The sonographic features are radial diameter, antiradial diameter, anteroposterior diameter, background tissue echo texture, mass shape, mass orientation, mass margin, lesion boundary, echo pattern, posterior acoustic features, calcifications within mass, special cases (as defined by the BI-RADS lexicon: clustered microcysts, complicated cysts, mass in or on skin, foreign body, intramammary lymph node, and axillary lymph node), and vascularity. The six features suggested by Stavros [16] are mass shape, mass margin, acoustic transmission, thin echo pseudocapsule, mass echogenicity, and calcifications. The four other sonographic mass descriptors are edge shadow, cystic component, and two mammographic BI-RADS descriptors applied to sonography—mass shape (oval and lobulated are separate descriptors) and mass margin (replaces sonographic descriptor angular with obscured). The three patient history features were family history, patient age, and indication for sonography [13].

The dataset was separated into two parts by using 5-fold cross-validation – 80% of the patterns were used for training and validation purposes using; 20% were used to test the model performance given that they were never presented to the network for training.

A specific problem with the dataset is the large amplitude of values due to different units of measurements of the data variables, e.g. mass size (0 – 75) vs. calcification (0 – 3). Such an inconsistency in values could affect the predictive abilities of the neural networks by making some variables be more 'influential' than others. Moreover, the ARTMAP models require inputs within the unit hypercube, i.e. values are between 0 and 1. A natural approach of meeting this requirement could be a linear transformation that divides all input values by the dataset maximum, however most of the input values would fall very close to zero, and the model would perform poorly. A better approach is to process each data variable (data column) separately. We did so by using the transformation

$$ x_i^{new} = \frac{x_i^{old} - \min_i}{\max_i - \min_i} , \tag{1} $$

which scales down the each variable x_i separately within the unit hypercube.

We also explored how presence or absence of variables presented to the model for training and testing affects the performance. Removing most irrelevant and redundant features from the data usually helps to alleviate the effect of the curse of dimensionality and to enhance the generalization capability of the model, yet to speed up the learning process and to improve the model interpretability. The exhaustive search approach that considers all possible subsets is best for datasets with small cardinality, but impractical for large number of features as in our case. By using stepwise feature selection method, Jesneck et al. [8] proposed a feature subset set of 14 descriptors (*set14*). In [15] was suggested a set of 17 descriptors (*set17*) for MLPs, which was derived by using an extension of the linear forward selection, proposed by Guetlien et al. [3]. There is no guarantee, however, that an optimal variable selection for one classification models will be optimal for another. We considered several feature selection algorithms, both feature ranking and subset selection. Comparing best first, genetic search, subset size forward selection, race search, and scatter search we found that the best is genetic search [9] combined with a set evaluation technique that considers individual predictive ability of each feature along with the degree of redundancy between them [11]. The feature set we obtained consists of the following 21 descriptors (*set21*): patient age, family history, mass margin, architectural distortion, associated findings, comparison with prior examinations, anteroposterior diameter, mass shape, mass orientation, mass margin, lesion boundary, calcification within mass, special cases, mass shape, mass margin, thin echo pseudocapsule, mass echogenicity, edge shadow, cystic component, mass shape, and mass margin.

4   Experiments and Analysis

We used fuzzy, distributed, default and ic ARTMAP neural network simulators trained and tested by using five-fold cross-validation. Results were grouped in four categories: true positives (TP), true negatives (TN), false positives (FP), and false negatives (FN). The model performance was estimated by the metrics accuracy (Acc), true positive rate (TPR), and false positive rate (FPR), where

$$Acc=(TP+TN)/(TP+TN+FP+FN), \qquad (2)$$

$$TPR=TP/(TP+FN), \qquad (3)$$

$$FPR=FP/(FP+TN). \qquad (4)$$

Exploring how variance of the network parameters affect the model performance we found that only one parameter plays significant role - the vigilance. Figure 2 shows the prediction accuracy of each of the models with each of the four feature sets and 42 vigilance values from 0 to 1 with step of increment 0.025. In the figures only the set with maximal accuracy is depicted as a line. Results show that set s21 is best performer for the fuzzy and distributed models, whilst the default and ic perform well with s14. Certain vigilance values give higher accuracy than MLP gives with the same dataset (e.g. 86.3% of default vs. 82.5% of MLP [15]). Accuracy details are shown in Table 1.

   In machine learning, the accuracy is the most common estimator, but sometimes it can be misleading because it depends on the model parameters and does not estimate different outcome cost (e.g. misclassified malignant and benign lesions have different practical consequences / cost). In such cases the receiver operating characteristics (ROC) analysis is better model estimation technique [8]. ROC space can be defined by the FPR and TPR as x and y axes respectively, which depicts relative trade-off between true positive (benefits) and false positive (costs). Being crisp classifiers, the ARTMAP models appear as point aggregations where each point corresponds to a classifier with certain vigilance value (Figure 3). The analysis shows that the distributed and ic ARTMAP models tend to position close to the left-hand side, which means that they are 'conservative' in prognoses with few false positives. Unfortunately, the TPR (sensitivity) is too low for the current clinical practices, which makes these two models unacceptable for the application domain. The fuzzy model well balances sensitivity and specificity but shows moderate results being not very close to the ideal classification (top-left corner). The default ARTMAP is closer to the top line showing itself 'liberal' or providing positive classifications with less evidences for that, which makes it a bit risky (but not riskier than the fuzzy one), at the same time very sensitive. These characteristics are sought in the domain.

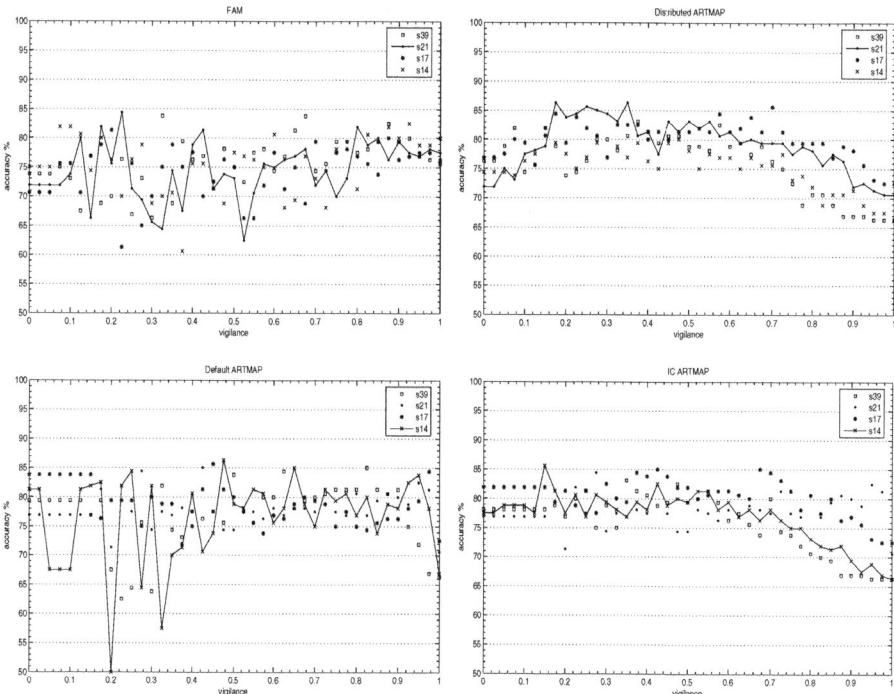

Fig. 2. Prediction accuracy of fuzzy distributed, default, and ic ARTMAP models with four feature sets: *set39*, *set21*, *set 17*, and *set14* and 41 vigilance parameter values from 0 to 1 with step of increment 0.025

We quantified our observation by ROC metrics such as area under the convex hull (AUH), partial area under the convex hull (pAUH), and specificity at a very high level of sensitivity (98%). The AUH / pAUH calculations for crisp classifiers involve trapezoidal approximation (5).

$$AUH = \frac{1}{2} \sum_{i \in ROCCH} (TPR_i + TPR_{i+1})(FPR_{i+1} - FPR_i) \qquad (5)$$

Further details can be found in Table 1. According to the figures, best overall performer (max AUH) is the default ARTMAP (0.931), followed by distributed (0.854), fuzzy (0.851), and ic (0.833).

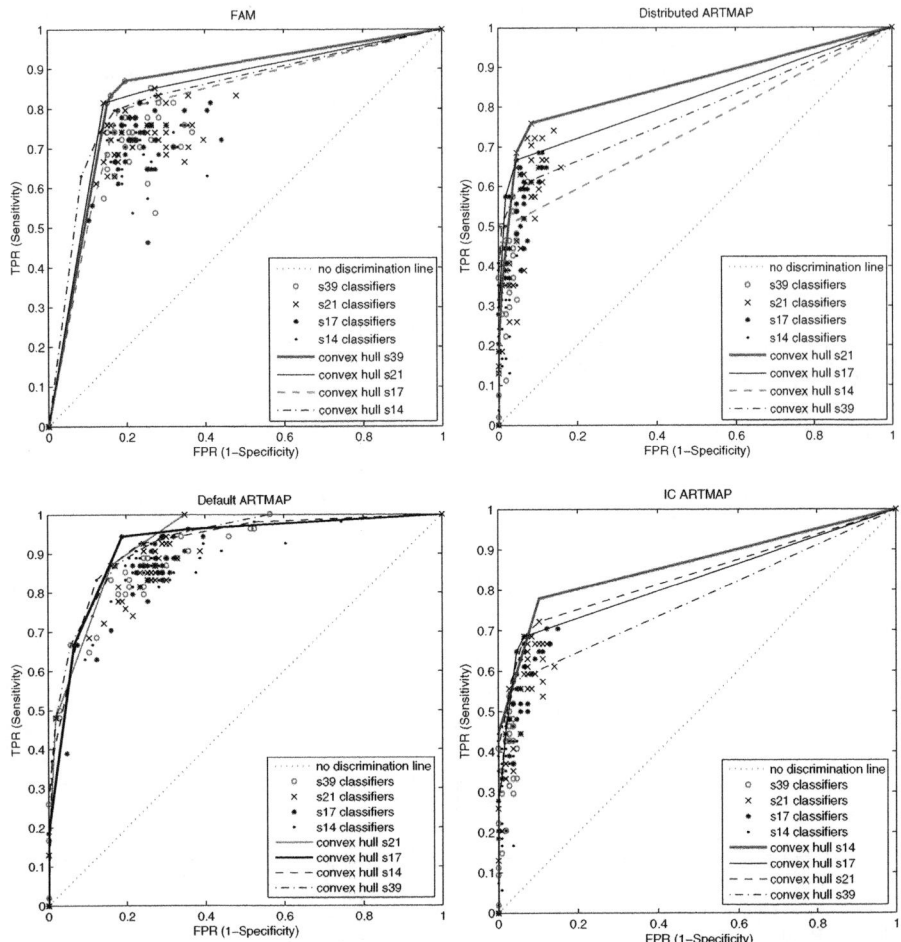

Fig. 3. ROC analysis of fuzzy distributed, default, and ic ARTMAP models with four feature sets: *set39*, *set21*, *set 17*, and *set14* and 41 vigilance parameter values from 0 to 1 with step of increment 0.025

As long as AUH provides an overall estimation of a model, the pAUH is more relevant from practitioners' viewpoint as they use to operate at high sensitivity (above 90%). The $_{0.90}$AUH values in Table 1 show that best performer is again the default ARTMAP (0.809), followed by fuzzy (0.586), ic (0.384), and distributed (0.361). The default ARTMAP outperforms all other prediction models used so far, and even the radiologist assessment (0.92) [13, 15].

We also estimated the models by specificity at 98% sensitivity – another clinically relevant metric. The table shows that the default ARTMAP significantly outperforms the other three models, and all other techniques used with the same dataset and even the radiologist assessment (0.52) [13, 15].

Table 1. Performance metrics of fuzzy, distributed, default, and ic ARTMAP neural networks with feature sets s39, s21, s17, and s14

Fuzzy ARTMAP performance	s39	s21	s17	s14
AUH	0.851	0.840	0.815	0.838
$_{0.90}AUH$	0.586	0.465	0.393	0.413
Spec / 98% sens	0.099	0.099	0.082	0.091
Acc_{max}	0.838	0.844	0.813	0.819

Distributed ARTMAP performance	S39	s21	s17	s14
AUH	0.786	0.854	0.819	0.744
$_{0.90}AUH$	0.226	0.361	0.272	0.187
Spec / 98% sens	0.047	0.075	0.057	0.039
Acc_{max}	0.831	0.863	0.856	0.819

Default ARTMAP performance	s39	s21	s17	s14
AUH	0.927	0.929	0.931	0.918
$_{0.90}AUH$	0.725	0.789	0.809	0.778
Spec / 98% sens	0.686	0.680	0.649	0.690
Acc_{max}	0.85	0.85	0.856	0.863

IC ARTMAP performance	S39	s21	s17	s14
AUH	0.776	0.833	0.821	0.860
$_{0.90}AUH$	0.215	0.306	0.282	0.384
Spec / 98% sens	0.045	0.064	0.059	0.081
Acc_{max}	0.831	0.850	0.850	0.856

5 Conclusions

This study explores and compares four neural networks based on the adaptive resonance theory: fuzzy, distributed, default, and ic ARTMAP as predictors of malignant breast masses. The networks were trained using data that contains a combination of 39 mammographic, sonographic, and other descriptors proposed by [13] which is novel for the field. To the best of our knowledge, the ARTMAP neural networks have not been studied for the purpose of the task before. Our findings show that the four models outperform the most popular MLP neural ` networks given that they are setup properly and used with appropriate selection of data variables.

We did ROC analysis in order to identify characteristics of the models and utilised metrics, such as area under the convex hull, partial area under the convex hull, and specificity at 98% sensitivity which are important from a practical viewpoint. Our results show that the distributed and ic ARTMAP tend to be conservative in their prediction which reduces the sensitivity and thus makes the models inappropriate for the task. The fuzzy model shows itself neutral by balancing the sensitivity and specificity. Best performer, however, is the default ARTMAP, which tends to be a bit risky in the prognoses, but not riskier that the fuzzy one, at the same time achieving a very high sensitivity and specificity. We find this model most appropriate for the task because such characteristics are sought in the domain, due to the low efficiency in biopsies (about 30% of all suspected cases). In terms of those characteristics the default ARTMAP outperforms all other techniques used so far with the same dataset. A limitation of all the four models, is that they require a very careful tuning.

In conclusion, we find that the default ARTMAP neural networks can be a promising technique for computer-aided detection and diagnosis tools that can help the practitioners in the field.

References

1. American College of Radiology: BI-RADS: ultrasound, 1st ed. In: Breast Imaging Reporting and Data System: BI-RADS atlas, 4th, VA: Am. College of Radiology (2003)
2. Carpenter, G., Grossberg, S., Reynolds, J.: ARTMAP: Supervised Real-Time Learning and Classification of Non-stationary Data by a Self-Organizing Neural Network. Neural Networks 6, 565–588 (1991)
3. Carpenter, G., Grossberg, S., Markuzon, N., Reynorlds, J., Rosen, D.: Fuzzy ARTMAP: A Neural Network Architecture for Incremental Supervised Learning of Analog Multidimensional Maps. IEEE Transaction on Neural Networks 3(5), 698–713 (1992)
4. Carpenter, G.: Distributed Learning, Recognition, and Prediction by ART and ARTMAP Neural Networks. Neural Networks 10(8), 1473–1494 (1997)
5. Carpenter, G., Markuzon, N.: ARTMAP-IC and Medical Diagnosis: Instance Counting and Inconsistent Cases. Neural Networks 11(2), 323–336 (1998)
6. Carpenter, G.: Default ARTMAP. In: Proceedings of the International Joint Conference on Neural Networks (IJCNN 2003), Portland, Oregon, pp. 1396–1401 (2003)
7. Chen, S., Hsiao, Y., Huang, Y., Kuo, S., Tseng, H., Wu, H., Chen, D.: Comparative Analysis of Logistic Regression, Support Vector Machine and Artificial Neural Network for the Differential Diagnosis of Benign and Malignant Solid Breast Tumors by the Use of Three-Dimensional Power Doppler Imaging. Korean J. Radiol. 10, 464–471 (2009)
8. Fawcett, T.: An introduction to ROC analysis. Pattern Recognition Letters 27(8), 861–874 (2006)
9. Goldberg, D.: Genetic Algorithms in Search, Optimization, and Machine Learning. Addison-Wesley, Reading (1989)
10. Grossberg, S.: Adaptive Pattern Recognition and Universal Encoding II: Feedback, expectation, olfaction, and illusions. Biological Cybernetics 23, 187–202 (1976)
11. Hall, M.: Correlation based feature selection for machine learning. PhD dissertation, Dept. of Computer Science, The University of Waikato, Hamilton, New Zealand (1999)
12. Jesneck, J., Nolte, L., Baker, J., Floyd, C., Lo, J.: Optimized Approach to Decision Fusion of Heterogeneous Data for Breast Cancer Diagnosis. Med. Phys. 33(8), 2945–2954 (2006)
13. Jesneck, J., Lo, J., Baker, J.: Breast Mass Lesions: Computer-Aided Diagnosis Models with Mamographic and Sonographic Descriptors. Rad. 244(2), 390–398 (2007)
14. Lacey, J., Devesa, S., Brinton, L.: Recent Trends in Breast Cancer Incidence and Mortality. Environmental and Molecular Mutagenesis 39, 82–88 (2002)
15. Nachev, A., Stoyanov, B.: An Approach to Computer Aided Diagnosis by Multi-Layer Preceptrons. In: Proc. of International Conference Artificial Intelligence, Las Vegas (2010)
16. Stavros, A., Thickman, D., Rapp, C., Dennis, M., Parker, S., Sisney, G.: Solid Breast Modules: Use of Sonography to Destinguish between Benign and Malignant Lesions. Radiology 196, 123–134 (1995)

An Adaptive Approach for Integration Analysis of Multiple Gene Expression Datasets

Veselka Boeva and Elena Kostadinova

Technical University of Sofia, branch Plovdiv
Computer Systems and Technologies Department, 4400 Plovdiv, Bulgaria
vboeva@tu-plovdiv.bg, elli@tu-plovdiv.bg

Abstract. In recent years, microarray gene expression profiles have become a common technique for inferring the relationship or regulation among different genes. While most of the previous work on microarray analysis focused on individual datasets, some global studies exploiting large numbers of microarrays have been presented recently. In this paper, we investigate how to integrate microarray data coming from different studies for the purpose of gene dependence analysis. In contrast to a meta-analysis approach, where results are combined on an interpretative level, we propose a method for direct integration analysis of gene relationships across different experiments and platforms. First, the algorithm utilizes a suitable metric in order to measure the relation between gene expression profiles. Then for each considered dataset a quadratic matrix that contains the interrelation values calculated between the expression profiles of each gene pair is constructed. Further a recursive aggregation algorithm is used in order to transform the set of constructed interrelation matrices into a single matrix, consisting of one overall inter-gene relation value per gene pair. At this stage a matrix of overall inter-gene relations obtained from previous data can be added and aggregated together with the currently constructed interrelation matrices. In this way, the previously generated integration results can, in fact, be updated with newly arriving ones studying the same phenomena. The obtained overall inter-gene relations can be considered as trade-off values agreed between the different experiments. These values express the gene correlation coefficients and therefore, may directly be analyzed in order to find the relationship among the genes.

Keywords: gene expression analysis; data integration; biomarker identification.

1 Introduction

Gene expression microarrays are the most commonly available source of high-throughput biological data. Each microarray experiment is supposed to measure the gene expression levels of a set of genes in a number of different experimental conditions or time points. Microarray experiments are often performed over many months, and samples are often collected and processed at different laboratories. Therefore, it is sensible to think of integrating related results from several laboratories in order to draw robust conclusion from microarray experiments. However, such integration

D. Dicheva and D. Dochev (Eds.): AIMSA 2010, LNAI 6304, pp. 221–230, 2010.
© Springer-Verlag Berlin Heidelberg 2010

analysis of microarray datasets is challenging, because of differences in technology, protocols and experimental conditions across datasets. Thus any microarray integration approach must be robust to such differences and should easily adjust to new data.

Different microarray combination techniques and meta-analysis studies have been published in bioinformatics literature. For instance, [4, 6, 13] all used meta-analysis to integrate microarray experiments for the detection of differential gene expression. A method which merges multiple cDNA microarrays and computes the correlations across all arrays was presented in [5]. However, this merging technique is hard to generalize to include other microarray platforms, because their expression values are typically incomparable. Another approach uses an idea to average the Pearson correlations over all datasets [7], but as it was shown in [17], averaging does not capture the realistic correlation structure. Zhou et al. also addressed the microarray integration issue in [23] by proposing a technique for functional classification of genes, called second-order correlation analysis, which utilizes the pairwise correlations of gene expression across different datasets. Other integration techniques, considered in [8, 11], present models that use inter-gene information.

In this work, in contrast to the meta-analysis approach, where results are combined on an interpretative level, we introduce a method for direct integration analysis of gene dependences across different experiments. Initially, for each dataset the proposed algorithm builds an interrelation matrix that contains the inter-gene relation values calculated between the expression profiles of each gene pair. Then applying a recursive aggregation algorithm the obtained interrelation matrices are transformed into a single matrix, which contains one overall value per gene pair. The values of this matrix can be interpreted as consensus inter-gene relations supported by all the experiments. These values as it was shown in [3, 18, 19] capture more realistically the gene correlation structure than the corresponding average values. Further instead of analyzing the spatial relation among the points in a multi-dimensional space, which is the case in [8, 11], the gene correlation coefficients are immediately expressed by the overall inter-gene relation values. The later ones may directly be analyzed and help find the relationship among the genes.

Microarray analysis is mostly split into two categories: functional classification of genes through clustering analysis and disease classification of samples by biomarker identification. Therefore, the performance of the proposed integration method is demonstrated and evaluated on the clustering of cell cycle involved genes and on the identification of marker genes for Dilated cardiomyopathy detection.

The used procedures are implemented in Perl, MATLAB and C++ and the corresponding programs are available upon request.

2   An Adaptive Gene Expression Data Integration Approach

Assume that a particular biological phenomenon is monitored in a few high-throughput experiments under n different conditions. Each experiment is supposed to measure the gene expression levels of m genes in a number of different experimental conditions or time points. Thus a set of n different data matrices $E_1, ..., E_n$ will be produced, one per experiment. The proposed integration algorithm consists of two distinctive phases: 1) construction of inter-gene relation matrices; 2) calculation of overall inter-gene relation values.

First, a suitable metric has to be chosen for the purpose of constructing an inter-gene relation matrix G_i for each considered gene expression matrix E_i ($i = 1, 2, ..., n$). Each value in the constructed interrelation matrix G_i presents a relation (correlation) between the expression profiles of the corresponding genes. Consequently, n matrices are obtained as a whole $G_1, ..., G_n$. Then at the second phase a recursive aggregation algorithm, discussed in [3] for time series data, can be applied in order to transform these n matrices into a single matrix $\mathbf{G}$, consisting of one overall interrelation value per gene pair. This aggregation algorithm has been inspired by a work on non-parametric recursive aggregation, where a set of aggregation operators is applied initially over a vector of input values, and then again over the result of the aggregation, and so on until a certain stop condition is met [19]. This process defines an aggregation operator that acts as a trade-off between the conflicting behaviour of the used operators. For instance, a set of p different aggregation operators $A_1, ..., A_p$ can be considered. The matrices $G_1, ..., G_n$ can initially be combined in parallel with these p aggregation operators. Consequently, p new matrices $G_1^0, ..., G_p^0$ (one per aggregation operator) are generated as follows: $G_i^0 = A_i(G_1, ..., G_n)$. The new matrices can be aggregated again, generating again a set of p matrices $G_1^1, ..., G_p^1$, where $G_i^1 = A_i(G_1^0, ..., G_p^0)$. In this way, each step is presented via p parallel aggregations applied over the results of the previous step, *i.e.* at step r ($r = 1, 2, ...$) matrices $G_1^r, ..., G_p^r$ are obtained and $G_i^r = A_i(G_1^{(r-1)}, ..., G_p^{(r-1)})$. Thus the final result is obtained after passing a few aggregation layers. In [19], it has been shown that any recursive aggregation process, defined via a set of continuous and strict-compensatory aggregation operators, following the algorithm described herein is convergent.

Notice that the matrix of overall inter-gene relation values $\mathbf{G}$ can easily be updated when a new study examining the same phenomenon is available. The idea behind the proposed updating procedure is that the currently produced matrix of overall inter-gene relations can be adjusted to newly arrived data by adding the matrix to the set of interrelation matrices constructed on the new datasets and then applying the recursive aggregation algorithm. The values of the resulting matrix $\mathbf{G}$ can be interpreted as the consensus interrelation values supported by all the experiments. These values can be further used to analyze and find the relationship among the genes. For example, a high inter-gene relation value $\mathbf{G}_{pr}$ will imply that genes p and r are involved in the same biological activity. Further a close distance between interrelation vectors $\mathbf{G}_p$ and $\mathbf{G}_r$ will suggest that genes p and r have high similarity with respect of their correlation with the other studied genes. Thus on one hand, gene clustering can easily be done by directly applying some cluster algorithm as *e.g.*, hierarchical clustering or k-means. For instance, if gene i is a cluster center then the corresponding cluster can easily be obtained by selecting all genes, whose inter-gene relation values in row vector $\mathbf{G}_i$ are above a given cutoff. On the other hand, traditional gene clustering analysis can be extended to take into account gene's correlation with other studied genes. Namely, genes will be grouped into a cluster if they have high similarity with respect to both their expression values and their relationships with other genes.

The proposed integration model has several advantages. First, the construction of the matrix of overall inter-gene relations is independent of the subsequent analysis. Second, the adjustment to newly available data is naturally built into the design of the integration algorithm. Third, instead of analyzing the spatial relation among the points in an n-dimensional space, which is the case in [8, 11], the gene correlation coefficients are immediately expressed by the inter-gene relation values, supported by all the experiments. The latter values as it was shown in [3, 18, 19] capture more adequately the gene correlation structure than the corresponding average values. In addition, as it was explained above, traditional gene clustering analysis can easily be extended to gene's correlation with other genes. Moreover, applying some information about the quality of microarrays, weights may be assigned to the experiments and further used in the aggregation procedure.

It is important to mention that the computational complexity of our integration algorithm on extremely large datasets may turn out very computationally intensive. We must first calculate the pairwise distance between the gene expression profiles in each given expression matrix, which implies $O(nm^2K)$ complexity for n matrices of m rows (genes). K is the cost of the applied distance algorithm ($m^2 >> K$) and it may significantly vary depending on the selected distance metric. Then at the second phase the recursive aggregation algorithm employs a set of p different aggregation operators in order to reduce the individual interrelation matrices into a single matrix, *i.e.* its computational complexity will be in the range of $O(pm^2)$. Thus the total cost of the proposed integration algorithm will be approximately $O(nm^2K)$, assuming that $n >> p$.

3   Experimental Setup

Most microarray computational studies widely fall into two groups: functional classification of genes through clustering analysis and disease classification of samples by biomarker identification. In view of this, the performance of the proposed integration algorithm has been evaluated and demonstrated on the clustering of cell cycle involved genes and on the identification of Dilated cardiomyopathy (DCM) marker genes. Therefore two types of gene expression data have been used: time series and non-time series.

3.1   Time Series Data

The time series data are microarray datasets coming from two independent studies both examining the global cell-cycle control of gene expression in fission yeast *Schizosaccharomyces pombe* [12, 15]. We have included in our test corpus all nine datasets from Rustici *et al.* [15] and the elutriation experiments from Oliva *et al.* [12]. Thus eleven different expression matrices have been used in the validation phase. In the pre-processing phase the rows with more than 25% missing entries have been filtered out from each expression matrix and any other missing expression entries have been imputed by the DTWimpute algorithm [20]. In this way eleven complete matrices have been obtained. Further, the set of overlapping genes has been found across all eleven datasets. Subsequently, the time expression profiles of these genes have been extracted from the original data matrices and thus eleven new matrices have been constructed.

3.2   Non - Time Series Data

In order to further evaluate the proposed integration model, three different non-time series gene expression matrices are included in our test corpus. These datasets were designed to identify a unique disease-specific gene expression that exists between end-stage Dilated cardiomyopathy (DCM) of different etiologies and non-failing (NF) human hearts [2, 9]. DCM is a leading cause of congestive heart failure and cardiac transplantation in Western countries [14], so that the high morbidity and mortality underscore the need for a better understanding of the underlying molecular events. However, the biological heterogeneity associated with the use of human tissue as well as the differences in platform technologies, experimental design and small sample sizes make difficult to compare the results obtained by different studies in heart failure.

To demonstrate the potential of integrating different datasets together, we have used three different microarrays which were generated from two independent laboratories using two different array platforms and experimental designs [2, 9]. The cDNA dataset from Barth *et al.* (dataset A) consists of 28 septal myocardial samples obtained from 13 DCM hearts at the time of transplantation and 15 NF donor hearts which were not transplanted. The second dataset (dataset B) was based on 7 end-stages DCM and 5 NF donor heart samples collected from the left ventricular and hybridized to Affymetrix U133A arrays [2]. The third expression matrix (referred to as dataset C) was generated by an independent laboratory [9]. It contains 37 samples representing 21 DCM, 10 Ischemic cardiomyopathy (ICM) and 6 NF human hearts and the experiments were performed with Affymetrix U133A GeneChip platform. In this work, we have focused on genome-wide expression analysis in end-stage human DCM compared to NF hearts and thus, the 10 ICM samples are not included in the present study. In summary, we have used for data analysis a total number of 67 (41 DCM and 26 NF) samples from the three matrices.

All microarray datasets are deposited in Gene Expression Omnibus database (GEO) [1]. The normalized matrices of three experiments have been downloaded and then the expression profiles of duplicated genes in each dataset have been fused by estimating their average. Further the set of overlapping genes have been found and a total of 5491 common probes have remained for further analysis. The expression profiles have been standardized by applying z-transformation across the experiments.

4 Results and Discussions

4.1 Clustering of Cell Cycle Involved Genes

The gene function classification refers to computationally predicting the function of a gene, whose function is not known yet. This is typically performed by clustering genes according to their expression similarity. Therefore, we have designed an experiment, which extracts gene clusters from the individual and integrated matrices, in order to investigate the performance of the proposed integration method. The Dynamic Time Warping (DTW) distance [16] has been chosen for the purpose of building inter-gene relation matrices, referred to as DTW distance matrices.

As benchmark sets we have used three gene clusters (*hht1*, *TF2-7* and *cdc22*), which were identified in [18] on the base of Rustici *et al.* data [15]. These clusters have been selected since the corresponding cluster center genes have some known cell cycle involvement. We have specially developed and implemented in C++ a relevant clustering algorithm based on the DTW distance. This algorithm is applied to identify a set of genes all at a maximum R DTW distance from the profile of the cluster center gene. *R* is preliminary determined as a percentage of the mean DTW distance to the gene profile in question. The performance of our method in the identification of genes from the benchmark sets is measured by calculating the coverage between the identified and benchmark clusters. A higher coverage value implies better performance of the underlying integration algorithm on the corresponding matrix [3].

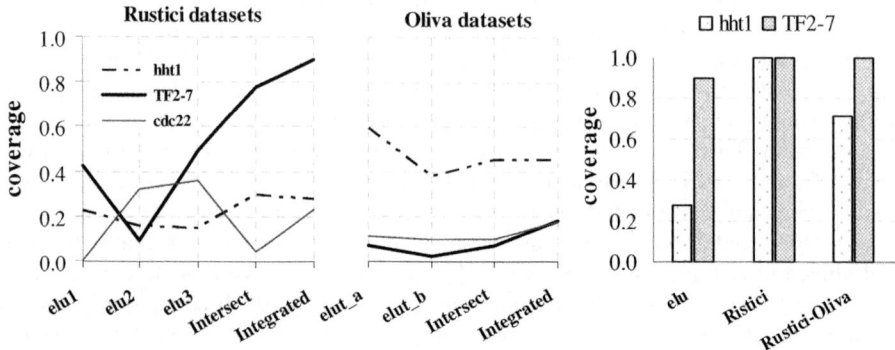

Fig. 1. Comparison of the coverage results generated on the individual and integrated matrices

Fig. 1 benchmarks the calculated coverage values for the identified gene clusters on individual and integrated DTW distance matrices. The results have been obtained for R equal to 50% of the mean DTW distance to the profile of the considered cluster center. It can easily be seen that the coverage values on the integrated DTW distances outperform the results obtained by the intersection of the individually generated clusters for elutriation experiments in both studies and in the majority of cases those given on the corresponding individual DTW matrices. In fact, the integrated results are comparable to (in most cases even better than) those given on the best performing individual dataset. This is possibly due to the fact that the proposed integration method is based on an aggregation procedure, which generates overall interrelation values adequately reflecting the values calculated in the different individual experiments.

Next the overall DTW matrix generated on elutriation datasets of Rustici *et al.* [15] has been updated by the other six datasets by adding this matrix to the DTW matrices constructed on these datasets and then applying the recursive aggregation. Further in the same fashion the obtained result has been extended to the whole test corpus including all eleven matrices. Notice that the coverage result generated on all test datasets is equal to that given on Rustici *et al.* data for *TF2-7* cluster, while the corresponding coverage value for *hht1* cluster is worse than that generated on the integrated datasets of Rustici *et al.* This may be due to the quality of Oliva *et al.* data.

4.2 Identification of DCM Marker Genes

Although, each microarray platform is possible to determine genes that can be used as bio-markers for prognosis and diagnosis, several studies, *e.g.* [10, 22], have demonstrated that these potential gene expression profiles may agree poorly, and that the sets of genes identified as differentially expressed may be very different. Therefore, the goal herein is to identify marker genes for DCM detection by integration analysis of datasets coming from two different microarray platforms. The proposed integration algorithm first translates the multiple-experiment expression values into overall inter-gene comparison indicators and then computes each gene pair's discriminative effectiveness by the distance of end-stage DCM defined groups and NF group. The most discriminative gene pairs can be then selected and further used as the marker genes for classifying new samples. In the present work, the Euclidean distance has been selected as a metric for gene similarity estimation. First, the input data have been grouped into end-stage DCM and NF samples. Thus for each considered expression matrix we have obtained two new matrices: one containing all samples from the end-stage DCM hearts and another including all samples coming from the donor hearts. In this way two groups of expression matrices (six matrices in total) have been obtained. Then an inter-gene relation matrix has been generated for each considered expression matrix and the recursive aggregation algorithm has been applied in order to transform each group of matrices into a single matrix consisting of one overall interrelation value per gene pair. In this way two integrated matrices have been produced, one per each group. Finally, these two inter-gene relation matrices have been transformed into a single one, which contains at each position the difference between the corresponding overall interrelation values of two matrices. The latter matrix can be used to identify the most discriminative gene pairs.

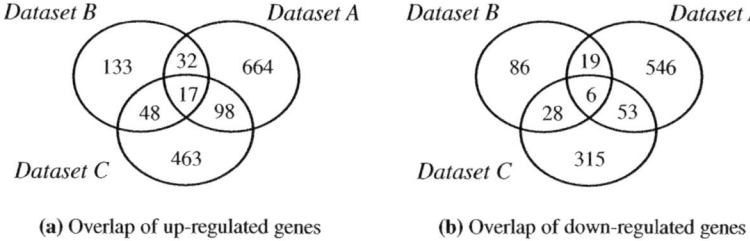

(a) Overlap of up-regulated genes   (b) Overlap of down-regulated genes

Fig. 2. Venn diagrams illustrating the number of identified (a) up- and (b) down-regulated differentially expressed genes in human DCM on each single matrix based on SAM analysis.

To validate the robustness of the proposed integration algorithm we have performed a series of experiments for identifying genes that are differentially expressed in human DCM. For instance, the well-established statistical procedure Significance Analysis of Microarray (SAM) [21] has been applied on the individual matrices. In this way three different gene lists containing significantly induced and significantly repressed expression profiles in human DCM have been generated. Considering a q-value < 0.05 as a threshold for statistical relevance and a fold-change minimum of 1.2 (similarly to the original report in [2]), we have identified 1210 differentially expressed genes in dataset A, 219 genes in B and 778 in C. The Venn's diagrams in Fig. 2 show the number of up- and down-regulated differentially expressed genes on each matrix, the number consistently identified across each pair of matrices and the number identified on all three datasets. It is obvious that only a small number of 23 genes (commonly for (a) and (b)) have been found to be consistently deregulated in the three datasets representing about 3% overlap at a single gene level.

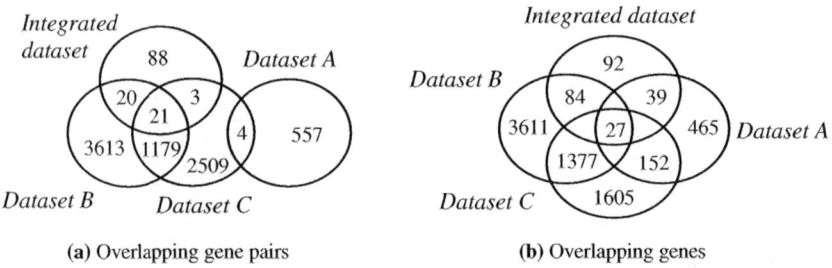

(a) Overlapping gene pairs   (b) Overlapping genes

Fig. 3. Venn diagrams illustrating the number of identified R-radius scoring gene pairs and individual genes in human DCM on the individual and integrated matrices

In order to demonstrate the advantages of the proposed data integration method, we have developed a relevant selection algorithm based on the gene pair discriminative effect. This algorithm, referred to *R-Radius Scoring Pairs* (RRSP), is applied to identify a set of pairs, which have at least R discriminative value. R is preliminary determined as a percentage of the discriminative score of the top scoring gene pair. The RRSP algorithm has been applied on the individual and integrated matrices. Thus we have selected gene pairs with most prominent changes in expression for each individual matrix and for the integrated matrix, as well. The results have been obtained for R equal to 65% of the discriminative value of the top scoring gene pair. The Venn's diagram in Fig. 3 (a) depicts the number of selected gene pairs on each individual matrix and the number of consistently identified across each pair of matrices. As one can see the agreement between the datasets coming from two different microarray platforms is remarkably poor. Namely, only a very small number of 4 common marker gene pairs between experiments A and C (versus none shared in A and B) have been identified. In contrast to this lack of consistency, the proposed integration method can always identify a set of marker gene pairs, which are supported by all the experiments. Thus although, the three individual matrices have an empty intersection, 88 most discriminative gene pairs have been found on the base of integration analysis of the individual experiments. The half of these gene pairs has also been identified as bio-markers either by dataset B either by C, but only a quarter of the selected pairs are supported by both experiments.

The Venn's diagram in Fig. 3 (b) shows the number of selected individual genes on each individual matrix, the number consistently identified across each pair of matrices and the number identified on all datasets. In this context, the intersection is not empty. However, it can again be observed that only a quarter of the genes selected on the base of the integration analysis (92 genes) are as well identified simultaneously by the individual experiments (27 genes). The potential of our integration approach has further been investigated by comparing the list of selected individual genes for R equal to 80% of the top scoring gene pair's discriminative value to those identified by applying SAM analysis on Barth *et al.* data [2]. We have found that 87.5% of the genes selected by the integrated matrix are, in fact, not presented in the intersection of the gene's lists obtained by single-set analysis performed in [2]. Evidently, the identification of discriminative gene expression signatures of DCM across different microarray platforms and independent studies may significantly be improved by applying the integration approach rather than relying only on the intersection of the single-set results. In addition, once the marker gene pairs are selected, a second-order gene analysis can be performed by additionally studying the most frequently appearing individual genes, *i.e.* those that participate in more gene pairs. Such genes may be assigned a higher importance by the classification algorithm.

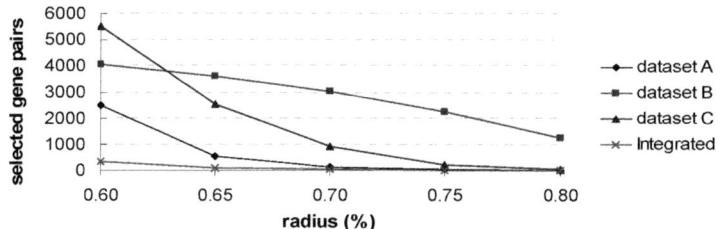

Fig. 4. Number of selected marker gene pairs for different R

Recall that the RRSP algorithm has been used to identify a set of gene pairs, which have at least R discriminative score. Therefore, we have investigated the number of selected marker gene pairs for a list of different R on the integrated matrix and on the individual ones, as well. As it can be noticed in Fig. 4 the number of the selected gene pairs varies significantly between the individual matrices. Moreover, this number is very sensitive to the different values of R on a particular matrix. In contrast to this sensitivity the results generated by the integrated matrix are more stable and balanced with respect to the different values of R.

5 Conclusion

In this paper, we have introduced a method for integration analysis of gene dependence across different experiments and platforms. It has been evaluated and validated on two microarray studies: the clustering of cell cycle involved genes and the identification of marker genes for DCM detection. The proposed method has been shown to be a robust data integration procedure, which is a flexible and independent of the subsequent analysis.

Acknowledgment

This research is supported by U904-17/2007 Bulgarian National Science Fund grant.

References

1. Barrett, T., et al.: NCBI GEO: mining tens of millions of expression profiles - database and tools update. Nucleic Acids Research 35(Database issue), 760–765 (2007)
2. Barth, A.S., et al.: Identification of a Common Gene Expression Signature in Dilated Cardiomyopathy Across Independent Microarray Studies. Journal of the American College of Cardiology 48(8), 1610–1617 (2006)

3. Boeva, V., Kostadinova, E.: A Hybrid DTW based Method for Integration Analysis of Time Series. In: Data. Proc. ICAIS 2009, Austria, pp. 49–54 (2009)

4. Choi, J.K., et al.: Combining multiple microarray studies and modeling interstudy variation. Bioinformatics 19, i84–i90 (2003)

5. Eisen, M., Spellman, P., Brown, P., Bostein, D.: Cluster analysis and display of genome-wide expression patterns. PNAS 95(25), 14863–14868 (1998)

6. Hu, P., et al.: Integrative analysis of multiple gene expression profiles with quality-adjusted effect size models. BMC Bioinformatics 6, 128 (2005)

7. Jansen, R., Greenbaum, D., Gerstein, M.: Relating whole-genome expression data with protein-protein interactions. Genome Res. 12(1), 37 (2002)

8. Kang, J., et al.: Integrating heterogeneous microarray data sources using correlation signatures. In: Ludäscher, B., Raschid, L. (eds.) DILS 2005. LNCS (LNBI), vol. 3615, pp. 105–120. Springer, Heidelberg (2005)

9. Kittleson, M.M., Minhas, K.M., Irizarry, R.A., et al.: Gene expression analysis of ischemic and nonischemic cardiomyopathy: shared and distinct genes in the development of heart failure. Physiological Genomics 21, 299–307 (2005)

10. Kuo, P.W., et al.: Analysis of matched mRNA measurements from two different microarray technologies. Bioinformatics 18, 405–412 (2002)

11. Lin, K., Kang, J.: Exploiting inter-gene information for microarray data integration. In: Proc. SAC 2007, Korea, pp. 123–127 (2007)

12. Oliva, A., et al.: The cell cycle-regulated genes of *Schizosaccharomyces pombe*. PLOS 3(7), 1239–1260 (2005)

13. Rhodes, D.R., et al.: Large-scale meta-analysis of cancer microarray data identifies common transcriptional profiles of neoplastic transformation and progression. Proc. Natl. Acad. Sci. USA 101, 9309–9314 (2004)

14. Roger, V.L., Weston, S.A., Redfield, M.M., et al.: Trends in heart failure incidence and survival in a community-based population. JAMA 292, 344–350 (2004)

15. Rustici, G., et al.: Periodic gene expression program of the fission yeast cell cycle. Nat. Genetics 36, 809–817 (2004)

16. Sakoe, H., Chiba, S.: Dynamic programming algorithm optimization for spoken word recognition. IEEE Trans. on Acoust, Speech, and S. Proc. ASSP-26, 43–49 (1978)

17. Tornow, S.: Functional modules by relating protein interaction networks and gene expression. Nucleic Acids Res. 31(21), 6283–6289 (2003)

18. Tsiporkova, E., Boeva, V.: Fusing Time Series Expression Data through Hybrid Aggregation and Hierarchical Merge. Bioinformatics 24(16), i63–i69 (2008)

19. Tsiporkova, E., Boeva, V.: Nonparametric Recursive Aggregation Process. Kybernetika, J. of the Czech Society for Cyber. and Inf. Sci. 40(1), 51–70 (2004)

20. Tsiporkova, E., Boeva, V.: Two-pass imputation algorithm for missing value estimation in gene expression time series. J. of Bioinformatics and Computational Biology 5(5), 1005–1022 (2007)

21. Tusher, V.G., Tibshirani, R., Chu, G.: Significance analysis of microarrays applied to the ionizing radiation response. Proc. Natl. Acad. Sci. USA 98, 5116–5121 (2001)

22. Yuen, T., et al.: Accuracy and calibration of commercial oligonucleotide and custom cDNA microarrays. Nucleic Acids Res. 30, e48 (2002)

23. Zhou, et al.: Functional annotation and network reconstruction through cross-platform integration of microarray data. Nature Biotechnology 23(2), 238–243 (2005)

EVTIMA: A System for IE from Hospital Patient Records in Bulgarian

Svetla Boytcheva[1], Galia Angelova[2], Ivelina Nikolova[2], Elena Paskaleva[2], Dimitar Tcharaktchiev[3], and Nadya Dimitrova[4]

[1] State University of Library Studies and Information Technologies, Sofia, Bulgaria
svetla.boytcheva@gmail.com
[2] Institute for Parallel Processing, Bulgarian Academy of Sciences
25A Acad. G. Bonchev Str., 1113 Sofia, Bulgaria
{iva,galia,hellen}@lml.bas.bg
[3] Medical University, Sofia, Bulgaria
dimitardt@gmail.com
[4] National Oncological Hospital, Sofia, Bulgaria
dimitrova.nadia@gmail.com

Abstract. In this article we present a text analysis system designed to extract key information from clinical text in Bulgarian language. Using shallow analysis within an Information Extraction (IE) approach, the system builds structured descriptions of patient status, disease duration, complications and treatments. We discuss some particularities of the medical language of Bulgarian patient records, the architecture and functionality of our current prototype, and evaluation results regarding the IE tasks we tackle at present. The paper also sketches the original aspects of our IE solutions.

Keywords: biomedical natural language processing, knowledge-based aspects of information extraction from free text, applied AI systems.

1 Introduction

Patient Records (PRs) contain much textual information. Free text paragraphs communicate the most important findings, opinions, summaries and recommendations while clinical data usually supports the textual statements or provide clarification of particular facts. Thus the essence of PR content is communicated as unstructured text message which contains various words, terms and terminological paraphrases or abbreviations. In addition the clinical documents present only partial information about the patients, thus some kind of aggregation is needed to provide an integrated view to the patient health status. However, automatic analysis of biomedical text is a complex task which requires various linguistic and conceptual resources [1]. Despite the difficulties and challenges, however, there are industrial systems and research prototypes in many natural languages, which aim at knowledge discovery and extraction of features from patient-related texts. Therefore the application of language technologies to medical PRs is viewed as an advanced but standard task which is a must in health informatics.

D. Dicheva and D. Dochev (Eds.): AIMSA 2010, LNAI 6304, pp. 231–240, 2010.

In this paper we present an IE prototype that extracts important facts from hospital PRs of patients diagnosed with different types of diabetes. The system, called EV-TIMA, is under development in a running project for medical text processing in Bulgarian language. It should be classified as an ontology-driven IE system, following the classification in [1]. The article is structured as follows: section 2 briefly comments related approaches; section 3 discusses the prototype and its functionality; section 4 deals with the evaluation and section 5 contains the conclusion.

2 Related Work

When designing our system, its rules for shallow analysis and the training corpus, we have considered the CLEF project [2] approach. Other systems which process patient symptoms and diagnosis treatment data are: the Medical Language Extraction and Encoding System which was designed for radiology reports and later extended to other domains such as discharge summaries [3]; caTIES [4] which processes surgical pathology reports; the open-source NLP system Health Information Text Extraction HITEx [5], and the Clinical Text Analysis and Knowledge extraction system cTAKES [6]. Interesting and useful ideas about processing of medical terminology and derivation of terminological variants are given in [7]. Negative statements in Bulgarian patient-related texts are studied in [8] where is given an algorithm for negation processing by treating the negative phrases as holistic text units. Another attempt for negation processing in IE for Slavic languages is shown in [9]. Actually most systems use partial text analysis and aim at the construction of structured representations of the medical notions corresponding to text units.

The IE approach arose in the late 80s as a scenario for partial natural language understanding, i.e. extraction of entities and relations of interest without full text understanding (see e.g. [10]). Various IE applications work on free text and produce the so-called templates: fixed-format, unambiguous representations of available information about preselected events. The IE steps are implemented after text preprocessing (tokenisation, lemmatisation) and the task is usually split into several components [11]:

- *Named entity recognition* – finds entities (e.g. nouns, proper names) and classifies them as person names, places, organisations etc.;
- *Coreference resolution* – finds which entities and references (e.g. pronouns) are identical, i.e. refer to the same thing;
- *Template element construction* – finds additional information about template entities – e.g. their attributes;
- *Template relation construction* – finds relations between template entities;
- *Scenario template production* – fills in the event template with specified entities and relationships.

These five steps are convenient for performance evaluation which enables the comparison of IE systems (because one needs intermediate tasks where the performance results can be measured). Groups working actively for English report: more than 95% accuracy in Named entities recognition, about 80% in template elements construction and about 60% in scenario template production [11].

Evaluation results for our present prototype will be presented in Section 4. Currently our results are quite good but concern only discovery of isolated facts in the PR texts.

3 IE in the EVTIMA System

The system presented here deals with anonymised PRs supported by the Hospital Information System (HIS) of the University Specialised Hospital for Active Treatment of Endocrinology "Acad. I. Penchev" at Medical University, Sofia. The current HIS facilitates PR structuring since the diagnosis, encoded in ICD-10 (the International Classification of Diseases v. 10), is selected via menu. The drugs prescribed to the patient, which treat the illness causing the particular hospital stay, are also supported via HIS by the so-called Computerised Provider Order Entry. In this way some important information is already structured and easy to find. However, in the PR discussion of case history, the previous diseases and their treatments are described as unstructured text only. In addition in the hospital archive we find PRs as separated text files, and these PRs consist entirely of free text. Therefore EVTIMA needs to recognise the ICD terms, drug names, patient age and sex, family risk factors and so on. It is curious to note that the list of drugs and medications is supported with Latin names by the Bulgarian Drug Agency, even for drugs produced in Bulgaria [12], but in the PR texts the medications are predominantly referred to in Bulgarian language, so the drug-related vocabulary is compiled on the fly in two languages.

3.1 Medical Language in Hospital PRs

The length of PR texts in Bulgarian hospitals is usually 2-3 pages. The document is organised in the following zones: *(i)* document identification; *(ii)* personal details; *(iii)* diagnoses of the leading and accompanying diseases; *(iv)* anamnesis (personal medical history), including current complains, past diseases, family medical history, allergies, risk factors; *(v)* patient status, including results from physical examination; *(vi)* laboratory and other tests findings; *(vii)* consult with other clinicians and *(viii)* discussion (some of these zones are be omitted in the PRs, see Table 1). Despite the clear PR fragmentation, there are various problems for automatic text processing.

Bulgarian medical texts contain a specific mixture of Cyrillic and Latin terms transcribed with Cyrillic letters, and terminology in Latin. The terms occur in the text with a variety of wordforms which is typical for the highly-inflectional Bulgarian language. The major part of the text consists of short declarative sentences and sentence phrases without agreement, often without proper punctuation marks. There are various kinds of typos in the original text which cannot be properly tackled within our research project since the project objectives exclude spell checker development. The training corpus contains only normalised texts with standard abbreviations and without spelling errors, because we aim at a research study, but the evaluation is run on test corpus with original PR texts without normalisation and correction of spell errors. Another PR text particularity is that specific descriptions are often omitted since the medical experts consider them insignificant or implicitly communicated by some other description. As reported in [13], only 86% of the PRs in our corpus discuss explicitly the patient status regarding skin colour, 63% - the fat tissue, about

42% - skin turgor and elasticity, and 13% - skin hydration. So our medical experts have to assign default values to many IE template slots.

Our present training corpus consists of 197 PRs. The test corpus contains 1000 PRs with some 6400 words, about 2000 of them being medical terms.

Table 1. Numbers of PRs without certain zones in the test corpus of 1000 PRs

Zone	1	2	3	4	5	6	7	8
No. PRs without this zone	1	0	0	3	11	79	422	484

3.2   Text Analysis and Filling in Template Slots

The specific nature of the PR language does not allow for deep syntactic analysis; therefore we use only shallow rules for feature extraction at present. Acquisition of extraction rules for features like *age, sex, illness, illness duration*, and *patient status* was performed semi-automatically. Phrases containing the focal terms were selected with window up to 5 words, left and right. These phrases were clustered and common rules were created. However the features describing organ conditions have sometimes more complex description in the PR text hence more complicated rules for their extraction are needed. These rules first fix the area where the organ description starts and are applied until some feature signals the end of the organ description. In more details the text analysis in described in [13] and [14].

About 96% of the PRs in our corpus present organ descriptions as a sequence of declarative nominal phrases, organised into short sentences. In this way the shallow analysis of phrases and sentences by cascades of regular expressions is a successful approach for structuring the statements concerning patient status. For each particular body part of interest, there is a predefined default template, where text units are stored as conceptual entities during the domain interpretation phase. There is some default sequence of discussing the patient status attributes which is kept in all PRs; this specific feature of the genre also helps to design the IE templates where the text description of patient status is to be structured. However, sometimes the PR text considers specific details regarding complications for particular patients; therefore we try to develop a strategy for dynamic template extension using conceptual resources of declarative domain knowledge [15]. Evaluation results of our current prototype are presented in Section 4.

Here we briefly discuss the project-specific aspects of our IE approach. One of them concerns the usage of domain knowledge: at the moment when certain particular template is selected for filling in and the IE system looks for template elements, the ontology provides constraints and helps to determine a text fragment. Let us consider an example of PR fragments discussing the patient status:

Sample text 1: Глава – без патологични изменения. Очни ябълки – правилно положени в орбитите, без очедвигателни нарушения. Език - суховат, зачервен. Шия – запазена подвижност. Щитовидна жлеза не се палпира увеличена, пресен несекретиращ оперативен цикатрикс. Нормостеничен гръден кош, ...

Head – without pathological changes. Eyes correctly placed in the eye-sockets, without disturbances in the eye-movements. Tongue – dry, red. Neck – preserved *mobility. Thyroid gland does not palpate enlarged, fresh non-secreting operative cicatrix. Normostenic thorax, ...*

Here the IE system finds the term *"head"* in the first sentence and runs the IE process in order to extract *head's* status, which is to be structured in the slots of a predefined template. The second and third sentences contain terms referring to body parts, linked to *"head"* by the *"has-location"* and *"part-of"* relations - *"eye"*, and *"tongue"* which is located in the *head's* part *"mouth"*. Mapping these terms to the concepts and relations in the ontology, the IE system considers sentences 1-3 as status descriptions of *"head"*, *"eye"*, and *"tongue"* correspondingly. The fourth and fifth sentences contain the terms *"neck"* and *"thyroid gland"* which usually appear in consecutive sentences. The *"neck"* is not *head's* part and therefore, the IE system considers the fourth sentence as a beginning of new body part description. The *cicatrix* in the fifth sentence refers to *"neck"* despite the fact that it is mentioned together with the *"thyroid gland"* in the same sentence. The discussion continues by presenting the *thorax* status in the sixth sentence which signals focus shift to another body part. Usually new descriptions start in another sentence but all statements in the patient status zone are mixed into one paragraph. This text genre particularity requires special efforts for recognition of the subtopics boundaries. More details about our present empirical approach to template extension are given in [15]. Currently we work with a compilation of fragments of conceptual resources which are integrated for the first time with Bulgarian vocabulary. Our target is to compile a compact domain model covering concepts and relationships relevant to diabetes.

Figure 1 illustrates one template generated to capture the information of patient status in file 100046-08. We would like to note that in the classical IE approach, when event templates for message understanding are defined, the enumeration of fixed template slots seem to be a practical design solution [11]. However, discussions of medical patient status might be extended to various details concerning organ/tissue particularities; in our corpus we found up to 45 organs mentioned for complicated medical cases. Therefore we are exploring the idea of dynamic template extension given predefined slots for the 10-15 features which are available in almost all PRs.

Another original aspect of our IE approach is the normalisation of values which includes tracking of correlated attributes. By *normalisation* we mean the classification of status attributes (found in the text as various words and phrases) into predefined scales of *good*, *worse*, and *bad* status. This empirical procedure is motivated by the observation that filling in template slots by plenty of words (text units) is not very helpful for understanding the patient status. For instance, *skin colour* is characterised by 93 words or phrasal descriptions in the training corpus; they are entered into the *skin* template after analysis whether they signal serious deviations from the normal status. In addition the text analysis procedures learn correlated attributes from the PR texts. For instance, *skin turgor* and *elasticity* are interrelated; if one of them is *decreased* then the other one is also *decreased*. But only in 42% of the PRs we find explicit descriptions of the *turgor* and *elasticity* together; in most PRs at most one of them is mentioned. Applying the correlation learned from the PRs, EVTIMA is able to change the default values of template slots: if e.g. only the *turgor* is mentioned in a PR and it is *decreased*, then the default value for *elasticity* is switched to *decreased* too. In this way the system tries to elaborate the classical IE approach for text analysis and to tune its results to the specific data and requirements in the medical domain.

3.3 Present Functionality

The EVTIMA system is designed as a stand-alone desktop application to be used by clinicians and knowledge engineers who want to study the textual data in the patient records. It is created as a scientific prototype and does not have ambitions for an industrial application. Please note that due to medical privacy and confidentiality conditions, we refrain from making a publicly available web-demo of EVTIMA.

The present system is able to work on single and multiple PRs. The implemented functionalities are the following:

- exploring and updating of the terminology bank;
- automatic segmentation of file/s in zones (as described in section 3.1);
- feature and relation extraction from the separate zones by text analysis;
- side-by-side document comparison;
- retrieval of analised documents by given features;
- export of analisises.

EVTIMA is composed of several interrelated modules. It contains a dictionary and a *terminology bank* where all vocabularies are kept and used for further IE tasks. Several modules maintain the language and conceptual resources, the PRs database and their exploration and update.

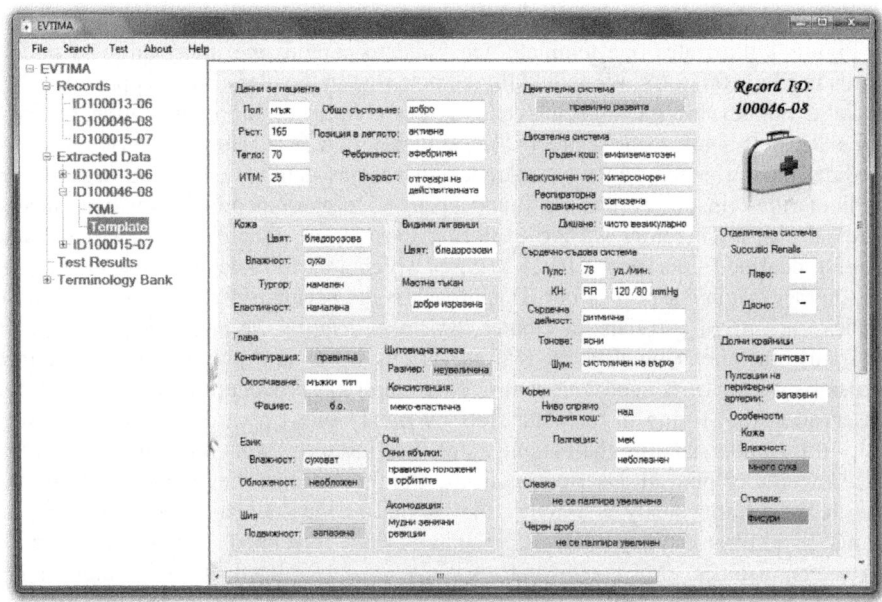

Fig. 1. An IE template filled with extracted features of the patient status

Another module is dealing with the text processing of the PRs. Once a file is loaded, the first task to be performed is automatic segmentation of the input in its semantic zones as described in section 3.1. The zones are to be found by keywords represented by regular expressions. Further text analysis operations are done separately on the so formed parts of the documents. Thus different rules are built for working on distinct areas of the text and for capturing various classes of features.

So far, recognition modules are available for the diagnosis, anamnesis and status zones. The user may run automatic recognition of the following features: *age, sex, diabetes type, diabetes term, diagnosis, skin condition, and limbs condition.*

The recognition can be done automatically and manually. When the automatic mode is chosen EVTIMA picks the most probable value for each feature. In manual mode the system offers for each predefined feature possible characteristics and their values, it is up to the user to chose which is the correct one to be stored in the analysis. Thus the system supports semi-automatic annotation of the text which is also used to validate the performance of the fully-automatic one. All saved analyses are available in XML format and could be later exported. Exports can be done in two different ways - they are either original texts enriched with annotation information or extracts of structured information captured during the analysis. An example of the second one is a template for patient status shown in Figure 1. A retrieval engine is available for document search within the repository of analysed PRs. Search criteria are the features and their characteristic values.

For each PR the system instantiates a new template object whose fields are to be filled after analysing the text. Each feature is presented as the relation of its characteristics and their values. In the future, the conjunction of the recognised features should represent the case summarisation and generalisation. The feature relations are binary. An example for this internal representation is shown in Figure 1. The feature characteristics preserved in the template are normalised to standard patient status conditions: good, worse, bad (shown correspondingly in white, yellow and red windows). The default values are shown in green windows.

4 Evaluation

Usually the Information Extraction performance is assessed in terms of three measures. The precision is calculated as the number of correctly extracted features, divided by the number of all recognised features in the test set. The recall is calculated as the number of correctly extracted feature descriptions, divided by the number of all available features in the test set (some of them may remain unrecognised by the particular IE module). Thus the precision measures the success and the recall - the recognition ability and "sensitivity" of the algorithms. The *F* measure (harmonic mean of precision and recall) is defined as

$$F = 2 * Precision * Recall / (Precision + Recall)$$

The segmentation and extraction of attributes were evaluated using a random excerpt of our test corpus of 1000 PRs. We have considered manually the IE results on 200 PRs from the test corpus.

4.1 Segmentation Task

The segmentation task even though looking as a more formal one, has been the first obstacle to overcome before analysing the PRs. Markers signaling the beginning of each section are rarely overlapping with other expressions in the PRs which allows for a high precision of the task. The PRs have been written by different clinicians without using a standard form and due to this fact often some of the sections are merged and this results in lower recall. The detailed precision and recall figures are shown in Table 2. The correct identification of these sections is of major importance for the IE tasks to be performed after the segmentation.

Both precision and recall are quite high for this task due to the comparatively strict zone mark-up in the PRs. It was interesting to notice that except from the standard keyterms, zone mark-up is strongly tight to the authors' writing style.

Table 2. Zone segmentation precision, recall and f-measure

Zone	Precision (%)	Recall (%)	F-measure (%)
#1	97.47	88.08	92.54
#2	97.43	99.48	98.44
#3	97.35	96.42	96.88
#4	98.49	99.50	98.99
#5	98.49	99.50	98.99
#6	98.66	98.28	98.47
#7	90.02	87.29	88.63
#8	90.19	99.80	94.75

4.2 Information Extraction Task

In a previous paper we have evaluated the recognition of skin characteristics [13]. These results are presented here, together with additional evaluation data [14], to allow for a more complete assessment of our IE task. Table 3 summarises the results.

Since we are using nested rules for capturing features, spelling errors in (or lack of) expressions which are to be matched on an upper level may prevent the further matching of the rules in the nested fields. We have noticed that this was one of the reasons for the comparatively low recall value of the diabetes duration, because it is dependent on the match of the diabetes acronym and diabetes type recognition. The recall for the feature sex is surprisingly low. This can be explained with the fact that there were too few samples in the anonymised training corpus and when testing on a new dataset, the available rules could not capture the relevant features. The main reason are the new author styles encountered in the enlarged corpus. These inconsistences are covered by continuous update and adjustment of the IE rules for shallow analysis.

The system design and current functionalities have been also tested by clinicians and they have given their positive opinions regarding the direction of development. Medical experts are attracted by the functionality to compare two templates side-by-side and explore dependencies which are not easy to find by reading the text of single documents. Actually, working with the EVTIMA prototype stimulates the medical

Table 3. Evaluation of IE performance for patient characteristics

Feature	Precision (%)	Recall (%)	F-measure (%)
Age	88.89	90.00	89.44
Sex	80.00	50.00	61.54
Diagnosis	98.28	96.67	97.47
Diabetes duration	96.00	83.33	89.22
Skin	95.65	73.82	81.33
Neck	95.65	88.00	91.67
Thyroid glant	94.94	90.36	92.59
Limbs	93.41	85.00	89.01

professionals to invent further functionality which can be useful in the context of their HIS. So EVTIMA is under development as a joint effort of computer scientists and medical users.

5 Conclusion

Our system is the first one which supports medical text mining for Bulgarian. The approach is language dependent and benefits from the strongly defined default discourse structure of the descriptions in some parts of the patient records. So far, we have evaluated its performance on several IE tasks dealing with the anamnesis and patient' status zones of the PRs. The results are promising and show that partial analysis is a successfull approach and needs to be developed further. EVTIMA serves as well as a tool for semi-automatic annotation of the recognised features. It supports different output formats which facilitate further processing of the obtained analysed structures. We envisage an extend of its IE capabilities to all PR zones and temporal information exctraction which will help building chronicles. Further exploration of the negative expressions and their semantic interpretation in the internal representation is also one of our future tasks. Our system will be successful if it can serve clinicians and knowledge engineers, by offering explicit and inferred knowledge and dependencies which are not directly obtainable from a single document in free text format.

There are many aspects of development and elaboration of our IE approach which have to be explored in the future. Medical patient records contain complex temporal structures which are connected to patient's case history and the family history. It would be useful for a hospital information system to monitor different hospital episodes of the patient thus supporting temporal information as well. Another interesting question is related to the automatic recognition of illness duration and the periods of drug admission. A target for our future work is to develop algorithms for discovering more complex relations and other dependences.

Acknowledgements. The research work presented here is partly supported by grant DO 02-292 "Effective Search of Conceptual Information with Applications in Medical Informatics", funded by the Bulgarian National Science Fund in 2009-2011.

References

1. Spasic, I., Ananiadou, S., McNaught, J., Kumar, A.: Text mining and ontologies in biomedicine: Making sense of raw text. Briefings in Bioinformatics 6(3), 239–251 (2005)
2. Clef Clinical E-Science Framework, Univ. of Sheffield (2008), http://nlp.shef.ac.uk/clef/
3. Friedman, C.: Towards a comprehensive medical language processing system: methods and issues. In: Proc. AMIA Annual Fall Symposium, pp. 595–599 (1997)
4. caTIES Cancer Text IE System (2006), https://cabig.nci.nih.gov/tools/caties
5. HITEx Health Information Text Extraction (2006), https://www.i2b2.org/software/projects/hitex/hitexmanual.html
6. Savova, G.K., Kipper-Schuler, K., Buntrock, J.D., Chute, C.G.: UIMA-based Clinical Information Extraction System. In: LREC 2008 Workshop W16: Towards Enhanced Interoperability for Large HLT Systems: UIMA for NLP (May 2008)
7. Valderrábanos, A., Belskis, A., Moreno, L.I.: 2002. Multilingual Terminology Extraction and Validation. In: Proc. LREC 2002 (3rd Int. Conf. on Language Resources and Evaluation), Gran Canaria (2002)
8. Boytcheva, S., Strupchanska, A., Paskaleva, E., Tcharaktchiev, D.: Some Aspects of Negation Processing in Electronic Health Records. In: Proceedings of the International Workshop Language and Speech Infrastructure for Information Access in the Balkan Countries, Held in Conjunction with RANLP 2005, Borovets, Bulgaria, pp. 1–8 (September 2005)
9. Mykowiecka, A., Marciniak, M., Kupść, A.: Rule-based information extraction from patients' clinical data. J. of Biomedical Informatics 42(5), 923–936 (2009)
10. Grishman, R., Sundheim, B.: Message understanding conference - 6: A brief history. In: Proc. 16th Int. Conference on Computational Linguistics COLING 1996, Copenhagen (July 1996)
11. Cunningham, H.: Information Extraction, Automatic. In: Encyclopedia of Language and Linguistics. Elsevier, Amsterdam (2005), http://gate.ac.uk/sale/ell2/ie/main.pdf (last visited June 2010)
12. BDA Bulgarian Drug Agency (2010), http://www.bda.bg/index.php?lang=en
13. Boytcheva, S., Nikolova, I., Paskaleva, E., Angelova, G., Tcharaktchiev, D., Dimitrova, N.: Extraction and Exploration of Correlations in Patient Status Data. In: Savova, G., Karkaletsis, V., Angelova, G. (eds.) Biomedical Information Extraction, Proc. of the Int. Workshop Held in Conjunction with RANLP 2009, Bulgaria, pp. 1–7 (September 2009)
14. Boytcheva, S., Nikolova, I., Paskaleva, E., Angelova, G., Tcharaktchiev, D., Dimitrova, N.: Structuring of Status Descriptions in Hospital Patient Records. In: The Proc. 2nd Int. Workshop on Building and Evaluating Resources for BioMedical Text Mining, Associated to the 7th Int. Conf. on Language Resources and Evaluation (LREC 2010), Malta, pp. 31–36 (May 2010)
15. Angelova, G.: Use of Domain Knowledge in the Automatic Extraction of Structured Representations from Patient-Related Texts. In: Croitoru, M., Ferre, S., Lucose, D. (eds.) Conceptual Structures: from Information to Intelligence, Proceedings of the 18th International Conference on Conceptual Structures ICCS 2010, Kuching, Malaysia, July 2010. LNCS (LNAI). Springer, Heidelberg (to appear 2010)

Direct Field Oriented Neural Control of a Three Phase Induction Motor

Ieroham S. Baruch[1], Irving P. de la Cruz[1], and Boyka Nenkova[2]

[1] CINVESTAV-IPN, Department of Automatic Control, Av. IPN No 2508,
Col. Zacatenco, A.P. 14-740, 07360 Mexico D.F., Mexico
{baruch,idelacruz}@ctrl.cinvestav.mx
[2] IIT-BAS, 1113 Sofia, Bulagaria
nenkova@riskeng.bg

Abstract. The paper proposed a complete neural solution to the direct vector control of three phase induction motor including real-time trained neural controllers for velocity, flux and torque, which permitted the speed up reaction to the variable load. The basic equations and elements of the direct field oriented control scheme are given. The control scheme is realized by nine feedforward and recurrent neural networks learned by Levenberg-Marquardt or real-time BP algorithms with data taken by PI-control simulations. The graphical results of modelling shows a better performance of the NN control system with respect to the PI controlled system realizing the same general control scheme.

Keywords: Angular velocity control, direct field oriented neural control, recurrent and feedforward neural networks, backpropagation and Levenberg-Marquardt learning, three phase induction motor.

1 Introduction

The application of Neural Networks (NN) for identification and control of electrical drives became very popular in last decade. In [1], [2], a multilayer feedforward and a recurrent neural networks are applied for a DC motor drive high performance control. In the last decade a great boost is made in the area of Induction Motor (IM) drive control. The induction machine of cage type is most commonly used in adjustable speed AC drive systems, [3]. The control of AC machines is considerably more complex than that of DC machines. The complexity arises because of the variable-frequency power supply, the AC signals processing, and the complex dynamics of the AC machine, [3], [4]. In the vector or Field-Oriented Control (FOC) methods, an AC machine is controlled like a separately excited DC machine, where the active (torque) and the reactive (field) current components are orthogonal and mutually decoupled so they could be controlled independently, [3]-[7]. There exist two methods for PWM inverter current control – direct and indirect vector control, [3]. This paper applied the direct control method, where direct AC motor measurements are used for field orientation and control. There are several papers of NN application for AC motor drive direct vector control. In [8] a feedforward NN is used for vector PW modulation,

D. Dicheva and D. Dochev (Eds.): AIMSA 2010, LNAI 6304, pp. 241–250, 2010.

resulting in a faster response. In [9] an ADALINE NN is used for cancellation of the integration DC component during the flux estimation. In [10] a fuzzy-neural uncertainty observer is integrated in a FOC system, using an estimation of the rotor time constant. In [11] an Artificial NN is used for fast estimation of the angle ρ used in a FOC system. In [12] a flux and torque robust NN observer is implemented in a FOC system. In [13], an ADALINE-based-filter and angular-velocity-observer are used in a FOC system. In [14], a NN velocity observer is used in FOC high performance system for an IM drive. In [15] a Feedforward-NN (FFNN)-based estimator of the feedback signals is used for IM drive FOC system. The paper [16] proposed two NN-based methods for FOC of IM. The first one used a NN flux observer in a direct FOC. The second one used a NN for flux and torque decoupling in an indirect FOC. The results and particular solutions obtained in the referenced papers showed that the application of NN offers a fast and improved alternative of the classical FOC schemes, [17]. The present paper proposed a neural solution of a direct FOC. The system achieved adaptation to a variable load applying real-time learned neural controllers of IM velocity, flux, and torque. In our early work, [17], the phase (a,b,c), the (q,d,0) model, and the coordinate transformation between them has been completely described, so here we may skip those parts.

2   Field Orientation Conditions and Flux Estimation

The flux and torque decoupling needs to transform the stator flux, current and voltage vectors from (a, b, c) reference frame into (q-d,s) reference frame and than to stationary and synchronous reference frames, [17]. In the next equations, the following notation is used: v –voltage, i-current, λ-flux, r-resistance, L-inductance, ω-velocity; the sub-indices are r- rotor, s-stator, q, d- components of the (q, d, 0) model; the upper index s means stator reference frame and e means synchronous reference frame; the prime means relative rotor to stator value. The Fig. 1a illustrates the current and voltage vector representations in stator and rotor synchronous frames and also the magnetic FO, where the rotor flux vector is equal to the d-component of the flux vector, represented in a synchronous reference frame ($\lambda'^e_{dr}=\lambda_r$), which is aligned with the d-component of the current in this frame. For more clarity, the current and flux orientation in the synchronous reference frame are shown on Fig. 1b. So, the FO conditions are:

$$\lambda'^e_{qr} = 0; \ p\lambda'^e_{qr} = 0; \ \lambda_r = \lambda'^e_{dr} \tag{1}$$

Taking into account that the rotor windings are shortcut, (the rotor voltage is zero), also the given up field orientation conditions, and the (q, d, 0) model, [17], we could write:

$$0 = r'_r i'^e_{qr} + (\omega_e - \omega_r)\lambda'^e_{dr}; 0 = r'_r i'^e_{dr} + p\lambda'^e_{dr} \tag{2}$$

Using the (q, d, 0) model, [17], for the q-component of the rotor flux, it is obtained:

$$\lambda'^e_{qr} = L_m i'^e_{qs} + L'_r i'^e_{qr} = 0; L'_r = L'_{1r} + L_m; i'^e_{qr} = -(L_m / L'_r)i'^e_{qs}; \tag{3}$$

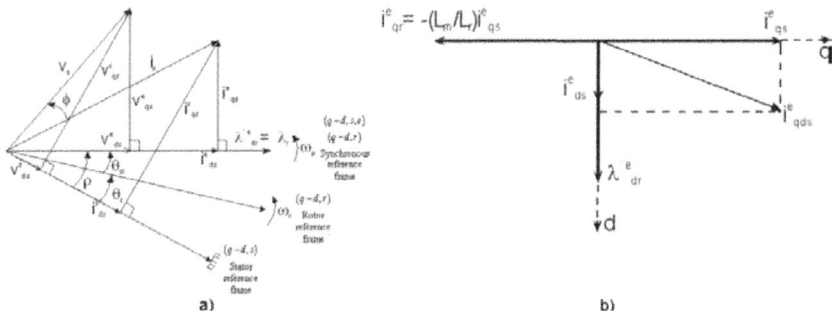

a)   b)

Fig. 1. Vector diagrams of the stator current, voltage and the rotor flux. a) The current and voltage vector representations in stator and in rotor synchronous reference frames. b) The stator current and the rotor flux vector representations in synchronous reference frame.

Using (1) and (3), the torque equation, [17], could obtain the form:

$$T_{em} = \frac{3}{2}\frac{P}{2}\frac{L_m}{L_r^{'}} \lambda_{dr}^{'e} i_{qs}^{e} \tag{4}$$

The equation (4) shows that if the flux of the rotor is maintained constant, so the torque could be controlled by the q-component of the stator current in synchronous reference frame. From the second equation of (2), taking into account (3) it is easy to obtain the slipping angular velocity as:

$$\omega_e - \omega_r = (r_r^{'}L_m / L_r^{'})(i_{qs}^{e} / \lambda_{dr}^{'e}) \tag{5}$$

The final equations (3), (4), (5) gives us the necessary basis for a direct decoupled field oriented (vector) control of the AC motor drive, where (see Fig. 1b) the q- component of the stator current produced torque and its d-component produced flux. Following [17], we could write:

$$\lambda_{qs}^{s} = (1/p)\left(v_{qs}^{s} - r_s i_{qs}^{s}\right); \lambda_{ds}^{s} = (1/p)\left(v_{ds}^{s} - r_s i_{ds}^{s}\right) \tag{6}$$

$$i_{qr}^{'s} = (\lambda_{qs}^{s} - L_s i_{qs}^{s})/L_m ; i_{dr}^{'s} = (\lambda_{ds}^{s} - L_s i_{ds}^{s})/L_m \tag{7}$$

$$\lambda_{qr}^{'s} = (L_r^{'}/L_m)\left(\lambda_{qs}^{s} - L_s^{'} i_{qs}^{s}\right); \lambda_{dr}^{'s} = (L_r^{'}/L_m)\left(\lambda_{ds}^{s} - L_s^{'} i_{ds}^{s}\right); L_s^{'} = [L_s - (L_m^2/L_r^{'})] \tag{8}$$

Now it is easy to compute the angle ρ needed for field orientation, the rotor flux, and the *sin, cos* - functions of this angle, needed for flux control, torque estimation, and coordinate transformations, which are:

$$\lambda_r^{'} = \sqrt{(\lambda_{qr}^{'s})^2 + (\lambda_{dr}^{'s})^2}; sin\rho = \lambda_{qr}^{'s}/\lambda_r^{'}; cos\ \rho = \lambda_{dr}^{'s}/\lambda_r^{'} \tag{9}$$

$$T_{em} = \frac{3}{2}\frac{P}{2}\frac{L_m}{L_r^{'}}\lambda_r^{'} i_{qs}^{e} \tag{10}$$

3   General Control Scheme and NN Realization of the IM Control

A general block diagram of the direct vector control of the Induction Motor drive is given on Fig. 2a. The direct control scheme contains three principal blocks. They are: G1, G2, G3 – blocks of PI controllers; block of coordinate (abc) to (q-d,s,e) transformation, [17]; block of vector estimation, performing the field orientation and the torque, flux and angle computations (see equations (9), (10)); block of inverse (q-d,s,e) to (a,b,c) transformation; block of the converter machine system and induction motor. The block of the converter machine system contains a current three phase hysteresis controller; a three phase bridge ASCI DC-AC current fed inverter; an induction motor model; a model of the whole mechanical system driven by the IM $((2/P)J(d\omega_r/dt)=T_{em}-T_L$, where J is the moment of inertia, T_L is the load torque). The block of vector estimation performed rather complicated computations. The Fig. 2b illustrates the flux and angle estimation for field orientation, computing (6), (8), (9). The rotor flux computations block (see Fig. 2b) performs computations given by (6), (8), illustrated by the Fig. 2c. The rotor flux, the angle, and the *sin, cos* -functions computations are given by equation (9). The torque estimation is computed by equation (10).

3.1   Neural Network Realization of the Control Scheme

The simplified block-diagram of the direct neural vector control system, given on Fig. 2a is realized by nine FFNNs. We will describe in brief the function, the topology and the learning of each FFNN. The main contribution here is the introduction of the neural P/PI velocity, flux and torque controllers which are capable to adapt the control system to load changes.

The FFNN1. The first NN1 is an angular velocity neural PI controller with two inputs (the velocity error, and the total sum of velocity errors) and one output (the torque set point). The weights learning is done in real – time using the Backpropagation (BP) algorithm. The FFNN1 function is given by the following equation:

$$T*(k) = \varphi[g_p(k)e_{vel} + g_i(k)e_{vel}^{sum}(k)] \tag{11}$$

Where: g_p and g_i are proportional and integral FFNN1 weights; φ is a *tanh* activation function; e_{vel} is a velocity error; $T*$ is the torque set point – output of the FFNN1. The integration sum of errors is:

$$e_{vel}^{sum}(k) = \sum_{k=0}^{n} e_{vel}(k) \tag{12}$$

Where n is the total number of iterations. The BP algorithm for this FFNN1 is:

$$g_p(k+1)=g_p(k)+\eta e_{vel}(k)[1-(T*(k))^2]e_{vel}(k)$$
$$g_i(k+1)=g_i(k)+\eta e_{vel}(k)[1-(T*(k))^2]e_{vel}^{sum}(k) \tag{13}$$

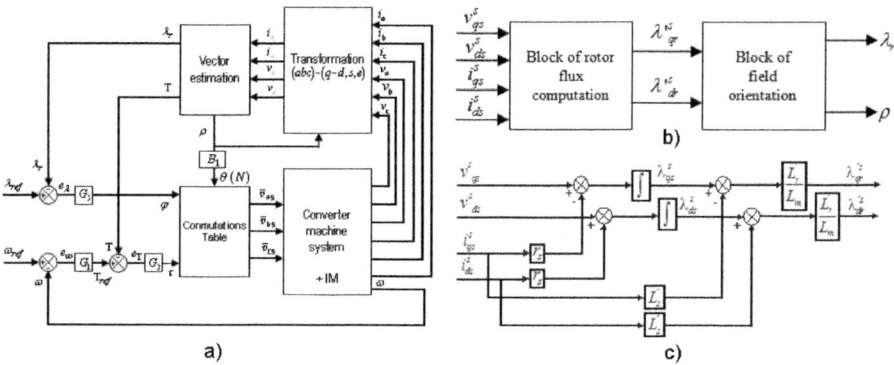

Fig. 2. Block diagrams. a) General BD of a direct IM vector control. b) BD of the vector estimation computations. c) BD of the flux estimation computations.

The FFNN2. The second FFNN2 is a torque neural P controller with one input and one output (the torque error and the stator q-current set point). The function and the real-time BP learning of this FFNN2 are given by:

$$i_{qs}^{e*}(k) = \phi[g_p(k)e_T(k)] \tag{14}$$

$$g_p(k+1) = g_p(k) + \eta e_T(k)[1-(i_{qs}^{e*}(k))^2]e_T(k) \tag{15}$$

Where: g_p is a proportional weight; Φ is a *tanh* activation function; e_T is a torque error; η is a learning rate parameter; $i_{qs}^{e}*$ is a current set point - output of FFNN2.

The FFNN3. The third FFNN3 is a flux neural PI controller with two inputs and one output (the flux error and its sum, and the stator d-current set point). The function and the real-time BP learning of this NN3 are given by:

$$i_{ds}^{e*}(k) = \varphi[g_p(k)e_{flux} + g_i(k)e_{flux}^{sum}(k)] \tag{16}$$

$$g_p(k+1) = g_p(k) + \eta e_{flux}(k)[1-(i_{ds}^{e*}(k))^2]e_{flux}(k)$$
$$g_i(k+1) = g_i(k) + \eta e_{flux}(k)[1-(i_{ds}^{e*}(k))^2]e_{flux}^{sum}(k) \tag{17}$$

Where: g_p and g_i are proportional and integral FFNN3 weights; φ is a *tanh* activation function; e_{flux} is a flux error; η is a learning rate parameter; $i_{ds}^{e}*$ is a current set point - output of FFNN3. The integration sum of errors during n iterations is:

$$e_{flux}^{sum}(k) = \sum_{k=0}^{n} e_{flux}(k) \tag{18}$$

The FFNN4. The fourth FFNN4 is a torque off-line trained neural estimator (realizing (10) equation computation) which has two inputs and one output (the rotor flux, the stator q-current, and the estimated torque). The topology of this FFNN4 is (2-10-1).

The FFNN5. The fifth FFNN5 performed a stator current (a,b,c) to (q-d,s,e) transformation, [17]. The FFNN5 topology has five inputs (three i_{as}, i_{bs}, i_{cs} —stator currents; $sin\rho$, $cos\rho$), two outputs (i_{qs}^e, i_{ds}^e – stator currents) and two hidden layers of 30 and 20 neurons each (5-30-20-2).

The FFNN6. The sixth FFNN6 performed an inverse stator current (q-d,s,e) to (a,b,c) transformation using the transpose of the transformation matrix, [17]. The FFNN6 topology is (4-30-10-3) (four inputs -two i_{qs}^e, i_{ds}^e —stator currents; $sin\rho$, $cos\rho$; three outputs- i_{as}, i_{bs}, i_{cs} – stator currents; two hidden layers of 30 and 10 neurons).

The FFNN7. The seventh FFNN7 performed rotor flux estimation using equation (9). The rotor (q-d,r) flux components $\lambda_{qs}^s, \lambda_{ds}^s$ are previously computed using equation (6) (see Fig. 2c), and they are inputs of FFNN7. The other two inputs are the stator currents: i_{qs}^s, i_{ds}^s. The FFNN7 output is the rotor flux: λ_r'. The FFNN7 topology is (4-30-10-1).

The FFNN8 and FFNN9. The FFNN8, and FFNN9 are similar to FFNN7 and performed separately the q and d rotor flux components estimation using equations (8). The FFNN8, FFNN9 topologies are: (2-10-5-2). The values of $sin\rho$, $cos\rho$, (9), needed for the coordinate transformations are obtained dividing the outputs of FFNN8, FFNN9 by the output of FFNN7.

All the FFNN4-FFNN9 are learned by 2500 input-output patterns (half period) and generalized by another 2500 ones (the other half period). The FFNN4-FFNN9 learning is off-line, applying the Levenberg-Marquardt algorithm [18], [19] during 61, 29, 32, 35, 47 and 49 epochs of learning, respectively. The final value of the MSE reached during the learning is of 10^{-10} for all the FFNN4-FFNN9.

4   Graphical Results of the Control System Modeling

The parameters of the IM used in the control system modelling are: power- 20Hp; nominal velocity – N = 1800 Rev.pm; pole number P = 4; voltage- 220 volts; nominal current – 75 A; phase number 3; nominal frequency 60 Hz; stator resistance r_s = 0.1062 Ohms; rotor resistance referenced to stator r_r'= 0.0764 Ohms; stator inductance L_s = 0.5689. 10^{-3} Henry; rotor inductance referenced to stator L_r' = 0.5689. 10^{-3} Henry; magnetizing inductance L_m = 15.4749. 10^{-3} Henry; moment of inertia J = 2.8 kg.m^2. The control system modeling is done changing the load torque in different moment of time. The Fig. 3a, b showed the angular velocity set point vs. the IM angular velocity in the general case of velocity control and particularly with load torque changes (PI and NN control). The results show that the angular velocity control system has a fast speed up response and satisfactory behaviour in the case of load change. The Fig. 4 a, b showed the flux graphics of control system with hysteresis control applying the PI control scheme and NNs. The results show a faster and better

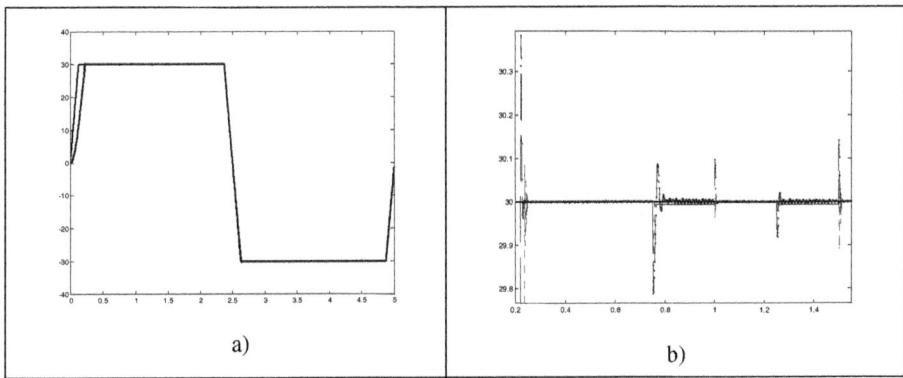

Fig. 3. Graphical results of the IM velocity control. a) General graphics of the angular velocity control; b) Graphical results of angular velocity control with load changes.

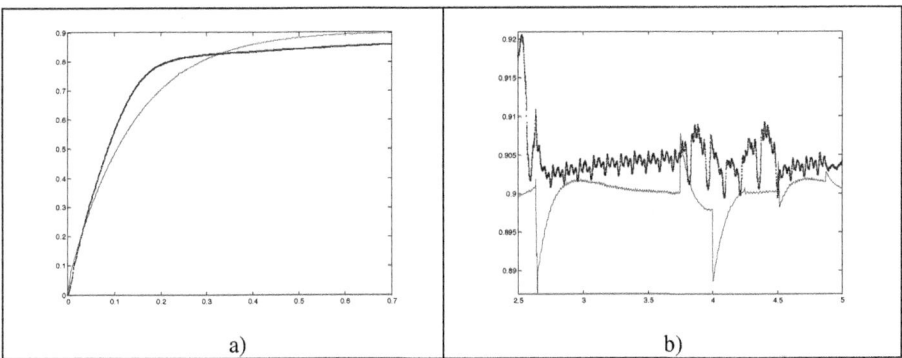

Fig. 4. Graphical results of the IM flux control. a) Graphics of the flux classical control vs. flux neural control; b) Graphics of both (classical vs. NN) flux control with load changes.

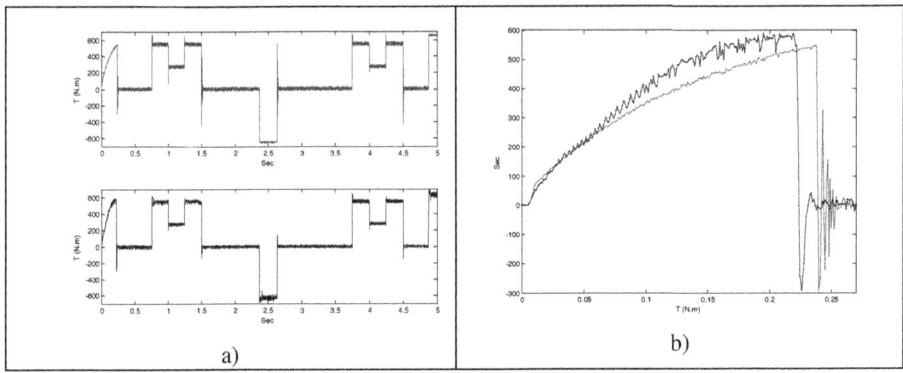

Fig. 5. Graphics of the torque control with load changes. a) Graphics of the torque classical and neural control; b) Graphics of both torque control (PI control vs. neural control) in the IM start.

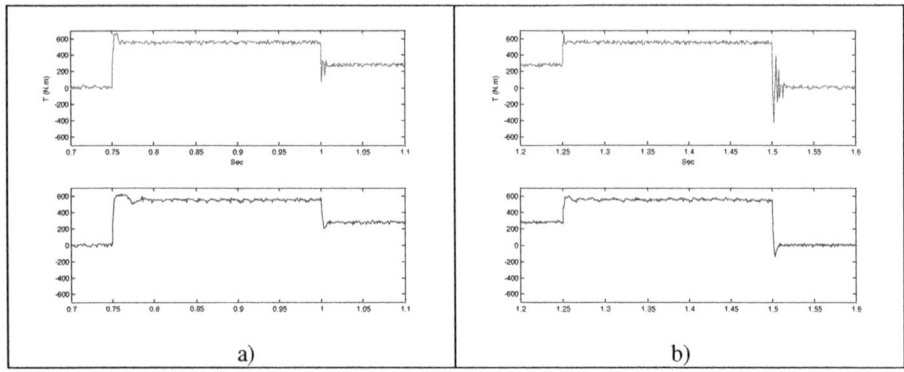

Fig. 6. Detailed graphics of the torque control using both control schemes and load changes. a) Processes from 0.7 sec. to 1.1 sec; b) Processes from 1.2 sec. to 1.6 sec.

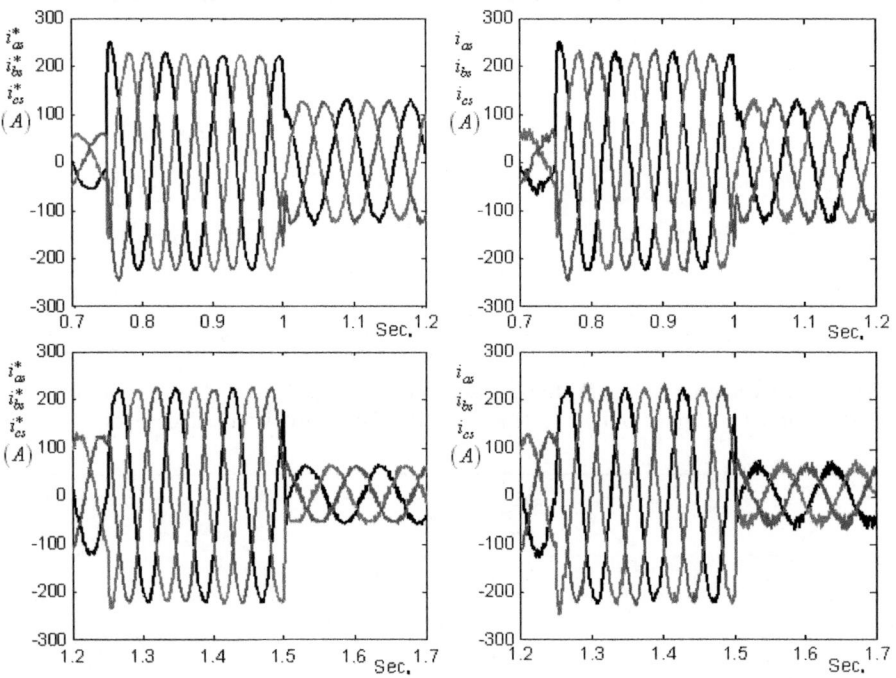

Fig. 7. Graphical results of (a,b,c) stator currents during neural control and load changes for periods of time of (0.7-1.2 sec.) and (1.2-1.7 sec.). a), c) Current set points; b), d) Currents.

response of the neural system which tried to maintain the flux constant in the case of load changes. The Fig. 5 a, b; Fig. 6 a, b; Fig. 7 a, b, c, d show the torque and current graphics with hysteresis control in the same cases and load changes.

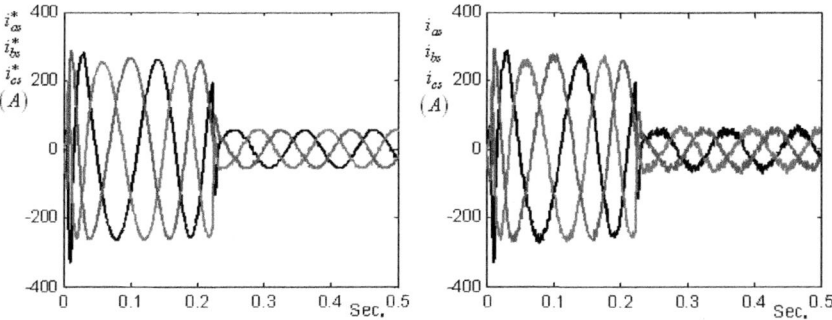

Fig. 8. Graphical Results of (a,b,c) Stator Currents During the Start of the IM. a) Current Set Points; b) Currents.

The Fig. 7 a, b, c, d shows the (a,b,c) stator currents set points and the stator currents of hysteresis controlled system using neural control schemes in load changes conditions for different time intervals. The Fig. 8 a, b shows the same stator current set points and currents during the start of the IM. The results show a good performance of the neural control system at all.

5 Conclusions

The paper proposed a complete neural solution to the direct vector control of three phase induction motor including real-time trained neural controllers for velocity, flux and torque, which permitted the speed up reaction to the variable load. The basic equations and elements of the direct FOC scheme are given. The control scheme is realized by nine feedforward neural networks learned with data taken by PI-control simulations. The NN PI or P adaptive neural controllers are learned on-line using the BP algorithm. The complementary blocks which realized coordinate and computational operations are learned off-line using the Levenberg-Marquardt algorithm with a 10^{-10} set up error precision. The graphical results of modelling shows a better performance of the adaptive NN control system with respect to the PI controlled system realizing the same computational control scheme with variable load.

Acknowledgements. The MS student Irving-Pavel de la Cruz A. is thankful to CONACYT, Mexico for the scholarship received during his studies in the Department of Automatic Control, CINVESTAV-IPN, Mexico City, Mexico.

References

1. Weerasooriya, S., El-Sharkawi, M.A.: Adaptive Tracking Control for High Performance DC Drives. IEEE Trans. on Energy Conversion 4, 182–201 (1991)
2. Baruch, I.S., Flores, J.M., Nava, F., Ramirez, I.R., Nenkova, B.: An Advanced Neural Network Topology and Learning. In: Sgurev, V., Jotsov, V. (eds.) Proc. of the 1-st Int. IEEE Symposium on Intelligent Systems, Applied for Identification and Control of a D.C. Motor, Varna, Bulgaria, pp. 289–295. IEEE, Los Alamitos (2002), ISBN 0-7803-7601-3

3. Bose, B.K.: Power Electronics and AC Drives, pp. 264–291. Prentice-Hall, Englewood Cliffs (1986)
4. Ortega, R., Loría, A., Nicklasson, P., Sira - Ramírez, H.: Passivity – Based Control of Euler – Lagrange Systems. Springer, Heidelberg (1998)
5. Ong, C.M.: Dynamic Simulation of Electric Machinery. Prentice Hall, New York (1998)
6. Novotny, D.W., Lipo, T.A.: Vector Control and Dynamics of AC Drives. Oxford University Press, New York (1996)
7. Woodley, K.M., Li, H., Foo, S.Y.: Neural Network Modeling of Torque Estimation and d-q Transformation for Induction Machine. Engng. Appl. of Artif. Intell. 18(1), 57–63 (2005)
8. Pinto, J.O., Bose, B.K., Da Silva, L., Borges, E.: A Stator – Flux – Oriented Vector – Controlled Induction Motor Drive With Space – Vector PWM and Flux – Vector Synthesis by Neural Networks. IEEE Trans. on Industry Applications 37, 1308–1318 (2001)
9. Cirrincione, M., Pucci, M., Capolino, G.: A New Adaptive Integration Methodology for Estimating Flux in Induction Machine Drives. IEEE Trans. on PE 19(1), 25–34 (2004)
10. Lin, F.J., Wai, R.J., Lin, C.H., Liu, D.C.: Decoupled Stator-Flux-Oriented Induction Motor Drive with Fuzzy Neural Network Uncertainty Observer. IEEE Trans. on Industrial Electronics 47(2), 356–367 (2000)
11. Keerthipala, W.L., Duggal, B.R., Chun, M.H.: Implementation of Field – Oriented Control of Induction Motors using Neural Networks Observers. In: Proc. IEEE International Conference on Neural Networks, vol. 3, pp. 1795–1800 (1996)
12. Marino, P., Milano, M., Vasca, F.: Robust Neural Network Observer for Induction Motor Control. In: Proc. 28-th Annual IEEE PE Specialists Conference, vol. 1, pp. 699–705 (1997)
13. Cirrincione, M., Pucci, M., Capolino, G.: A New TLS Based MRAS Speed Estimation UIT Adaptive Integration for High-Performance Induction Machine Drive. IEEE Trans. on Industry Applications 40(4), 1116–1137 (2004)
14. Cirrincione, M., Pucci, M., Capolino, G.: An MRAS Based Sensorless High Performance Induction Motor Drive with a Predictive Adaptive Model. IEEE Trans. on Industrial Electronics 52(2), 532–551 (2005)
15. Simoes, M.G., Bose, B.K.: Neural Network Based Estimation of Feedback Signals for a Vector Controlled Induction Motor Drive. IEEE Trans. on Ind. Appl. 31, 620–629 (1995)
16. Ba – Razzouk, A., Cheriti, A., Olivier, G., Sicard, P.: Field - Oriented Control of Induction Motors Using Neural Network Decouplers. IEEE Trans. on P. Elec. 12, 752–763 (1997)
17. Baruch, I.S., Mariaca-Gaspar, C.R., De la Cruz, I.P.: A Direct Torque Vector Neural Control of a Three Phase Induction Motor. In: Sossa-Azuela, J.H., Baron-Fernandez, R. (eds.) Research in Computer Science, Special Issue on Neural Networks and Associative Memories, vol. 21, pp. 131–140 (2006), ISSN 1870-4069
18. Hagan, M.T., Menhaj, M.B.: Training Feedforward Networks with the Marquardt Algorithm. IEEE Trans. on Neural Networks 5, 989–993 (1994)
19. Demuth, H., Beale, M.: Neural Network Toolbox User's Guide, version 4, The Math. Works, Inc. COPYRIGHT (1992-2002)

General Shape Analysis Applied to Stamps Retrieval from Scanned Documents

Dariusz Frejlichowski and Paweł Forczmański

West Pomeranian University of Technology, Szczecin, Faculty of Computer Science
and Information Technology, Żołnierska Str. 49, 71–210 Szczecin, Poland
{dfrejlichowski,pforczmanski}@wi.zut.edu.pl

Abstract. The main purpose of the paper is to present a method of
detection, localization and segmentation of stamps (imprints) in the
scanned document. It is a very actual topic these days since more and
more traditional paper documents are being scanned and stored on digital media. Such digital copy of a stamp may be then used to print a
falsified copy of another document. Thus, an electronic version of paper
document stored on a hard drive can be taken as a forensic evidence of
possible crime. The process of automatic image retrieval on a basis of
stamp identification can make the process of crime investigation more efficient. The problem is not trivial since there is no such thing like stamp
standard. There are many variations in size, shape, complexity and ink
color. It should be remembered that the scanned document may be degraded in quality and the stamp can be placed on relatively complicated
background. The algorithm consists of several steps: color segmentation
and pixel classification, regular shapes detection, candidates segmentation and verification. The paper includes also some results of selected
experiments on real documents having different types of stamps.

1 Introduction

Nowadays, when computer technology is present in various areas of life, the problem of computer crime is becoming more and more important. It covers both
strictly electronic and traditional types of law-breakings. On the other hand,
there are still many areas of life, where computers and digital media are employed only as tools and play just a supporting role. The most evident example
of such domain is an area associated with official documents, identity cards,
formal letters, certificates, etc. All these documents are being issued by formal
authorities and are often in a form of a paper letter consisting of several typical
elements: heading, body text, signatures and stamps which, from this historical
point of view confirm its official character. In business environments, they are
often used to provide supplemental information (date received/approved, etc).
In other words, its main purpose is to authenticate a document which in many
cases is a subject to forgery or tampering with help of modern computer means.
In general, the process of forgery consists of the following steps: obtaining the
original document, high resolution scanning, digital image manipulation and final printing. It is rather easy to recognize fake stamps, even if they are printed

D. Dicheva and D. Dochev (Eds.): AIMSA 2010, LNAI 6304, pp. 251–260, 2010.
© Springer-Verlag Berlin Heidelberg 2010

using ink-jet printers. This article addresses the problem, which is definitely not new, since the task of seal imprint identification on bank checks, envelopes, and transaction receipts have emerged from mid-1980s. On the other hand reliable recognition of stamps in the documents is not trivial and has not been solved till today [1,2,3]. The most advanced method found in the scientific literature is described in [2], where the authors present a stamp detection approach, which treats stamps as regions with analytically shaped contours, however these regions are limited to oval (round) shapes only. From the strictly technical point of view, rubber stamping, also called stamping, is a craft in which certain type of ink made of dye or pigment is applied to an image or pattern that has been carved, molded, laser engraved or vulcanized, onto a sheet of rubber. The ink coated rubber stamp is then pressed onto some type of medium such that the colored image has now been transferred to the medium. The medium is generally some type of fabric or paper. This kind of stamping has not changed for centuries (in fact it is as old as writing itself) and it is supposed that it will not change in the close future. The general motivation of the research presented in this paper is a need of semi-automatic computer software that is able to analyze an image and detect and localize different types of stamps in it. The application area of this kind of a system is broad, ranging form law-enforcement forces, law offices, official archives and any other institutions that utilize stamping technique. Instead of browsing large sets of images stored on digital media, the proposed software approach is able to retrieve images that contain objects recognized as stamps. Finally, it classifies them as one of the fixed classes associated with shape and color.

2   Stamp Detection and Localization

The algorithm of initial stamps processing is divided into several stages: detection, localization and candidates verification. Detailed descriptions of each stage are presented below.

2.1   Stamps Characteristics

It is a fundamental issue to define the features that can be employed to distinguish stamps from not-stamps and further between official and unofficial stamps. In this paper we focus on official stamps as it plays a meaningful role in practical tasks. Typical stamps which can be found on paper documents have specific characteristics which are derived from the process of stamping. These characteristics (e.g. shape, complexity, typical patterns) have evolved during many centuries into de-facto standards. The analysis of the problem shows that there are two main groups of stamps having its distinguishable properties:

 − official stamps met mostly on official documents,
 − unofficial stamps used as decoration.

Fig. 1. Sample official stamps, often regular and without decorations (left-hand side) and unofficial stamps, more complex, with many decorative motives (right-hand side)

The first group (see Fig. 1) consists of regularly-shaped objects (ovals, squares, rectangles) with clearly visible text and mere ornaments. They are often colored red or blue. On the other hand, stamps from the second group (see Fig. 1) are more fancy, irregularly-shaped, with decorative fonts and complex patterns. Thus, the features which can be used to describe stamps are divided into two classes:

1. spatial characteristics [4,5,6,7,8], including dimensions (proportions of dimensions), edge distributions, mean and variance of gradients, moment representation, low-frequency spectrum;
2. color characteristics [9,10], which include color distribution is HSV and YC_bC_r color spaces;

The stamps on official documents do not cover large area (often not more than 3%–5% of the total image area for A4 documents). It is worth noticing that we do not employ generalized Hough transform to detect circles or other regular shapes, since it is a very time-consuming. The approach is much more flexible and less complex than one presented in [2] and does not assume a stamp to have any specific shape. The color information is very important to us since we are looking for original documents which in most cases includes color stamps. The black ink is less frequently used for official stamping, and in many cases suggests a copy (not original document).

2.2 Stamp Detection

We assume that an input image of a document is stored in a file with possibly lossless compression, high spatial resolution and full color range (24-bit RGB).

Fig. 2. Sample document containing a stamp (left), C_r and C_b components of its YC_bC_r representation

First, the input image is converted into YC_bC_r color space (ITU-R BT.709 standard) in order to expose some very useful color properties:

$$\begin{cases} Y = 16 + (65.481 \cdot R + 128.553 \cdot G + 24.966 \cdot B) \\ C_b = 128 + (-37.797 \cdot R - 74.203 \cdot G + 112.0 \cdot B) \\ C_r = 128 + (112.0 \cdot R - 93.786 \cdot G - 18.214 \cdot B) \end{cases} \quad (1)$$

where: R, G, B and Y, C_b, C_r are appropriate color components.

It is worth noticing that most popular stamps are often blue or red colored, thus we select C_b and C_r components only for further analysis (see Fig. 2). In the case presented below (Official radio amateur license), several potential areas are detected and passed to the verification/classification stage.

For each matrix which represents C_b an C_r channel we perform spatial convolution filtering in order to blur the image, which eliminates small objects and

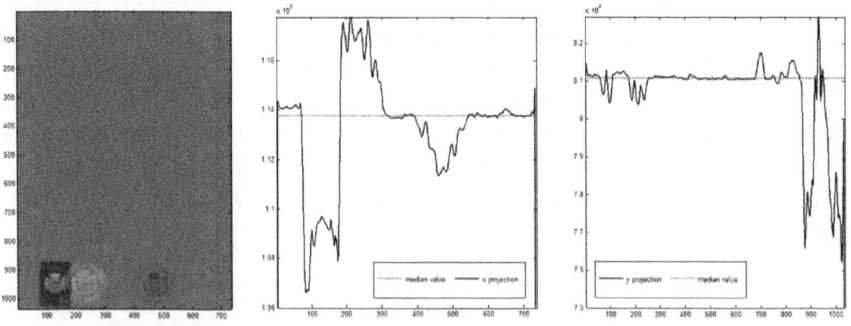

Fig. 3. Blurred C_r component (left), its column projection (middle) and row projection (right). Median values (dashed line) used for thresholding are shown.

noise. We investigated several masks of the convolution and came to the conclusion that the most optimal results concerning time of the processing and the final effect gives averaging filter which convolution size equals to 3% of the smaller side of the image.

Then we project the filtered image in horizontal and vertical directions in order to find areas of high intensity. Sample projections for one of the test images are presented below, in Fig. 3. As it can be seen, areas occupied by possible stamps are represented by values higher than median value (assumed to be a background value). This is an obvious restriction, making detection of black-colored stamps impossible, however it is coherent with our assumptions regarding detection of color stamps.

2.3   Candidates Verification

Next, the candidates for stamps are segmented and passed to the stage where two simple features are calculated: width to height proportion and standard deviation of pixel intensities. In the case of stamps collected for the experimental purposes, the proportion of width to height should be not less than 1/3 and not more that 3. This prevents the situation where relatively narrow objects are accepted. The second condition promotes objects with relatively high deviation of pixel intensities, ranging from 0.25 to 0.45 (see Fig. 4).

Sample stamps extracted from test documents are presented in Fig. 5 (blue stamps) and Fig. 6 (red stamps). As it can be seen, the detection accuracy is not influenced by text or signatures placed under a stamp.

The next stage is a coarse classification of extracted stamps. It is performed on monochromatic images taken from adequate color channel (C_r or C_b).

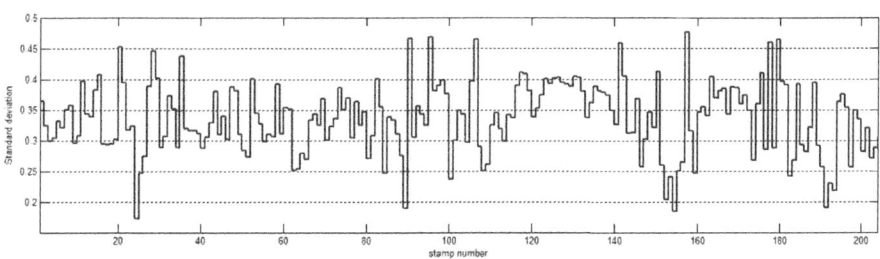

Fig. 4. The distribution of standard deviation of pixel intensities of 204 stamps from the experimental database

Fig. 5. Sample blue colored stamps retrieved from test documents

Fig. 6. Sample red colored stamps retrieved from test documents

3 Stamp Classification Based on the General Shape Analysis

3.1 Application of the General Shape Analysis to the Coarse Stamp Classification

The coarse classification of shapes is one of the most useful applications of the *General Shape Analysis* (*GSA*, [11]). In this case the number of templates is significantly limited, for example to ten instances. Those are the most general objects in the group of shapes, e.g. circle, square, triangle, star, ellipse, etc. The object under processing does not belong usually to any of the base classes. Therefore, it is possible to conclude the general information about it, e.g. how round, square, triangular, star-alike, oval, etc. it is.

The above properties of the *GSA* allow us to apply them to the evaluation of the overall stamp shape. In this case we can make a decision about this property. We can for example conclude how round or rectangular is particular stamp. It is important, because official, governmental stamps have precisely established shapes, and there are not many of them. The analysis of an extracted stamp can end here or it can be later more precisely classified within the pre-selected general group. In the work presented in this paper we are stopping on the general analysis, which indicates only the overall information about a shape, e.g. round are supposed to be governmental stamps, rectangular — doctoral, etc.

3.2 Selection of the Shape Description Algorithm for the Problem

The problem of *General Shape Analysis* was explored for various types of shapes, using seven shape descriptors, and described in [11]. Here, the algorithms are applied to particular problem of stamp analysis. In order to select the best method for this application, five shape descriptors from [11] with the best experimental results were experimentally compared: *2D Fourier Descriptors* ([12]), *Point Distance Histogram* ([13]), *Roundness* ([14]), *Moment Invariants* ([15]) and *UNL-Fourier* ([16]). For this purpose a collection of sixty various stamps of various size and rotation, extracted from documents was prepared and used. The number of templates was equal to 5 and covered the most popular stamp shapes, i.e. circle, ellipse, rectangle, triangle and hexagon. As it turned out the best results were achieved using *Point Distance Histogram* (see Table 1) and this method will be used in the final approach to the problem.

Table 1. Result of the experiment with five shape description algorithms applied to the general shape analysis of stamps

Shape descriptor	Percentage of proper analysis (1st place)
1. FD	95%
2. PDH	97%
3. Roundness	93%
4. MI	75%
5. UNL-F	67%

The *Point Distance Histogram* (*PDH*) is a combination of the histogram and the transformation of contour points into polar coordinates. As usual in the case of polar shape descriptors is starts with the calculation of the centroid ([13]):

$$O = (O_x, O_y) = \left(\frac{1}{n} \sum_{i=1}^{n} x_i, \frac{1}{n} \sum_{i=1}^{n} y_i \right), \tag{2}$$

where:
n — number of points in the contour,
x_i, y_i — a point in the contour, $i = 1, \ldots, n$.

The polar coordinates are now calculated and put into two vectors Θ^i for angles (in degrees) and P^i for radii ([13]):

$$\rho_i = \sqrt{(x_i - O_x)^2 + (y_i - O_y)^2}, \qquad \theta_i = atan \left(\frac{y_i - O_y}{x_i - O_x} \right). \tag{3}$$

The resultant values in the vector Θ^i are converted into nearest integers ([13]):

$$\theta_i = \begin{cases} \lfloor \theta_i \rfloor, & if \ \theta_i - \lfloor \theta_i \rfloor < 0.5 \\ \lceil \theta_i \rceil, & if \ \theta_i - \lfloor \theta_i \rfloor \geq 0.5 \end{cases} . \tag{4}$$

Later the elements in the vectors Θ^i and P^i are rearranged according to the increasing values in the first of them. The resultant vectors are denoted as Θ^j and P^j. For equal values in Θ^j only the one with highest corresponding value in P^j is left and the other ones are rejected. This process is performed in order to eliminate repeating angular values by selecting only the one the most distant from the centroid. Thanks to this the outer points of the contour are only selected and the method is insensitive to the internal structure of a stamp. In result we achieve two vectors with at most 360 elements, one for each integer angle (in degrees). The vector Θ^j is not required for further work. Only the new vector P^k is used, where $k = 1, 2, \ldots, m; m \leq 360$. The elements in P^k are normalized according to the maximal one ([13]):

$$M = \max_{k} \{\rho_k\}, \qquad \rho_k = \frac{\rho_k}{M}. \tag{5}$$

Finally, the histogram is built — elements in P^k are assigned to r bins in histogram, ρ_k to l_k ([13]):

$$l_k = \begin{cases} r, & if \ \rho_k = 1 \\ \lfloor r\rho_k \rfloor, & if \ \rho_k \neq 1 \end{cases}. \tag{6}$$

In order to evaluate the dissimilarity measure between two received shape descriptions (denoted as h_1 and h_2) the L_2 norm was used ([17]):

$$L_2(h_1, h_2) = \sqrt{\sum_i ((h_1(i) - h_2(i))^2}. \tag{7}$$

In the experiments described earlier in this subsection the parameter r was equal to 25. This value was pre-established experimentally. The selected results for *PDH* descriptor in the general analysis of stamps are provided in Fig. 7.

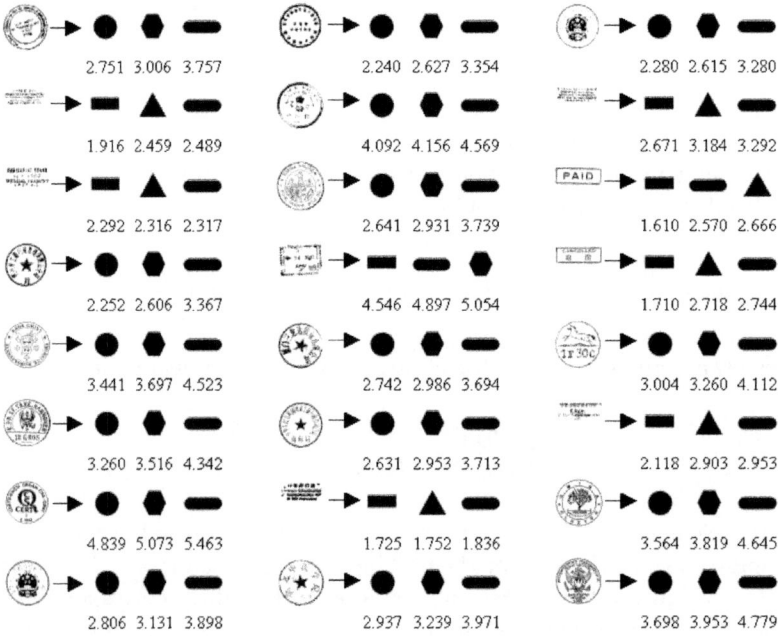

Fig. 7. Exemplary results of applying the PDH algorithm to the general shape analysis of stamps. Three the most similar templates (general shapes) are presented, along with the dissimilarity measures achieved for them.

4   Summary

In the paper the problem of stamp extraction and coarse classification was considered. According to our best knowledge this problem has not been discussed in the scientific literature (except [2]). In comparison to the mentioned approach, our algorithm is more flexible, since it detects not only oval (round) stamps. Firstly, the stamp detection was performed, using the YC_bC_r color channels as well as the horizontal and vertical projections. The coarse classification was later performed by means of the *General Shape Analysis*. Five shape descriptors were experimentally compared and amongst them the best result was achieved using the *Point Distance Histogram*. Our experiments were conducted on 170 different scanned color documents and showed that in more than 82% cases the stamp was successfully segmented and then in 97% successfully classified. This can be considered very promising since in some cases the scanned documents were of poor quality and sometimes a color was different from the fixed standards. Nevertheless, the above problem can be easily solved using more detailed color separation based on custom color model and Gram-Schmidt orthonormalization method [18]. Moreover, the classification stage can be performed more precisely. However, in that case, the comparison with the coarse classification presented in this paper is necessary. Also the more precise analysis of stamps can be performed. Not only the shape but some other characteristics can be analyzed, e.g. the text on a stamp.

References

1. Ueda, K., Nakamura, Y.: Automatic verification of seal impression patterns. In: Proc. 7th. Int. Conf. on Pattern Recognition, pp. 1019–1021 (1984)
2. Zhu, G., Jaeger, S., Doermann, D.: A robust stamp detection framework on degraded documents. In: Proceedings - SPIE The International Society For Optical Engineering, vol. 6067 (2006)
3. Zhu, G., Doermann, D.: Automatic Document Logo Detection. In: The 9th International Conference on Document Analysis and Recognition (ICDAR 2007), pp. 864–868 (2007)
4. Pham, T.D.: Unconstrained logo detection in document images. Pattern Recognition 36, 3023–3025 (2003)
5. Zhang, D., Lu, G.: Review of shape representation and description techniques. Pattern Recognition 37, 1–19 (2004)
6. Loncaric, S.: A survey on shape analysis techniques. Pattern Recognition 31, 983–1001 (1998)
7. Mehtre, B.M., Kankanhalli, M.S., Lee, W.F.: Shape measures for content based image retrieval: a comparison. Information Proc. & Management 33, 319–337 (1997)
8. Wood, J.: Invariant pattern recognition: a review. Pattern Recognition 29, 1–17 (1996)
9. Deng, Y., Manjunath, B.S., Kenney, C., Moore, M.S., Shin, H.: An Efficient Color Representation for Image Retrieval. IEEE Transactions on Image Processing 10(1), 140–147 (2001)
10. Manjunath, B.S., Ohm, J.-R., Vasudevan, V.V., Yamada, A.: Color and Texture Descriptors. IEEE Transactions on Circuits and Systems for Video Technology 11(6), 703–715 (2001)
11. Frejlichowski, D.: An Experimental Comparison of Seven Shape Descriptors in the General Shape Analysis Problem. In: Campilho, A., Kamel, M. (eds.) Image Analysis and Recognition. LNCS, vol. 6111, pp. 294–305. Springer, Heidelberg (2010)

12. Kukharev, G.: Digital Image Processing and Analysis. SUT Press (1998) (in Polish)
13. Frejlichowski, D.: The Point Distance Histogram for Analysis of Erythrocyte Shapes. Polish Journal of Environmental Studies 16(5b), 261–264 (2007)
14. Nafe, R., Schlote, W.: Methods for Shape Analysis of two-dimensional closed Contours — A biologically important, but widely neglected Field in Histopathology. Electronic Journal of Pathology and Histology 8(2) (2002)
15. Rothe, I., Süsse, H., Voss, K.: The method of normalization to determine invariants. IEEE Trans. On Pattern Anal. and Mach. Int. 18, 366–375 (1996)
16. Rauber, T.W.: Two-dimensional shape description. Technical Report: GR UNINOVA-RT-10-94, Universidade Nova de Lisboa (1994)
17. Miklasz, M., Aleksiun, P., Rytwinski, T., Sinkiewicz, P.: Image Recognition Using the Histogram Analyser. Computing, Multimedia and Intelligent Techniques: Special Issue on Live Biometrics and Security 1(1), 74–86 (2005)
18. Borawski, M., Forczmanski, P.: Orthonormalized color model for object detection. Computing, Multimedia and Intelligent Techniques: Special Issue on Live Biometrics and Security 1(1), 125–132 (2005)

Selection of Foundational Ontology for Collaborative Knowledge Modeling in Healthcare Domain

Farooq Ahmad and Helena Lindgren

Department of Computer Science, Umeå University,
SE 90187 Umeå, Sweden
{farooq,helena}@cs.umu.se

Abstract. Ontology design is an important process for structuring knowledge to be reused in different projects in the health domain. In this paper, we describe an ontology design for the collaborative knowledge building system ACKTUS to be used for developing personalized knowledge applications for different domains. Different foundational ontologies were compared with respect to selected criteria considered vital for the project, such as modularity and descriptiveness.

1 Motivation

The objective of the work presented in this paper is to propose a foundational ontology selection process for developing decision support systems in the health domain. Involving health professionals in the modeling of knowledge is a challenging task, since the use of formal frameworks and systems requires knowledge in formalization and is often time consuming. ACKTUS is being developed as a common architecture and framework for facilitating knowledge modeling in the health domain [1]. The resulting knowledge systems that are created will be web applications built on semantic web techniques that enable users to contribute with their experiences and knowledge. The purpose of the initial prototype was to provide the expert physicians participating in knowledge engineering an intuitive framework and system for modeling, validating and maintaining knowledge integrated in a decision support system for dementia care [1]. However, the ontology of the framework is currently being generalized to provide a common base for additional domains such as monitoring of health in work environments and rehabilitation. In this paper we draw a comparison to select a foundational ontology for OntACKTUS, a generalized ontology for the purpose. OntACKTUS should provide a base for the acquisition and management of medical and health related knowledge in a collaborative knowledge engineering process.

2 Results and Conclusions

Ontology organizes knowledge about some specific domain in a structured system of concepts, properties and their relationships. To overcome problem of ontology design methods, Guarino and Welty proposed a method for ontology analysis and cleaning [2]. With few extensions, a method was proposed to explicitly dividing ontology into

D. Dicheva and D. Dochev (Eds.): AIMSA 2010, LNAI 6304, pp. 261–262, 2010.

the layers *foundational* ontology, *core* and *domain* ontology for reusability across different domains and within the domain [3]. There is large number of foundational ontologies available [3], [4], [5] such as DOLCE, GFO, BFO, Cyc and SUMO. [4], [5] provide extensive analyses about selection of foundational ontologies by giving multipoint criteria for selection. Different options described by [4] that are considered important for our work, are *descriptiveness, multiplicative* and *perdurantism*. There are also some other qualities that are relevant like modularity, availability in a recommended language by the W3C and modeling flexibility for information objects. Table 1 shows the criteria selected for comparison of different features available in foundational ontologies [4], [5]. The comparison made shows that DOLCE sufficiently fulfills our requirements (Table 1). In addition, DOLCE provides careful formation, strong logical axiomatisation for the removal of conceptual ambiguities, availability of lighter versions for minimal use and further extendibility.

Table 1. Comparison of ontologies (NA *information not available*; - *not present*; **X** *present*)

	DOLCE	BFO	GFO	Cyc	SUMO
Descriptiveness	X	-	X	X	X
Multiplicative	X	-	NA	NA	X
Perdurantism	X	X	X	NA	X
Modularity	X	X	X	-	-
OWL Availability	X	X	X	X	X
Modeling of information objects	X	-	NA	-	-
Availability of domain ontologies	X	-	-	X	X

OntACKTUS is being further developed using DOLCE and is evaluated and modified in an iterative process as a part of the application developments. In our ontology design, the basic theme is re-usability and portability of available resources, and designing of new resources in such a way that it should be reusable and modifiable.

References

1. Lindgren, H., Winnberg, P.: Collaborative and Distributed Guideline Modeling in the Dementia Domain – An Evaluation Study of ACKTUS. In: Poster Proc. MEDINFO 2010, Kapetown, South Africa (2010)
2. Guarino, N., Welty, C.: Towards a methodology for ontology-based model engineering. In: ECOOP 2001 Workshop on Model Engineering, Cannes, France (2000)
3. Temal, L., Dojat, M., Kassel, G., Gibaud, B.: Towards an ontology for sharing medical images and regions of interest in neuroimaging. J. of Biomedical Informatics 41(5), 766–778 (2008)
4. Oberle, D.: Semantic Management of Middleware. In: The Semantic Web and Beyond. vol. I, Springer, New York (2006)
5. Mascardi, V., Cordì, V., Rosso, P.: Comparison of Upper Ontologies. In: Baldoni, M., Boccalatte, A., De Paoli, F., Martelli, M., Mascardi, V. (eds.) Conf. on Agenti e industria: Applicazioni Tecnologiche Degli Agenti Software, pp. 55–64 (2007)

Towards Ontological Blending

Joana Hois, Oliver Kutz, Till Mossakowski, and John Bateman

SFB/TR 8 Spatial Cognition, University of Bremen, Germany
{joana,okutz,till}@informatik.uni-bremen.de,
bateman@uni-bremen.de

Abstract. We propose ontological blending as a new method for 'creatively' combining ontologies.

Keywords: Ontologies, Creativity in AI, Blending, Algebraic Semiotics.

In contrast to other combination techniques that aim at integrating or assimilating categories and relations of thematically closely related ontologies, blending aims at 'creatively' generating new categories and ontological definitions on the basis of input ontologies whose domains are thematically distinct but whose specifications share structural or logical properties. As a result, ontological blending can generate new ontologies and concepts and allows a more flexible technique for ontology combination than existing methods. The approach is inspired by conceptual blending in cognitive science, and draws on methods from ontological engineering, algebraic specification, and computational creativity in general.

Well-known techniques directed towards unifying the semantic content of different ontologies, namely techniques based on matching, aligning, or connecting ontologies, are ill-suited for generating new conceptual schemas from existing ontologies as suggested by the general methodology of conceptual blending introduced by Fauconnier and Turner [3]: here, the blending of two thematically rather different *conceptual spaces* yields a new conceptual space with emergent structure, selectively combining parts of the given spaces whilst respecting common structural properties. A classic example for this is the blending of the theories of *house* and *boat* yielding as blends the theories of *houseboat* and *boathouse*, but also the blended theory of *amphibious vehicle* [6].

Conceptual blending inspires a structural and logic-based approach to 'creative' ontological engineering which allows the creation of new ontologies with emergent structure. Ontologies developed this way can be used, e.g., for applications in the area of computational creativity or analyses of artistic processes [2]. We believe that the principles governing ontological blending are quite distinct from the rather loose principles employed in blending phenomena in language or poetry, or the rather strict principles ruling blending in mathematics.

Our approach to ontological blending follows a line of research in which blending processes are primarily controlled through mappings and their properties [5,4,12]. By introducing blending to ontology languages, we propose a technique to combine two thematically different ontologies to create the *blendoid*, an ontology describing a newly created domain. The blendoid creatively mixes information from input ontologies on the basis of their structural commonalities and combines their axiomatisation, raising the following challenges: (1) when combining the terminologies of two ontologies, the

D. Dicheva and D. Dochev (Eds.): AIMSA 2010, LNAI 6304, pp. 263–264, 2010.

shared semantic structure is of particular importance to steer possible combinations; this shared semantic structure leads to the notion of base ontology and the problem of computing it. (2) Having established a shared semantic structure, there is typically still a huge number of blending possibilities: here, optimality principles for selecting blends take on a central role. We approach these challenges as follows: we

- differentiate alignment, matching, analogical reasoning, and conceptual blending, vis-à-vis ontological blending;
- give an abstract definition of ontological blendoids capturing the basic intuitions of conceptual blending in the ontological setting;
- provide a structured approach to ontology languages, in particular to OWL-DL, by defining the language hOWL. This combines the simplicity and good tool support for OWL with the more complex blending facilities of OBJ3 [7] or Haskell [8];

The tool HETS, the HETCASL [11] language, and in particular hOWL, provide an ideal starting point for developing the algorithmic side of the theory further. They (1) support various ontology language and their heterogeneous integration, and allow the specification of theory interpretations and other morphisms between ontologies [9]; (2) support the computation of colimits as well as the approximation of colimits in the heterogeneous case [1]; (3) provide (first) solutions for automatically computing a base ontology through ontology intersection [10].

These issues constitute almost completely new research questions in ontology research.

References

1. Codescu, M., Mossakowski, T.: Heterogeneous colimits. In: Proc. of MoVaH 2008 (2008)
2. Colton, S.: Towards ontology use, re-use and abuse in a computational creativity collective. In: Kutz, O., Hois, J., Bao, J., Grau, B.C. (eds.) Modular Ontologies – Proceedings of the Fourth International Workshop (WoMO'2010), pp. 1–4. IOS Press, Amsterdam (2010)
3. Fauconnier, G., Turner, M.: The Way We Think: Conceptual Blending and the Mind's Hidden Complexities. Basic Books, New York (2003)
4. Forbus, K., Falkenhainer, B., Gentner, D.: The structure-mapping engine. Artificial Intelligence 41, 1–63 (1989)
5. Gentner, D.: Structure mapping: A theoretical framework for analogy. Cognitive Science 7(2), 155–170 (1983)
6. Goguen, J.A., Harrell, D.F.: Style: A Computational and Conceptual Blending-Based Approach. In: The Structure of Style: Algorithmic Approaches to Understanding Manner and Meaning. Springer, Heidelberg (2009)
7. Goguen, J.A., Malcolm, G.: Algebraic Semantics of Imperative Programs. MIT Press, Cambridge (1996)
8. Kuhn, W.: Modeling the Semantics of Geographic Categories through Conceptual Integration. In: Egenhofer, M.J., Mark, D.M. (eds.) GIScience 2002. LNCS, vol. 2478, pp. 108–118. Springer, Heidelberg (2002)
9. Kutz, O., Lücke, D., Mossakowski, T.: Heterogeneously Structured Ontologies—Integration, Connection, and Refinement. In: Proc. KROW 2008. CRPIT, vol. 90, pp. 41–50. ACS (2008)
10. Kutz, O., Normann, I.: Context Discovery via Theory Interpretation. In: Workshop on Automated Reasoning about Context and Ontology Evolution, ARCOE 2009, IJCAI 2009 (2009)
11. Mossakowski, T., Maeder, C., Lüttich, K.: The Heterogeneous Tool Set. In: Grumberg, O., Huth, M. (eds.) TACAS 2007. LNCS, vol. 4424, pp. 519–522. Springer, Heidelberg (2007)
12. Pereira, F.C.: Creativity and Artificial Intelligence: A Conceptual Blending Approach. In: Applications of Cognitive Linguistics (ACL), vol. 4. Walter de Gruyter, Berlin (2007)

Integration of Ontology with Development of Personalized E-Learning Facilities for Dyslexics

Tatyana Ivanova[1], Rumen Andreev[2], and Valentina Terzieva[2]

[1] College of Energetics and Electronics, TU-Sofia, Blvd. Kl. Ohridski 8, 1000 Sofia, Bulgaria
[2] Inst. of Comp. & Comm. Systems - BAS, Acad. G. Bonchev Str. Bl. 2, 1113 Sofia, Bulgaria
`tiv72@abv.bg`, `{rumen,valia}@isdip.bas.bg`

Abstract. The paper describes a framework of adaptable e-learning environment for development of personalized e-learning facilities for dyslexic pupils. One of its basic components is reuse platform that is necessary for effective production and usage of e-learning resources. An ontology-based approach to design of this platform is represented by an open semantic model.

Keywords: Ontology, Personalization, E-Learning, Dyslexia, Agent approach.

1 Framework of Adaptable E-Learning Environment

Dyslexic people have learning difficulties in reading, writing, mathematics, orientation (in time and/or space) and coordination [1]. Their learning difficulties are due to different violations of sense channels in receiving and processing environment signals, i.e. they have problems with their cognitive abilities. That is why they are determined as people with specific learning (cognitive) disabilities that need of special education. It could be achieved by development of e-learning system that has to ensure *partnership* between educators and learners in learning and *adaptation* of learning facilities to specific cognitive abilities of individual dyslexic learner. The development of personalized e-learning facilities requires design of adaptable e-learning environment that supports production and delivery of learning resources. We suggest a framework of such adaptable e-learning environment – see Fig. 1 [2].

For effective development of personalized facilities the flexible realization of reuse strategy is needed. It has to support collaboration of teachers at different stages of learning resources authoring, as well as learners in dynamic automated personalized finding and selecting additional resources. An approach to its achievement is usage of machine-processable semantic representation of models of dyslexic pupil, teaching, subject domain, contextualization, pedagogical methods, i.e. application of ontologies for reuse implementation. For representation of complex learning resources that consist of several media types and have clear structure, the advantages of domain ontologies (*Precise Domain Ontology, Upper Domain Ontology, Etalon Ontologies, Dyslexia Document Structure ontology*), pedagogical ontologies, collaborative web 2.0 like ontologies (*Collaborative Learning Ontology, Tag ontology, MOAT, FOAF* for web 2.0 tagging and collaboration semantic) are taken. The adaptation and personalization of resources demand the use of *Dyslexia Profile Ontology and Learning Object Context Ontology*. Ontologies propose possibilities to avoid natural language

D. Dicheva and D. Dochev (Eds.): AIMSA 2010, LNAI 6304, pp. 265–266, 2010.

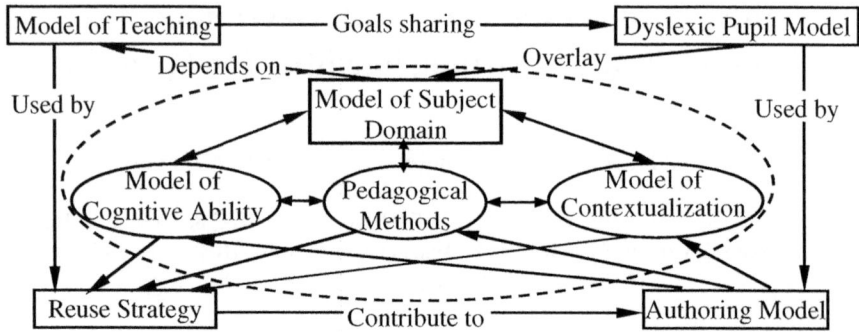

Fig. 1. Framework of adaptable e-learning environment

ambiguity and to use computers effectively in knowledge-management, exchange and reasoning. They ensure semantic search for specialized resources.

2 Agent-Based Framework for Resource Reuse

We present an open agent-based framework that supports automated ontology-based learning facilities production, composition and use, as well as management, (re)use and retrieval of external resources. It includes *Knowledge expert agent, Teacher agent, Dyslexic learner personal agent, Collaboration agent, Evaluator agent, Ontology search agent, Web search agent, Mapping, Annotation agent* and *External resource agent.* The above mentioned specialized ontologies, collaborative and goal-oriented multi agent system ensure a flexible adaptive environment for collaborative development of e-learning facilities for personalized education of dyslexic learners.

3 Conclusion

The design of a reuse platform for supporting the development of personalized e-learning facilities for dyslexic learners is a step to ensure effective and efficient adaptable e-learning environment. It uses capabilities of semantic and agent-based technologies to achieve flexibility and semi-automation. The knowledge-management and exchange during development of e-learning resources enables innovation.

References

1. Davis, R.: The Gift of Learning. The Berkley Publishing Group (2003)
2. Andreev, R., Terzieva, V., Kademova-Katzarova, P.: An Approach to Development of Personalized E-learning Environment for Dyslexic Pupils' Acquisition of Reading Competence. In: Rachev, B., Smrikarov, A. (eds.) CompSysTech 2009. ACM International Conference Proceeding Series, vol. 433. ACM, New York (2009)

A Meta Learning Approach: Classification by Cluster Analysis

Anna Jurek, Yaxin Bi, Shengli Wu, and Chris Nugent

School of Computing and Mathematics, University of Ulster, Jordanstown campus,
Shore Road, Newtownabbey, Co. Antrim, UK
BT37 0QB
jurek-a@email.ulster.ac.uk, {Y.Bi,S.Wu1,CD.Nugent}@ulster.ac.uk

Abstract. This paper describes a new meta-learning technique of combining multiple classifiers based on cluster analysis.

Keywords: Combining Classifiers, Stacking, Clustering, Meta-Learning.

1 Introduction

One of the most popular methods to building a combination of classifiers is referred to as Stacking and is based on a meta-learning approach. In this approach predictions of base classifiers (level-0) are used to train another algorithm (level-1 meta-classifier), which is responsible for making the final decision. In a study by Ting and Witten [2], it was demonstrated that MLR was the only good candidate for level-1 combining function. This was more successful than C4.5, IB1 and NB. In [3], the concept of Meta Decision Tree was proposed. This approach outperformed Stacking with ordinary J4.8, but did not provide better results than Stacking with MLR. In [1], Multi-Response Model Tree was applied, instead of MLR, as a level-1 combining function. The improved method was very successful and outperformed the old version of it. In our work, we propose a collection of clusters, that were built based on base classifiers' outputs, as a meta-classifier in the Stacking framework.

2 Classification by Cluster Analysis (CBCA)

The proposed method is applied with different learning methods. All models provide a class label as an output. The main aim is therefore to generate a collection of clusters containing similar instances according to their classification results. This means that one cluster should contain objects that were correctly/incorrectly classified to the same class by the same group of base classifiers. In the first step a data set D is divided into 3 disjunctive subsets: D_1, D_2, and D_3. D_1 is used for the training of base classifiers, D_2 is used for the purposes of building the clusters and D_3 is used as a testing set. The process of generating a collection of clusters with some technical implementation details are presented in Fig 1.

D. Dicheva and D. Dochev (Eds.): AIMSA 2010, LNAI 6304, pp. 267–268, 2010.

Settings	Algorithm
IGAE-weka.attributeSelection.InfoGainAttributeEval-Evaluates the worth of an attribute by measuring the information gain with respect to the class (autimaticly discretizes numeric data). Kmeans-weka.clusterers.SimpleKMeans-Clusters data using the k-means algorithm (Euclidean distance, Nominal and Numeric attributes accepted) $A[M,N]$-matrix with M rows and N columns $F(x)$-value of the feature F of the instance x **Inputs**:D_1-training set, $D_2 = (x_1,...,x_M)$-validation set, $L_1,...,L_N$ – N different learning methods, K-no. of most significant features to be selected, Z – no. of clusters to be generated **Output**: $(x_1^*, \alpha_1),...,(x_Z^*, \alpha_Z)$-centres of the cluster with their class labels	*Build N base classifiers:* $\quad$ For i=1→N $\quad$ $C_i=L_i(D_1)$ *Select K most significant features* $\quad$ IGAE(D_1,D_2)→$F_1,...,F_K$ *Build matrix $A[M, N+K]$* $\quad$ For i=1→M $\quad\quad$ For j=1→N $\quad$ $A(i,j)=C_j(x_i)$ $\quad\quad$ For k=1→K $\quad$ $A(i,N+k)=F_k(x_i)$ *Cluster all rows of matrix A, no. of clusters=Z* $\quad$ Kmeans$([A(1,1),...,A(1,N+K)],...,[A(M,1)...A(M,N+K)], Z)$ $\quad\quad$ →$P_1,...,P_Z$ For i=1→Z $\quad$ {Identify centoid of P_i→ x_i^* $\quad$ Read class label of instance x_i^*→α_i } Return $(x_1^*,\alpha_1),..., (x_Z^*,\alpha_Z)$

Fig. 1. The clustering algorithm with some implementation details

In the following steps, all instances in D_2, denoted by$\{x_1,...,x_M\}$, are classified by all base classifiers. As a result of this process, we obtain an MxN matrix, where rows represent instances from D_2 and columns represent base classifiers. For example, cell $\{a_{ij}\}$ of the matrix, contains the label of the class, where x_i was classified by C_j. Additionally, K columns that represent the K most significant features, were added to the matrix. Subsequently, all rows were clustered according to the values in the columns. Centroids of the clusters were identified. For an unseen instance, the distance between it and all centroids are calculated using the Euclidean metric and consider only the K most significant features. The class where the nearest centroid belongs is subsequently considered as a final decision. To establish the best performing model, we evaluated different values of parameters K and Z.

3 Conclusion

To evaluate the new approach, we conducted experiments with 14 data sets. The proposed method was tested with each of the data sets and then compared with individual classifiers and the Majority Voting method. The results show that CBCA is an effective combining tool, since it outperformed all base classifiers in most of the cases. It also outperformed MV for 12 out of 14 data sets. Beside this, our method had a more significant advantage over all base classifiers than MV.

References

1. Dzeroski, S., Zenko, B.: Is Combining Classifiers with Stacking Better than Selecting the Best One? Machine Learning, 255–273 (2004)
2. Ting, K.M., Witten, I.H.: Stacked generalization: when does it work? In: 15th International Joint Conference on Artificial Intelligence, pp. 866–871 (1997)
3. Todorovski, L., Dzeroski, S.: Combining Multiple Models with Meta Decision Trees. In: Zighed, D.A., Komorowski, J., Żytkow, J.M. (eds.) PKDD 2000. LNCS (LNAI), vol. 1910, pp. 54–64. Springer, Heidelberg (2000)

Mapping Data Driven and Upper Level Ontology

Mariana Damova, Svetoslav Petrov, and Kiril Simov

Ontotext AD, Tsarigradsko Chosse 135, Sofia 1784, Bulgaria
{mariana.damova,svetoslav.petrov,kivs}@ontotext.com

Abstract. The Linking Open data (LOD) [1] initiative aims to facilitate the emergence of a web of linked data by publishing and interlinking open data on the web in RDF. The access to these data becomes increasingly a challenge. This paper presents an innovative method for retrieving facts from vast amounts of data by using an upper level ontology (PROTON) [2] as an access interface. FactForge, the largest and most heterogeneous body of general factual knowledge that was ever used for logical inference served to develop and test the method of mapping ontologies.

Keywords: Semantic Web, Linking Open Data, Upper Ontology, FactForge, Ontology Mapping, Linked Data, Inference.

This effort originates from a series of projects within Ontotext AD on design and implementation of a reason-able view over the data integrated within Linking Open Data (LOD) [1]. Reason-able views are built according to several design principles:

    (a) All the datasets in the view represent linked data;
    (b) Single reasonability criteria is imposed on all datasets;
    (c) Each dataset is connected to at least one of the others.

We present a new version of the reason-able view on a subset of LOD data, Fact-Forge, which adds an additional layer to it in order to provide a better integration between the heterogeneous datasets and a unified access to them. This additional layer consists of an upper level ontology mapped to the ontologies of the datasets of Fact-Forge. The unified access simplifies the process of querying the datasets, as a single ontology is used instead of predicates from a multitude of ontologies requiring in depth knowledge of each of them. It also allows to obtain information from many datasets via one single predicate. For example, one ontology predicate can cover and retrieve several data driven predicates from different datasets. FactForge (http://factforge.net) is a collection of 8 of the central LOD datasets (DBPedia, Geonames, Wordnet, Musicbrainz, Freebase, UMBEL, Lingvoj and the CIA World Factbook into a single repository (BigOWLIM – http://www.ontotext.com/owlim) containing 1.2 billion explicit and 0.8 billion implicit statements. PROTON is a modular, lightweight, upper-level ontology, encoded in OWL Lite, and a minimal set of custom entailment rules, built with a basic subsumption hierarchy. An upper level ontology is a model of the common objects that are applicable across a wide range of domains. It contains generic concepts that can serve as a foundation of other more specific ontologies.

D. Dicheva and D. Dochev (Eds.): AIMSA 2010, LNAI 6304, pp. 269–270, 2010.
© Springer-Verlag Berlin Heidelberg 2010

The PROTON ontology set was extended to cover the available data within Fact-Forge. This extension was governed by two main principles:

(1) to provide a coverage for the available data; and
(2) to reflect the best approaches in the design of ontologies such as OntoClean methodology [4].

The first principle was met by performing several steps:

(a) analysis of the structure and the content of the data within FactForge;
(b) analysis of the structure of PROTON;
(c) analysis of the related classes within the ontology.

The mapping consisted in establishing of equivalence and subclass relations between the classes and properties of the ontologies in FactForge and the classes and properties of PROTON. In cases of missing explicit concepts new classes were created and used to do the mapping. The process of identification of new classes and properties is being performed several times until a reasonably large set of data is covered.

The outcomes of this work can be summarized as follows:

(1) a new layer of unified semantic knowledge over FactForge;
(2) an original approach to providing similar layers to other datasets;
(3) a new version of PROTON ontology to be used in other projects.

The new PROTON contains 140 new classes and 14 new properties for a total of 430 classes and 114 properties. After extensions and mappings loaded the datasets increased by 67 million statements when only 2600 explicit statements and 1400 new entities were added. The mapping was tested with 32 SPARQL queries. 16 of them were constructed with various predicates from several datasets of FactForge and 16 of them were (their equivalents) constructed with PROTON only predicates. The two sets of queries returned the same results, which proves the validity of the approach.

References

1. World Wide Web Consortium (W3C): Linking Open Data. W3C SWEO project home page as of January 2010 (2010), http://esw.w3.org/topic/SweoIG/TaskForces/CommunityProjects/LinkingOpenData
2. Terziev, I., Kiryakov, A., Manov, D.: D.1.8.1 Base upper-level ontology (BULO) Guidance, Deliverable of EU-IST Project IST – 2003 – 506826 SEKT (2005)
3. Omitola, T., Koumenides, C.L., Popov, I.O., Yang, Y., Salvadores, M., Szomszor, M., Berners-Lee, T., Gibbings, N.: Put in Your Postcode, Out Comes the Data: A Case Study. In: Aroyo, L., Antoniou, G., Hyvönen, E., ten Teije, A., Stuckenschmidt, H., Cabral, L., Tudorache, T. (eds.) ESWC 2010. LNCS, vol. 6088, pp. 318–332. Springer, Heidelberg (2010)
4. Welty, C., Guarino, N.: Supporting Ontological Analysis of Taxonomic Relationships. Data and Knowledge Engineering (2001)

Data Sample Reduction for Classification of Interval Information Using Neural Network Sensitivity Analysis

Piotr A. Kowalski[1] and Piotr Kulczycki[2]

[1] Systems Research Institute, Polish Academy of Sciences,
ul. Newelska 6, PL-01-447 Warsaw, Poland
pakowal@ibspan.waw.pl,
[2] Cracow University of Technology,
Department of Automatic Control and Information Technology
ul. Warszawska 24, PL-31-155 Cracow, Poland
kulczycki@pk.edu.pl

Abstract. The aim of this paper is present a novel method of data sample reduction for classification of interval information. Its concept is based on the sensitivity analysis, inspired by artificial neural networks, while the goal is to increase the number of proper classifications and primarily, calculation speed. The presented procedure was tested for the data samples representing classes obtained by random generator, real data from repository, with clustering also being used.

Keywords: classification, interval information, data sample reduction, artificial neural networks, sensitivity method.

1 Introduction and Main Results

Recently, interest in interval analysis has grown notably in many practical applications [2]. Fundamental here is the assumption that the only available information about the investigated quantity $x \in \mathbb{R}$, is that it fulfills the condition $\underline{x} \le x \le \overline{x}$, and consequently can be treated as the interval $[\underline{x}, \overline{x}]$. The multidimensional case $x \in \mathbb{R}^n$ was also examined. The main subject of the research presented here is a reduction data sample for the classification task. The tested element is of interval form, but elements consisting of patterns of particular classes are defined uniformly. A classification procedure [3] was worked out for the removal from samples of those elements having negligible or even negative influence on the correctness of classification. Its concept is based on the sensitivity method [1], inspired by neural networks, while the goal is to increase the number of proper classifications as well as, primarily, calculation speed. The concept of classification is based on the Bayes approach, ensuring a minimum of potential losses arising from misclassification. For a such-formulated problem, the methodology of statistical kernel estimators [4] is used, which frees the investigated procedure from arbitrary assumptions concerning shapes of samples.

2 Numerical Experiments

As an illustrative example, consider the one-dimensional case with two classes, represented by 50-elements samples given by Gaussian generators N(0,1) and N(2,1).

D. Dicheva and D. Dochev (Eds.): AIMSA 2010, LNAI 6304, pp. 271–272, 2010.

The interval-type elements subjected to classification were calculated by generating intervals' centers and then their lengths: 0.1; 0.25; 0.5; 1.0; 2.0. Using the concept of the classification method without reduction, 16.21; 16.38; 16.42; 16.43; 16.45 percent were misclassifications, with respect to interval length. Next, by applying the data reduction procedure, the number of misclassifications decreased to 14.76; 14.87; 14.89; 14.96; 15.00 percent, respectively, when sample size was also significantly reduced. The common occurrence of the results in both the precision and calculation speed aspects, is worth underlining: about 10% improvement of classification accuracy with around 40% reduction of sample sizes were simultaneously obtained. Obviously, following a reduction in sample size, calculation speed was also significantly reduced. In the case when the samples had been obtained by the clustering k-means method, the classification algorithm led to the following results: 16.60; 16.34; 16.32; 16.33; 16.33 percent of misclassifications before the reduction procedure, and 14.67; 14.55; 14.51; 14.51; 14.50 after it. The results obtained here were better than in the basic case described above, thanks to a more effective reduction in atypical elements of patterns incorrectly treated during the clustering procedure.

3   Comments and Final Remarks

The investigated method is two-phased in its nature – the time-consuming procedures for defining the classifier and reduction data take place only once at the beginning, however, the interval classification procedure itself is performed in a relatively short period, mainly thanks to analytical forms of formulas obtained. The developed reduction algorithm was compared with simple and natural random reduction as well as with the k-NN method. For all these cases for reduction, the concept worked out here produced much better results. Furthermore, there is no need for any arbitrary assumption concerning an algorithm's parameters, which is another positive aspect of the procedure presented in this paper.

References

1. Engelbrecht, A.P.: Sensitivity Analysis for Selective Learning by Feedforward Neural Networks. Fundamenta Informaticae, vol. 46(3), pp. 219--252 (2001)
2. Jaulin, L., Kieffer, M., Didrit, O., Walter, E.: Applied Interval Analysis. Springer, Berlin (2001)
3. Kowalski, P.A.: Bayesian Classification of Imprecise Interval-Type Information (in Polish). Systems Research Institute, Polish Academy of Sciences, Ph.D. Thesis (2009)
4. Kulczycki, P.: Kernel Estimators in Industrial Applications. In: Soft Computing Applications in Industry, B. Prasad (ed.), pp. 69--91, Springer-Verlag, Berlin, (2008)

Individualized Virtual Humans for Social Skills Training

H. Chad Lane

University of Southern California
lane@ict.usc.edu

Abstract. Virtual humans are now being used as role players for a variety of domains that involve social skills. In this paper, we discuss how such systems can provide individualized practice through dynamically adjusting the behaviors of virtual humans to meet specific learner needs.

Keywords: virtual humans; social skills; pedagogical experience manipulation.

1 Introduction

Pedagogical agents are most often designed to play the role of tutor or peer [1] in virtual learning environments. Over the last decade or so, a new breed of pedagogical agents have emerged that act as the *object* of practice – i.e., it is the interaction itself (with the agent) that is intended to have educational value. Here, the agent is usually a *virtual human* who is playing some defined social role in an interaction that requires application of specific communicative skills. In this paper, we briefly explore some methods for providing adaptive guidance *through* the virtual human role players.

2 Towards Individualized Practice with Virtual Humans

From supporting anti-bullying and intercultural learning with children [2] to negotiation skills for professionals [3], a number of mature systems now exist for the practice and learning of social skills. Given the richness of human interactions, it should be no surprise that the space of adjustability in virtual humans is vast. *Nonverbal behaviors*, such as gaze, nodding, and gestures, play a key role in the expression of emotion and thus it is possible to convey a great deal of implicit feedback through them (examples are shown figure 1). The information conveyed in the *content* of an utterance represents another critical dimension in the space of configurability. Finally, many virtual humans perform task-based reasoning and behave based on *underlying representations* of the dialogue, their intentions, desires, the task domain, and their emotions [2-3]. Manipulation of these underlying models for pedagogical purposes represents a third category. Given these dimensions, some broad categories of support that are possible through tailoring of behaviors include:

1. support *recognition* when errors are committed or ideal actions taken
2. provide an *explanation* for observed reactions and emotional state changes
3. suggest a *repair* for how a learner might revise their beliefs

D. Dicheva and D. Dochev (Eds.): AIMSA 2010, LNAI 6304, pp. 273–274, 2010.
© Springer-Verlag Berlin Heidelberg 2010

Fig. 1. Expressions of anger, skepticism, appreciation, and umbrage by virtual humans [3]

The goal is to provide feedback while maintaining the narrative context and not detracting from the perceived realism of the experience. Some examples of how a character might achieve these goals include:

1. amplification of virtual human response behavior, such as the intensity of facial expressions or use of emotionally charged vocabulary (recognition)
2. description of a causal link between a user action and a negative (or positive) result via additional content (e.g., "Why would you say...?"; explanation)
3. clarification of a relevant domain concept by including it in the content of an utterance ("I can tell you that..."; explanation; repair).
4. highlighting of an alternative communicative action that would have produced a better outcome (e.g., "If I were you, I'd ..."; repair)

The central idea behind all of these strategies is to build on the existing feedback already coming from the virtual human, but alter it to achieve a pedagogical goal. We have implemented strategies for amplification and clarification in a prototype system.

An important question is whether such adaptations threaten fidelity and the implications of that. If learners figure out the characters are secretly "helping", does it impact learner affect and motivation to engage? Future studies will need to address these questions as well as determining if support from pedagogical experience manipulation can be as effective or extend help delivered by a tutoring system.

References

1. Johnson, W.L., Rickel, J., Lester, J.C.: Animated Pedagogical Agents: Face-to-Face Interaction in Interactive Learning Environments. International Journal of Artificial Intelligence in Education 11, 47–48 (2000)
2. Aylett, R., Vannini, N., Andre, E., Paiva, A., Enz, S., Hall, L.: But that was in another country: agents and intercultural empathy. In: Proceedings of The 8th International Conference on Autonomous Agents and Multiagent Systems. International Foundation for Autonomous Agents and Multiagent Systems, Budapest, Hungary, vol. 1, pp. 329–336 (2009)
3. Swartout, W., Gratch, J., Hill, R.W., Hovy, E., Marsella, S., Rickel, J., Traum, D.: Toward virtual humans. AI Magazine 27, 96–108 (2006)

Signal Classification with Self-organizing Mixture Networks

Piotr Lipinski

Institute of Computer Science,
University of Wroclaw, Wroclaw, Poland
lipinski@ii.uni.wroc.pl

1   Introduction

This paper proposes a new method of signal classification with discrete wavelet transformations and self-organizing mixture networks (SOMN) [3] being an extension of popular self-organizing maps (SOM) [1]. While SOM try to describe the sample data with a single distribution of a single parametric form, SOMN use a mixture of different distributions of different parametric forms.

2   Signal Classification Framework

In recent research, two main approaches to feature extraction and signal representation usually exist: Discrete Fourier Transformations (DFT) and Discrete Wavelet Transformations (DWT). Although both approaches are comparable in energy preservation, we focus on Discrete Wavelet Transformations, with the Daubechies wavelets, $D4$, because they are usually faster to compute and offer the multi-resolution decomposition.

Once feature extraction done, signals are represented by d-dimensional signal vectors $\mathbf{x}$, where d is the number of DWT coefficients chosen, which allows to use SOMN to classify them. SOMN deal with a network consisting of M nodes representing clusters. Each node $i = 1, 2, \ldots, M$ has assigned an independent gaussian distribution described by a parameter vector $\boldsymbol{\theta}_i$, consisting of the mean vector $\mathbf{m}_i$, the covariance matrix $\boldsymbol{\Sigma}_i$ and a prior probability P_i.

Each signal vector $\mathbf{x}$ is assigned to one of the M nodes with a joint probability distribution

$$p(\mathbf{x}|\boldsymbol{\Theta}) = \sum_{i=1}^{M} p_i(\mathbf{x}|\boldsymbol{\theta}_i) \cdot P_i, \tag{1}$$

where $p_i(\mathbf{x}|\boldsymbol{\theta}_i)$ denotes the i-th node conditional distribution, being the gaussian distribution with the mean vector $\mathbf{m}_i$ and the covariance matrix $\boldsymbol{\Sigma}_i$, $\boldsymbol{\theta}_i$ denotes the parameter vector for the i-th conditional distribution and P_i denotes the prior probability of the i-th node.

Parameter vectors are set in an iterative training process, as in the case of SOM, on a training data set. Each iteration starts with randomly drawing a training vector $\mathbf{x}$. Next, a winner node is chosen according to its kernel output multiplied by its prior probability, and within a neighborhood of the winner, parameters are updated, based on

D. Dicheva and D. Dochev (Eds.): AIMSA 2010, LNAI 6304, pp. 275–276, 2010.

the Kullback-Leibler information metric, [3], by adding to the mean vector $\mathbf{m}_i$ and the covariance matrix Σ_i a noise $\Delta\mathbf{m}_i$ and $\Delta\Sigma_i$, respectively,

$$\Delta\mathbf{m}_i = \alpha \cdot P(i|\mathbf{x})(\mathbf{x} - \mathbf{m}_i), \tag{2}$$

$$\Delta\Sigma_i = \alpha \cdot P(i|\mathbf{x})((\mathbf{x} - \mathbf{m}_i)(\mathbf{x} - \mathbf{m}_i)^T - \Sigma_i), \tag{3}$$

where $\alpha \in (0, 1)$ is a learning rate decreasing in successive iterations.

3   Experiments and Conclusions

Experiments concerned fault detection in mechanical systems, based on measurement of vibration signals and their classifications, on synthetic data generated using the mathematical model described in [2] for three levels of stiffness degradation. Figure 1 presents numbers of wrong classified signals in each cluster for one experiment with 1500 vibration signals, 500 per class, represented by 64-dimensional signal vectors, classified by SOM (a) and SOMN (b). Experiments proved that SOMN outperforms SOM in classification of vibration signals, especially when applied to signals derived from significantly different sources, due to using mixtures of different distributions in different parametric forms (in 40 experiments, the average error rate was below 5%).

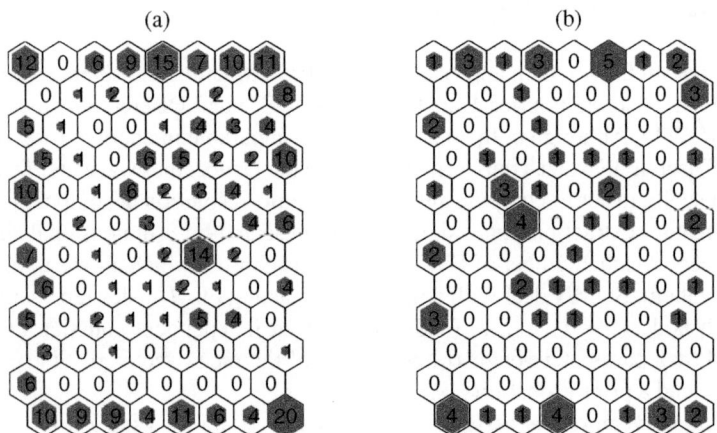

Fig. 1. Numbers of wrong classified signals in each cluster for one experiment with 1500 vibration signals, 500 per class, represented by 64-dimensional signal vectors, (a) SOM, (b) SOMN

References

1. Kohonen, T.: Self-Organizing Maps. Springer, Heidelberg (1995)
2. Sobczyk, K., Trebicki, J.: Stochastic dynamics with fatigue-induced stiffness degradation. Probabilistic Engineering Mechanics 15(1), 91–99 (2000)
3. Yin, H., Allinson, N.: Self-organizing mixture networks for probability density estimation. IEEE Transactions on Neural Networks 12(2), 405–411 (2001)

Simulation of the Autonomous Agent Behavior by Autoregressive Models

Vanya Markova

ICSR-BAS
2 Acad G. Bontchev Str., 1113, Sofia, Bulgaria
markovavanya@yahoo.com

Abstract. Agent's behaviour can be described as time series of agent's parameters and actions.The comparative analysis shows AR and ARMA methods are preferable in low volatility of the environment while GARCH methods have better predictive characteristics in case of heteroskedasticity.

Keywords: predictive model, behaviour, autonomous agent, autoregressive GARCH, time series.

1 Introduction

Autonomous Mobile Sensor Agent (AMSA) is intended to collect and transfer data in hardly accessible, inaccessible or hazardous locations. Its behaviour and environment parameters could be presented as time series[1]. It is of great importance to make reliable short term forecasts in order to perform in autonomous manner.The Autoregressive (AR) Models are widely used for short term forecasting. The ARCH and GARCH are further extensions of AR models. These methods deals well with heteroskedasticity time series. The term means "differing variance" and comes from the Greek "hetero" ('different') and "skedasis" ('dispersion'). However GARCH are rarely used outside of econometrics[2] [3] . The aim of this research is to apply and evaluate the potential of predictive analysis and autoregressive models as a means to further understanding the impact of previous states and changing environment on agent's behaviour. Comparison between GARCH and AR/ARMA methods is also the purpose of this paper.

2 Methods and Results

We claim that agent's behaviour could be described as time series of agent's parameters and actions. AMSA perform several actions: making_of_a_plan, making_of_a_route, self_localisation and movement. The agent omits some actions to achieve better performance. However due to such optimisation agent could stray from its route, especially in case of "volatile" way shape. It is desirable to have reliable short time forecasts in order to evaluate risk of these deviations. Owing to such forecasts the agent could omit more actions in linear parts of the route and perform more actions and checks in more complex and volatile parts of its way. Hence it would be of great

D. Dicheva and D. Dochev (Eds.): AIMSA 2010, LNAI 6304, pp. 277–278, 2010.

use to make a reliable forecast for agent's behaviour. The autoregressive models are useful in these applications. However in case of retrospective time series the result from conventional AR methods could be unreliable. To avoid this we should to apply GARCH methods. The software modules are built for the implementation of the three autoregressive models. Separate software module is developed for evaluation the adequacy of the prediction of the models. AR, ARMA and GARCH methods are nearly similar in case of low volatility (table 1). AR/ARMA methods are preferable in this case since they work faster than GARCH methods. In the case of heteroskedasticity GARCH methods are more accurate than AR/ARMA methods.

Table 1. Estimators of agent's behaviour prediction

	Time series from Map I			Time series from Map II		
	R^2	MSE	R_{SE}	R^2	MSE	R_{SE}
AR	0.92	0.18	16.84	0.72	0.29	4.26
ARMA	0.91	0.17	17.80	0.73	0.27	3.41
AR/GARCH	0.95	0.15	21.10	0.87	0.19	18.12

From table 1 one can see that for case of low volatility AR, ARMA and AR/GARCH are nearly similar. But because GARCH method is slower one should prefer AR/ARMA methods. On the other hand in cases of heteroskedasticity GARCH outperforms the AR/ARMA methods.

3   Conclusion

This paper presents the application of three autoregressive models (AR, ARMA and GARCH) for simulation the autonomous agent behaviour. The analysis of the results shows that AR, ARMA and GARCH demonstrate similar accuracy when time series aren't heteroskedastic. GARCH methods indisputably reveal their high abilities in case of dynamic environments and can be used with success for adaptive behaviour of autonomous agent.

References

1. Markova, V., Shopov, V., Onkov, K.: An Approach for Developing the Behaviour Models of Automomous Mobility Sensor Agent, Bulgaria, pp. 97–105 (2008)
2. Wong, F., Galka, A., Yamashita, O., Ozaki, T.: Modelling non-stationary variance in EEG time series by state space GARCH model. J. Comp. Biology and Medicine 36, 1327–1335 (2006)
3. Zhou, B., He, S.: Network Traffic Modeling and Prediction with ARIMA/GARCH, West Yorkshire, UK, July 18-20 (2005)

Semi-partitioned Horn Clauses: A Tractable Logic of Types

Allan Ramsay

School of Computer Science,
University of Manchester,
Manchester M13 9PL, UK

Abstract. Reasoning about types is a common task, and there are a number of logics that have been developed for this purpose [1,2]. The problem is that type logics with efficient proof procedures are not very expressive, and logics with more expressive power are not tractable. The current paper presents a logic which is considerably more expressive than using a simple taxonomy, but for which the proof procedure is nonetheless linear.

1 Introduction

There are two major issues in reasoning with types: determining whether one type subsumes another, and deciding whether two types are compatible. The majority of efficient logics of types concentrate on the former, but it is arguable that in many applications the latter is of equal importance (*e.g.* in the use of type constraints on slot-fillers in natural language processing). The logic of types outlined below provides efficient machinery for deciding both kinds of relation. This logic depends on two separate notions. In the first, we represent partitioned type lattices in a very compact form which supports extremely fast comparison of types. The algorithm for comparing types exploits the use of [3]'s notion of 'watched literals'. In the second we show how to use watched literals to implement Dowling & Gallier (D & G)'s [4] algorithm for reasoning with propositional Horn clauses, and show that this can be smoothly integrated with the machinery for managing type lattices. The result is an expressive logic of types in which subsumption and consistency of types can be checked in linear time.

The inference engine for this logic exploits two major ideas:

1. Nodes in a partitioned typ hierarchy such as the one in Fig. 1 can be represented as open lists of `<type>=<value>` lists, so that comparison of types becomes a matter of unification of such lists.
   From Fig. 1, for instance, we can see that human lies at the end of the path `[living, animal, ape, human]`. We could therefore take this list, or better take the open list `[living=yes, animal=yes, ape=yes, human=yes | _]`, to be the representation of this type. Then checking whether two types are compatible is simply a matter of unifying their representations, and checking whether a type T1 subsumes another T2 is just a matter of checking whether T1's representation subsumes T2's.

D. Dicheva and D. Dochev (Eds.): AIMSA 2010, LNAI 6304, pp. 279–280, 2010.

```
living >> [animal, plant, bacterium].
animal >> [bird, reptile, fish, insect, mammal].
mammal >> [cat, dog, ape].
ape >> [monkey, orangutang, human].
human >> [man, woman, boy, girl].
plant >> [fruit, vegetable, grass].
```

Fig. 1. Simple natural kinds hierarchy

2. If we replace concrete truth values in these lists by Prolog variables, then we can attach delayed constraints to these variables. These delayed constraints allow us to provide efficient indexing for a variant on [4]'s algorithm for reasoning with Horn clauses. This implementation allows us to carry out D & G's algorithm without enumerating the truth values of every clause, and provides a very efficient ways of indexing clauses.

2 Conclusions

The logic of types described above is tractable in theory and very fast in practice. The key innovations are the use of sorted lists of path descriptors for managing semi-partitioned type lattices, and the use of coroutining operations to allow for negated types and intersective subsets. The logic is more expressive than most other logics with similar complexity: it is thus suitable for use in contexts where speed is critical and expressive power is desirable. We currently use it, for instance, to help guide the parser in a general purpose NLP system. To make reliable choices between competing analyses of a natural language sentence can require complex inference based on large amounts of background knowledge [5]. Carrying out inference of this kind during parsing is likely to slow the parsing process down very substantially. We therefore use the type logic to guide choices about potential slot fillers during parsing, and then carry out full scale inference to choose between those analyses that survive this process. In this context, we want reasoning about whether some entity is of the right kind to play a given role to take about the same amount of time as the basic unification that we carry out on syntactic features. The inference algorithms here, which are after all driven largely by unification of paths, have the required level of performance.

References

1. Aït-Kaci, H., Boyer, R., Lincoln, P., Nasr, R.: Efficient implementation of lattice operations. ACM Transations on Programming Languages 11(115), 115–146 (1989)
2. Fall, A.: Reasoning with taxonomies. PhD thesis, Simon Fraser University (1990)
3. Moskewicz, M., Madigan, C., Zhao, Y., Zhang, L., Malik, S.: Chaff: Engineering an efficient SAT solver. In: 39th Design Automation Conference, Las Vegas (2001)
4. Dowling, W.F., Gallier, J.H.: Linear-time algorithms for testing the satisfiability of propositional horn formulae. Journal of Logic Programming 1(3), 267–284 (1984)
5. Hirst, G.: Semantic interpretation and the resolution of ambiguity. Studies in Natural Language Processing. Cambridge University Press, Cambridge (1987)

Feed Rate Profiles Synthesis Using Genetic Algorithms

Olympia Roeva

Centre of Biomedical Engineering, Bulgarian Academy of Sciences
105 Acad. G. Bonchev Str., Sofia 1113, Bulgaria

Abstract. In the paper a genetic algorithm for feed rate profiles synthesis is proposed. An *E. coli* fed-batch fermentation process is considered. The feed rate profiles based on three different lengths of chromosomes are synthesized. A satisfactory result for the fermentation system due to economical effect and process effectiveness is achieved.

Keywords: Genetic algorithms, Chromosome, Feed rate, *E. coli* fermentation.

1 Introduction

Optimization of fed-batch fermentation processes has been a subject of research for many years. Currently, the feed rate synthesis is commonly solved by mathematical model based optimization methods. If an accurate model of the system is available optimization procedures can be used to calculate the feeding strategy [5]. However, fermentation processes are typically very complex and this makes processes difficult to control with traditional techniques. As an alternative the global optimization methods are used, for instance genetic algorithms (GA). The GA is already used for synthesis of feed rate profiles [2]. In this paper a genetic algorithm for feed rate profiles synthesis during an *E. coli* fed-batch fermentation process is proposed. The bacterium *E. coli* is the microorganism of choice for the production of the majority of the valuable biopharmaceuticals. Here an optimal state of microorganisms' culture is maintained by GA synthesized feed rate profiles.

2 Application of GA for Feed Rate Profiles Synthesis

The model of *E. coli* fed-batch fermentation is presented in [1, 3]. The numerical values of the model parameters and initial conditions of the process variable used in simulations are given in [1, 3]. The GA operators and parameters are based on [4]. Three chromosomes representations are proposed: (i) the profile is divided into equal 30 genes; (ii) into equal 60 genes and (iii) into equal 100 genes. Every gene in a chromosome is coded in range $F = 0\text{-}0.05 \ 1\cdot\text{h}^{-1}$ [1].

Using the model [1, 3] and objective function (1) three problems (30, 60 and 100 genes in chromosome) are running 50 executions with the GA. The results in the case of 60 genes in chromosome are depicted on Fig. 1.

D. Dicheva and D. Dochev (Eds.): AIMSA 2010, LNAI 6304, pp. 281–282, 2010.

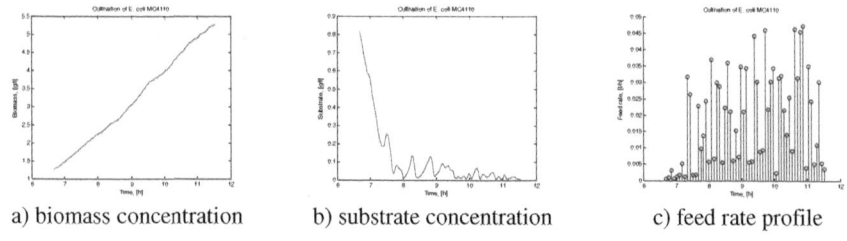

| a) biomass concentration | b) substrate concentration | c) feed rate profile |

Fig. 1. Resulting dynamics and feed rate profile in case of 60 genes in chromosome.

3   Results and Conclusion

The proposed genetic algorithm is found to be an effective and efficient method for solving the optimal feed rate profile problem. The results show that for all tests the required objective function has been achieved. However, the feed profile formed using chromosome with 60 genes is superior to the rest feeding trajectories. Based on obtained feed rate profile cell concentration has an ideal increase for the complete fermentation period, achieving final cell concentration of 5.26 g·l⁻¹ using 1.38 l feeding solution. This is a satisfactory result for the fermentation system due to the economical effect and process effectiveness.

Acknowledgements. This work is partially supported by the European Social Fund and Bulgarian Ministry of Education, Youth and Science under Operative Program "Human Resources Development", Grant BG051PO001-3.3.04/40, and by National Science Fund Grant DMU 02/4.

References

1. Arndt, M., Hitzmann, B.: Feed Forward/Feedback Control of Glucose Concentration during Cultivation of *Escherichia coli*. In: 8th IFAC Int. Conf. on Comp. Appl. in Biotechn., Quebec City, Canada, pp. 425–429 (2001)
2. Chen, L.Z., Nguang, S.K., Chen, X.D.: On-line Identification and Optimization of Feed Rate Profiles for High Productivity Fed-batch Culture of Hybridoma Cells using Genetic Algorithms. ISA Trans. 41(4), 409–419 (2002)
3. Roeva, O.: Parameter Estimation of a Monod-type Model based on Genetic Algorithms and Sensitivity Analysis. In: Lirkov, I., Margenov, S., Waśniewski, J. (eds.) LSSC 2007. LNCS, vol. 4818, pp. 601–608. Springer, Heidelberg (2008)
4. Roeva, O., Tzonkov, S.: A Genetic Algorithm for Feeding Trajectory Optimization of Fed-batch Fermentation Processes. Int. J. Bioautomation 12, 1–12 (2009)
5. Shin, H.S., Lim, H.C.: Maximization of Metabolite in Fed-batch Cultures Sufficient Conditions for Singular Arc and Optimal Feed Rate Profiles. Biochemical Eng. J. 37, 62–74 (2007)

An Ontology of All of Computing: An Update on Challenges and Approaches

Lillian N. Cassel[1], Siva Kumar Inguva[1], and Jan Buzydlowski[2]

[1] Villanova University,
Villanova, PA, USA
[2] Cabrini College,
Radnor, PA, USA
{Lillian.Cassel,Sivakumar.Inguva}@villanova.edu,
janb@cabrini.edu

Abstract. This poster provides a status update on the Computing Ontology project. A number of challenges are associated with constructing an ontology of all computing. These include deciding on boundaries for a topic area that touches nearly every domain. Several sources of content contribute to the list of topics. These, in turn, reflect the several potential application areas of the resulting ontology.

Keywords: Computing Ontology, Ontology, Education.

1 Introduction

The Computing Ontology is a collection of more than 3000 concepts that attempt to reflect the entirety of the computing disciplines. All areas, including computer science, computer engineering, software engineering, information systems and information technology appear in the ontology. In addition, more specialized subdomains such as security, networking, game programming, and others are part of the computing ontology. The terms in the ontology originally came primarily from documents related to computing education. A recent review of computer science publications adds another dimension to the collection.

2 The Computing Domain as Reflected in Curriculum Documents

Initial efforts to accumulate the topic list for the computing ontology started with curriculum recommendations and accreditation standards. [1-4] Challenges included melding the topic areas as represented in different curriculum reports to make a single, coherent presentation of the breadth and depth of the computing disciplines. Databases are important topics in both computer science and information systems, for example. However, the views of databases vary substantially between these areas. Merging the ideas from these aspects and other aspects of the computing disciplines required extensive discussion and negotiation. The result is a comprehensive collec-

D. Dicheva and D. Dochev (Eds.): AIMSA 2010, LNAI 6304, pp. 283–284, 2010.

tion of computing topics that does not distinguish among the subareas of computing. [5] This permits each area to identify its areas of specialty in a context of the whole of the discipline. This permits comparisons of separate areas to see easily what is covered and what is missing from each. The implications for curriculum design are important: a conscious choice is made about what a new curriculum will include and it is clear what will be excluded.

3   The Computing Domain as Reflected in Research Publications

Curriculum documents are a valuable source of computing concepts, but are not necessarily a complete representation of the discipline. The most recent additions to the computing ontology comes from an analysis of the significant terms extracted from all the publications of the ACM Digital Library for the last 10 years. The poster will describe the results of that analysis. Additional input to the ontology comes from consultations with experts with particular concerns with specific parts of the computing domain. Recent inputs include aspects of parallelism significant to the emergence of multi-core computers. These inputs illustrate the importance of constant updates to the ontology.

4   Summary

This poster provides an update on the computing ontology with its connections to computing education and its updates from research interests and the challenge of boundaries between computing and other disciplines.

References

1. Computing Curricula 2005: The Overview Report, p. 62 (2005)
2. Software Engineering 2004. Curriculum guidelines for undergraduate degree programs in software engineering, Computing Curricula Series (2004)
3. Computer Engineering 2004. Curriculum Guidelines for undergradutate degree programs in computer engineering. Computing Curricula Series (2004)
4. Computing curricula 2001. Journal on Educational Resources in Computing (JERIC) 1(3) (2001)
5. The computing ontology website, http://what.csc.villanova.edu/twiki/bin/view/Main/OntologyProject

Author Index